Building Production-Grade Web Applications with Supabase

A comprehensive guide to database design, security, real-time data, storage, multi-tenancy, and more

David Lorenz

Building Production-Grade Web Applications with Supabase

Copyright © 2024 Packt Publishing

Group Product Manager: Kaustubh Manglurkar

Publishing Product Manager: Chayan Majumdar

Book Project Manager: Sonam Pandey

Senior Editor: Hayden Edwards

Technical Editor: K Bimala Singha

Copy Editor: Safis Editing

Indexer: Pratik Shirodkar

Production Designer: Jyoti Kadam

DevRel Marketing Coordinator: Anamika Singh and Nivedita Pandey

First published: August 2024

Production reference: 1260724

Published by Packt Publishing Ltd.
Grosvenor House
11 St Paul's Square
Birmingham
B3 1RB, UK.

ISBN 978-1-83763-068-4

www.packtpub.com

To my spouse, Ella, for her endless support and love. To my mother and sister for their understanding. To Alexander Hachmann for the time provided. To Christoph Kolb for the support. To Kai Klostermann for being my best friend and supporter. To the Supabase team for providing me with their trust, help, and warm words.

– David Lorenz

Foreword

Six years ago, I joined Mercedes-Benz.io, the company where David worked, and I had the privilege of collaborating with him for four years. It quickly became evident that he was a "beacon" developer—a professional others admired for his exceptional skills and leadership.

David's passion for knowledge is unparalleled. He eagerly explores the latest technological advancements, conducting thorough research on each topic to form well-founded conclusions. His dedication to understanding the intricacies of his craft sets him apart as a truly exceptional professional.

As lead developers for our respective teams on the same project, we faced unique challenges. Our project had specific constraints, chief among them a commitment to the old-fashioned approach of vanilla JavaScript.

"So what?", you might ask. And it's a fair question.

In an industry where cutting-edge frontend frameworks reign supreme, our approach was akin to building a house with hand tools while others used power equipment. We often had to solve complex problems from the ground up, without the shortcuts and conveniences modern toolsets provide. This wasn't just about writing code—it was about rediscovering and applying core principles that many developers had begun to take for granted.

Moreover, our role extended beyond coding. We became mentors and educators, guiding our colleagues through the intricacies of framework-free development. This meant teaching not just how to solve problems, but why certain solutions worked, fostering a deeper understanding of fundamental concepts.

It was in this challenging environment that David's exceptional qualities truly shone. From the early days of my tenure, his ability to form and share strong, well-researched opinions on a wide range of topics became apparent. David didn't just know the how; he understood the why, and he had a knack for explaining both clearly and concisely. His enthusiasm for sharing these insights is a core part of who he is—a trait you, dear reader, are about to experience firsthand through this book.

In our fast-paced technological landscape, where the "flavor of the month" changes constantly and documentation often lags behind, individuals such as David are invaluable. They serve as anchors amidst the turbulent sea of evolving technologies, helping others navigate and understand the ever-changing currents. David's ability to provide clarity and insight where it's needed most isn't just useful—it's essential for staying ahead in our field.

This book offers you a glimpse into the same level of insight and enthusiasm that David brought to our team every day, distilled into a format that's accessible and practical for developers at all levels. Regardless of your current expertise with Supabase, this book promises to be a valuable resource. David provides the following:

- Clear explanations covering essential aspects of Supabase

- Practical guidance on integrating Supabase into your projects

- Strategies for building performant and scalable applications

- Insights on how Supabase can compete with more expensive solutions

Throughout these pages, you'll find David's knowledge translated into actionable insights. Whether you're new to Supabase or an experienced developer, this book aims to enhance your understanding and application of Supabase, empowering you to create robust, efficient, and cost-effective solutions.

Jorge Varandas

System Architect, Mercedes-Benz.io

Contributors

About the author

David Lorenz is a web software architect and lecturer who began programming at age 11. Before completing university in 2014, he had built a CRM system that automated an entire company and worked with numerous agencies through his own company. In 2015, he secured his first employment as a senior web developer, where he played a pioneering role in using cutting-edge technology and was an early adopter of progressive web apps. In 2017, he became the leading frontend architect and team lead for one of the largest projects at Mercedes-Benz.io, involving massive-scale architecture. Today, David provides valuable insights and guidance to clients across various industries, using his extensive experience and exceptional problem-solving abilities.

I want to thank the people who have been close to me and supported me, especially Ella and the team at Wahnsinn Design GmbH, as well as my mom and my sister.

About the reviewers

Kamil Pyrkosz is a multiplatform programmer who fell in love with Supabase the first time he used it, surprised by how easy and powerful it was. His passion for Supabase resulted in the creation of a YouTube channel, *Kamil the Supabase enjoyer*, where he makes Supabase tutorials and videos on other related topics.

Thor (Thorsten) Schaeff is a software developer, startup advisor, and Angel investor. Having grown up around the SAP headquarters in Germany, he started building websites back in high school, later studied computer science and media, and interned with Google in London. He joined early Stripe in Dublin, building out various user-facing engineering teams across Europe and Southeast Asia, contributing to open source software, and mentoring and investing in early-stage startups along the way.

Now settled in sunny Singapore, Thor works on DevRel and DX at Supabase, helping developers take advantage of the power of Postgres.

We would also like to thank **Jorge Varandas**, **Kai Klostermann**, and **Kushal Seth** for their help in reviewing the book.

Table of Contents

Preface   xvii

Part 1: Creating the Foundations of the Ticket System App

1

Unveiling the Inner Workings of Supabase and Introducing the Book's Project   3

Technical requirements (and some preamble)	4	Realtime – elevating the user experience	21
Understanding why Supabase is the stack you want	5	Storage – simple and scalable object storage	21
Demystifying the inner workings of Supabase with Postgres	8	Image Proxy – helping to transform images on the fly	22
Access logic within a route	9	Edge Functions – completing the optimization stack	23
Access logic as a central service	10		
How Supabase handles access control	13	pg-meta – an internal helper service for the database	24
How the access system works under the hood	14		
Supabase Studio – the convenient web dashboard	17	Kong – the overarching service orchestrator	25
Supabase Auth (GoTrue) – the authentication handler	18	Introducing the production-grade ticket system project	25
PostgREST – a REST and GraphQL API for your database	19	Summary	28

2

Setting Up Supabase with Next.js   29

Technical requirements 30
Getting ready with Next.js 31
Installing the Supabase CLI 31
Running your first Supabase instance on your machine 32
Initializing a new local Supabase instance 32
Starting your first Supabase instance 32

Managing multiple local Supabase instances 34
Option 1 – the start-stop technique 35
Option 2 – change ports 35

Connecting to Supabase with the Supabase JavaScript client 35

Initializing and testing the base Supabase JavaScript client within Next.js 36
Understanding the base Supabase client 40
Using the Supabase client with Pages Router and App Router 40

Connecting directly to the database 54
Using Supabase with TypeScript 55
Connecting Supabase to other frameworks 57
Nuxt 3 57
Python 58

Summary 59

3

Creating the Ticket Management Pages, Layout, and Components   61

Technical requirements 62
Setting up Pico.css with Next.js 62
Building the login form 63
Visualizing the Ticket Management UI 68
Creating a shared UI layout with navigation elements 69
Designing the Ticket List page 72

Constructing the Ticket Details page 75
Adding the comments section to the ticket details 78

Implementing a page to create a new ticket 83
Implementing a user overview 84
Enhancing the navigation component 86
Summary 88

Part 2: Adding Multi-Tenancy and Learning RLS

4

Adding Authentication and Application Protection 91

Technical requirements	92	Authenticating with magic links	114
Adding authentication protection with Supabase	92	Sending magic links with signInWithOtp() on the frontend	114
Creating users	93	Why I usually don't use signInWithOtp()	119
Preparing the middleware for authentication	95	Understanding a server-only magic link flow	121
Implementing the login functionality in our app	99	Implementing a server-only magic link flow with custom email content	123
Protecting access to the Ticket Management system	101	Adding password recovery	130
Adding a log out button	103	Learning about the Site URL and redirect URLs	132
Logging out using the frontend	104	How to configure site and redirect URLs	133
Logging out using the backend	107	Optional knowledge: adapting built-in templates	134
Understanding server authentication	109	Summary	136
Enhancing the password login	110		

5

Crafting Multi-Tenancy through Database and App Design 137

Technical requirements	137	Committing your database state (if you don't seed it, you lose it)	155
What kind of multi-tenancy do we need?	138	Making our Next.js application tenant-aware	158
Designing the database for multi-tenancy	139	Enhancing the middleware to safeguard dynamic routes	159
Planning our database	139	Fixing all static routes in the application	161
Creating the tenants table	141	Making the login tenant-based	165
Designing the users table	146	Summary	168
Designing the permission structure	152		

6

Enforcing Tenant Permissions with RLS and Handling Tenant Domains   169

Technical requirements	170	Keeping custom claims in sync with the table data	202
Learning to work with RLS	170		
Fetching tenant data with the restrictive Supabase client	171	**Making the authentication process tenant-based**	**203**
Defining RLS policies to access tenants based on permissions	173	Preventing password login on a foreign tenant	204
Creating a permission-based RLS policy	178	Preventing the magic link login for foreign tenants	205
Understanding and solving RLS implications	185	Rejecting to visit invalid and forbidden tenant URLs when signed in	206
Shrinking RLS policies based on the implications	187		
Learning about RLS implications	189	**Matching a tenant per domain instead of per path**	**207**
Minimizing RLS complexity with custom claims	**190**	Adding custom domains via the hosts file	208
		Mapping domains in our application	209
Extending app_metadata with tenant permissions	192	Bringing back localhost with mapped domains	219
		Summary	**220**

7

Adding Tenant-Based Signups, including Google Login   221

Technical requirements	222	Handling account creation	232
Understanding the impact of disabling signups	**222**	Adding the service user and permission rows	234
Disabling signups generally	223	Sending the activation email	236
Disabling specific signup methods	223	Redirecting the user to a success page	240
Implementing the registration page	**224**	**Enabling OAuth/Sign-in with Google**	**242**
Processing the registration with a Route Handler	**228**	Obtaining Google OAuth credentials	242
Reading and validating the form data	228	Configuring our Supabase instance with the OAuth credentials	247
Rejecting registration	230		

Adding a "Sign in with Google" option
triggering the OAuth process   247

Solving the crypto/HTTPS security problem   249

Building a verification route
to finalize the registration   250

**Dealing with invalid
user registration   254**

Summary   255

Part 3: Managing Tickets and Interactions

8

Implementing Dynamic Ticket Management   259

Technical requirements   259

**Creating the tickets table
in the database**   260

Creating tickets and using triggers   266

Implementing the ticket creation logic   267

Using triggers to derive and set the user ID 273

Improving loading behavior after
adding a ticket   279

Enforcing checks on the database columns   279

Viewing the ticket details   281

Caching the author's name with a trigger   284

Improving the date and status view   287

Listing and filtering tickets   288

Enabling paging   290

Sorting tickets   295

Creating a ticket filter   297

Deleting tickets   303

Summary   306

9

Creating a User List with RPCs and Setting Ticket Assignees   307

Technical requirements   307

Adding a user list with an RPC   308

Ensuring there are enough users to test   308

Enhancing the table structure   308

Fetching the users with an RPC   309

Using the function with an RPC   316

**Allowing the setting and
editing of an assignee to a ticket**   317

Adding assignee columns in the tickets table 318

Creating the trigger function to cache the name 319

Adding an assignee at ticket creation   322

Showing the assignee in the details   324

Updating the assignee   326

Summary   329

10

Enhancing Interactivity with Realtime Comments   331

Technical requirements	332	Implementing Realtime comments	350
Creating the comments table	332	Enabling Realtime and subscribing to it	350
Adding a trigger to set the tenant automatically	335	Updating the UI with Realtime data	353
Adding and optimizing RLS policies	338	Triggering impersonated real-time updates with the Table Editor	354
Creating RLS helper functions	338	Embracing additional Realtime insights and learning about potential pitfalls	359
Creating the policies	342		
Implementing comment creation	344	Summary	361
Listing existing comments from the server	346		

11

Adding, Securing, and Serving File Uploads with Supabase Storage   363

Technical requirements	364	Uploading files to storage	377
Creating and understanding Storage buckets	364	Connecting uploaded files with the written comment	382
Examining public buckets	367	Showing the connected files	385
Exploring files within a bucket programmatically	369	Serving image attachments directly in the UI	390
Learning how a basic RLS policy can be added to your bucket	370	Using Image Transformations	390
Understanding private buckets and revising our bucket choice	372	Building a pseudo-CDN for private buckets	391
Choosing a private or a public bucket?	374	Using the pseudo-CDN inside our UI	393
Enabling the addition of comments with file attachments	376	Writing RLS policies directly on buckets and objects table	395
Preparing the UI with file upload possibility	376	Diving into advanced storage restrictions	396
		Summary	399

Part 4: Diving Deeper into Security and Advanced Features

12

Avoiding Unwanted Data Manipulation and Undisclosed Exposures

Technical requirements	404	Enabling column-level security/working with roles	417
Understanding PostgREST's OpenAPI Schema exposure	404	Understanding security on views and manually created tables	418
Preventing schema exposure	407		
Removing schemas from usage via API	408	Changing the max_rows configuration	419
Specifically exposing a schema to the API	409		
		Understanding safe-guarded API updates or deletion	419
Being careful with current_user usage and understanding auth.role()	409	Adding middleware inside Postgres for each API request	420
Generating new Anonymous Keys, Service Role Keys, and database passwords	412	Adding middleware for PostgREST	423
		Using the Security Advisor	426
Benefiting from Supabase Vault	413	Allowing a listing of IPs for database connections	427
Creating secrets in the Vault and reading them	414		
Using the secret in the business logic/within your application	415	Enforcing SSL on direct database connections	429
		Summary	429
Utilizing silent resets to avoid data manipulation	416		

13

Adding Supabase Superpowers and Reviewing Production Hardening Tips

Technical requirements	432	Grasping the importance of extra_search_path	436
Making sense of search_path	432		
Comprehending search path in Postgres	433	Familiarizing yourself with database extensions	437

Installing an extension in the default
extensions schema   437

Installing extensions in their own schema   439

Using the programmatic installation
of extensions versus using the UI   441

Adding an AI-based semantic ticket search   442

Deciding on an embeddings provider   443

Creating the embeddings column
in the table   444

Creating embeddings with OpenAI   444

Comparing embeddings to find
matching search results   446

Using anonymous sign-ins   449

Transforming external APIs into tables with foreign data wrappers   452

Using webhooks   457

Creating webhooks with dynamic URLs per
environment   460

Understanding Edge Functions   463

Understanding when to use Edge Functions   463

Creating an Edge Function that runs for new
rows   463

Triggering the Edge Function   465

Using cronjobs to notify about due tickets   471

Using pg_jsonschema for JSON data integrity   475

Testing the database with pgTAP   476

Setting the auth.storageKey to avoid migration problems   477

Extending supabase.ts with custom typings   478

Improving RLS and query performance   479

Identifying database performance problems and bloat   481

Working with complex table joins   483

Reviewing the underestimated benefit of using an external database client   485

Understanding migrations   485

Utilizing database branching   487

Disabling GraphQL or PostgREST (if you don't need it)   488

Using a dead-end built-in mailing setup   489

Retrieving table data with the REST API and cURL   490

Summary   491

Index   493

Other Books You May Enjoy   504

Preface

Technology and web development have always fascinated me. Over the past 23 years, I've navigated the evolving landscape of coding, starting from my self-taught beginnings at the age of 11. Back then, the internet presented different challenges, such as achieving a proper box model in Internet Explorer or using border-radius and transparent PNGs.

As a web app architect, I carefully choose technology stacks. While I stay at the cutting edge of web development and closely observe new hypes, I don't like to adopt them immediately. Instead, I evaluate them meticulously. My extensive JavaScript experience has given me a broad, well-informed perspective and taught me to trust my gut feeling, developed over years of diverse projects and challenges.

My journey with Supabase has been unique as it's one of the rare stacks I can use for nearly any project, no matter which additional stack surrounds it, thanks to its superpowered Postgres database with an "all batteries included" approach.

This book is a culmination of my experiences, insights, and countless hours spent exploring Supabase. My hope is that it will not only guide you through the technical aspects but also inspire you to see the potential and possibilities that Supabase offers. We are at an exciting juncture in web development, and together, we will explore how Supabase can transform your projects and reshape the way you think about web development.

Welcome to the world of Supabase!

Who this book is for

This book is perfect for developers looking for a hassle-free, universally applicable solution to build robust apps. By using the open source Supabase backend and its simple integration libraries, you can significantly accelerate your development process.

What this book covers

Chapter 1, Unveiling the Inner Workings of Supabase and Introducing the Book's Project, explains why Supabase is the stack you want for your next project, demystifies its inner workings, and explores its interconnected services. You will also be introduced to the Multi-Tenant Ticket System project that we will be working on throughout the book.

Chapter 2, Setting Up Supabase with Next.js, covers setting up and connecting to Supabase, running local instances, managing multiple instances, and integrating Supabase with your Next.js application, laying the groundwork for your ticket system project.

Chapter 3, *Creating the Ticket Management Pages, Layout, and Components*, explains how to build the foundational design of your ticket system project using mock data, creating solid-looking pages and components with Next.js only.

Chapter 4, *Adding Authentication and Application Protection*, integrates user authentication with Supabase, protecting your application, and you will also learn about sending customized authentication emails.

Chapter 5, *Crafting Multi-Tenancy through Database and App Design*, covers how to design your database and the application to support multi-tenancy, including defining permissions and making the application tenant aware.

Chapter 6, *Enforcing Tenant Permissions with RLS and Handling Tenant Domains*, focuses on implementing **row-level security** (**RLS**) to secure tenant-specific data, streamlining RLS with Custom Claims, and adapting your application to use domain-based tenant identification.

Chapter 7, *Adding Tenant-Based Signups, Including Google Login*, takes you through implementing tenant-based user registration, enabling OAuth sign-in with Google, and handling invalid user registrations to enhance the onboarding process.

Chapter 8, *Implementing Dynamic Ticket Management*, shows you how to create and manage tickets, implement ticket details and related data, and enhance the ticket list with paging, sorting, and searching features.

Chapter 9, *Creating a User List with RPCs and Setting Ticket Assignees*, explores how to create a user list using **remote procedure calls** (**RPCs**), add assignees to tickets, and implement UPDATE RLS policies for enhanced security.

Chapter 10, *Enhancing Interactivity with Realtime Comments*, guides you through creating a comments table, implementing real-time comment functionality, and optimizing RLS policies to enhance user interactivity and experience.

Chapter 11, *Adding, Securing, and Serving File Uploads with Supabase Storage*, explains how to implement file uploads within ticket comments, secure files with RLS policies, and serve images using Supabase Storage and Image Transformations

Chapter 12, *Avoiding Unwanted Data Manipulation and Undisclosed Exposures*, discusses advanced security techniques to protect your Supabase application, including managing roles, using Supabase Vault, implementing **column-level security** (**CLS**), and much more.

Chapter 13, *Adding Supabase Superpowers and Reviewing Production Hardening Tips*, unleashes the full potential of your Supabase application with powerful techniques. You'll discover how to integrate database extensions, optimize performance, implement AI-based features, and secure your workflows like never before.

To get the most out of this book

To get the most out of this book, you should have a solid understanding of at least one programming language and be familiar with JavaScript, as we will be building a Next.js app to explore Supabase's features. However, since the concepts of Supabase are framework independent, this book is also incredibly insightful for experienced developers from non-JavaScript backgrounds who want to enhance their development speed.

While familiarity with Postgres is also a plus, all SQL statements used in the book are thoroughly explained.

Software/hardware covered in the book	Operating system requirements
Docker Desktop or orbstack	Windows, macOS, or Linux
npm/node	
An account on supabase.com	
DBeaver (recommended, not necessary)	

If you are using the digital version of this book, we advise you to type the code yourself or access the code from the book's GitHub repository (a link is available in the next section). Doing so will help you avoid any potential errors related to the copying and pasting of code.

Also, take your time and read the book carefully - there's no rush!

Download the example code files

You can download the example code files for this book from GitHub at https://github.com/PacktPublishing/Building-Production-Grade-Web-Applications-with-Supabase. If there's an update to the code, it will be updated in the GitHub repository.

Conventions used

There are a number of text conventions used throughout this book.

Code in text: Indicates code words in text, database table names, folder names, filenames, file extensions, pathnames, dummy URLs, user input, and Twitter handles. Here is an example: "Change port to something unique, such as port=9100, and now, for this project, your Supabase Studio Service will run on localhost:9100."

A block of code is set as follows:

```
document = createNewTextDocument();
userId = getUserIdFromLoginSession();
getFgaClient()
    .subject({userId})
    .addPermissions({objectType: 'document', objectId: document.id})
    .relations(['save', 'delete', 'share'])
```

When we wish to draw your attention to a particular part of a code block, the relevant lines or items are set in bold:

```
CREATE POLICY only_own_rows_thor
  FOR SELECT TO thor ON documents
  USING (owner  = 42);
```

Any command-line input or output is written as follows:

```
npm install supabase --save-dev
```

Bold: Indicates a new term, an important word, or words that you see onscreen. For instance, words in menus or dialog boxes appear in **bold**. Here is an example: "Within your project, just click on **Authentication**, then **URL Configuration**."

> **Tips or important notes**
> Appear like this.

Get in touch

Feedback from our readers is always welcome.

General feedback: If you have questions about any aspect of this book, email us at customercare@packtpub.com and mention the book title in the subject of your message.

Errata: Although we have taken every care to ensure the accuracy of our content, mistakes do happen. If you have found a mistake in this book, we would be grateful if you would report this to us. Please visit www.packtpub.com/support/errata and fill in the form.

Piracy: If you come across any illegal copies of our works in any form on the internet, we would be grateful if you would provide us with the location address or website name. Please contact us at copyright@packtpub.com with a link to the material.

If you are interested in becoming an author: If there is a topic that you have expertise in and you are interested in either writing or contributing to a book, please visit authors.packtpub.com.

Hiring the author: If you'd like the author to consult you, it's best to reach out at https://activeno.de/.

Join Us on Discord

If you encounter any problems, need any help, or would just like to discuss the book with the author and other readers, you can join the book's Discord channel here: `supa.guide/discord`.

Share Your Thoughts

Once you've read *Building Production-Grade Web Applications with Supabase*, we'd love to hear your thoughts! Scan the QR code below to go straight to the Amazon review page for this book and share your feedback.

https://packt.link/r/1-837-63068-2

Your review is important to us and the tech community and will help us make sure we're delivering excellent quality content.

Download a free PDF copy of this book

Thanks for purchasing this book!

Do you like to read on the go but are unable to carry your print books everywhere?

Is your eBook purchase not compatible with the device of your choice?

Don't worry, now with every Packt book you get a DRM-free PDF version of that book at no cost.

Read anywhere, any place, on any device. Search, copy, and paste code from your favorite technical books directly into your application.

The perks don't stop there, you can get exclusive access to discounts, newsletters, and great free content in your inbox daily

Follow these simple steps to get the benefits:

1. Scan the QR code or visit the link below

https://packt.link/free-ebook/9781837630684

2. Submit your proof of purchase
3. That's it! We'll send your free PDF and other benefits to your email directly

Part 1:
Creating the Foundations of the Ticket System App

In the first part of this book, we will embark on the journey of building the foundations for your ticket management app. You'll explore the basics of Supabase, set up your development environment, and create a user-friendly interface. This part is all about laying the groundwork and getting you ready for more complex features ahead, so let's dive in and start building!

This part includes the following chapters:

- *Chapter 1, Unveiling the Inner Workings of Supabase and Introducing the Book's Project*
- *Chapter 2, Setting Up Supabase with Next.js*
- *Chapter 3, Creating the Ticket Management Pages, Layout, and Components*

1

Unveiling the Inner Workings of Supabase and Introducing the Book's Project

In 2000, I started playing around with HTML, which had been around for 10 years at the time. Google, which was then only two years old, was already becoming popular. Back then, the internet was slow and expensive, and my connection was a snail-paced 56 Kbps, miles behind today's lightning-fast 1 Gbps, which is 18,000 times faster.

However, it was a great time to start with web development as the field wasn't overly complex. Making dynamic websites typically meant using PHP and MySQL, and things such as "JavaScript Frameworks" did not even exist yet. At that time, authentication was simpler but not very secure, and HTTPS (which is represented by the padlock icon in your browser) was rare. To log in, you had to enter a username and password on a website, and the server created a session cookie to connect you to your account on the backend.

Fast forward to today, when starting out with web development can be confusing. There are tons of libraries, frameworks, and tech choices to make. However, as most of them are some kind of interactive service, there are underlying needs such as file storage, databases, and authentication that are always the same for each project. That's where **Platform as a Service/Backend as a Service (PaaS/BaaS)** comes in. Supabase is a standout option for this. Choosing Supabase will save you lots of time, as I can personally attest—usually multiple weeks or even months of work. Even if all you need is a database to store data in, PaaS solutions such as Supabase will allow you to iteratively add the capabilities of the PaaS solution to your application.

Embarking on a new web project is like setting out on a thrilling adventure. In this chapter, I'll guide you through the gates to Supabase, your gateway to a seamless, powerful, and scalable web development experience. Here, I'll equip you with knowledge of, and insight into, each layer of Supabase. This will allow you to confidently choose Supabase as the foundation for your next web application. By unraveling the intricacies of the Supabase stack and delving into its internal workings, you'll not only grasp what sets it apart but also gain a deeper knowledge for the book project that lies ahead. On top of that, from a software architectural standpoint, it will allow you to make better decisions since you will have a deeper understanding of the stack.

So, in this chapter, we will first cover the following topics:

- Understanding why Supabase is the stack that you want
- Demystifying the inner workings of Supabase with Postgres

After that, and after having seen a systems diagram of how Supabase is built, you'll dive into learning more about its interconnected services:

- Supabase Studio – the convenient web dashboard
- Supabase Auth (GoTrue) – the authentication handler
- PostgREST – A REST and GraphQL API for the database
- Realtime – elevating the user experience
- Storage – simple and scalable object storage
- Image Proxy – helping to transform images on the fly
- Edge Functions– completing the optimization stack
- pg-meta – an internal helper service for the database
- Kong – the overarching service orchestrator

In the final section of this chapter, you'll then get an overview of the project that you are going to create.

By the end of this chapter, you won't just know how Supabase works; you'll understand why it's the stack that you'll want to rely on!

Technical requirements (and some preamble)

This chapter contains code samples. However, you don't need to remember them, as they simply serve the purpose of helping you understand the context of what's explained better. Just expose yourself to their meaning and let go of trying to map it to a real implementation.

In the upcoming chapters, the code *will* be real code samples, which can be found, sorted by branches, at `https://github.com/PacktPublishing/Building-Production-Grade-Web-Applications-with-Supabase`.

Now for the preamble. In this chapter, we'll dive into why Supabase is the ideal stack for building exceptional applications and explore its internal architecture.

Don't be intimidated by the length of this chapter—Supabase works seamlessly without you needing to remember all the details of this chapter. Approach this chapter like you would a novel: as something to read, enjoy, and have *aha*-moments with, rather than something that you need to implement immediately. This chapter is more theory-focused, while the next ones will be hands-on and practical.

It might be tempting to skip this theoretical section, but doing so means missing out on very essential background knowledge. This foundation is crucial for fully grasping why things work the way they do. Stick with it—it's worth your time!

Understanding why Supabase is the stack you want

Talking about an awesome product is one thing. Understanding its history of success is equally important, as this paves the way for the future that you will invest your time and effort into. Also, that same history is part of the reason why it resonates with the principles of many developers, as well as why it isn't just yet another **Software as a Service (SaaS)** with a short life span but one of the most promising tools that the web has seen in recent years.

In 2019, Paul Copplestone, the CEO of Supabase, worked on a Realtime extension for Postgres. This was part of a migration, in his former company, to move away from Google's Firebase to Postgres. As this gained popularity, he asked his friend Ant Wilson if he'd be open to building an open-source company with the Postgres database at its core.

Ant agreed, and they came up with the name Supabase, choosing this due to the planned "super database" nature of the tool (and with the potential for many memes based on the Nicki Minaj song *Super Bass*).

The goal of the company? I think that this quote from Supabase will give you a feeling about their underlying goals at the time:

> *At Supabase, we're building some amazing tools that make Postgres as easy to use*
> *as Firebase [...]. Why are database interfaces so hard to use? The Supabase team*
> *has built products for 70-year-olds, so we're confident we can make something*
> *easier for developers. (Supabase, June 2020)*

They were trying to establish an easy-to-use open-source Google Firebase alternative, going beyond just cloning it and coming up with their own innovative ideas.

They started actively fiddling with that idea in January 2020. The first video I know of where Paul shows a first promising Alpha release (May 2020) was published less than three months after they had started building it and can be found here: `https://www.youtube.com/watch?v=ck5MM_PD4Co`

At the time of writing this book, it has less than 800 views—apparently, not many people knew about Supabase in the early 2020s.

The Alpha release was a wonderfully preconfigured, hosted Postgres database with a dashboard. Not just that; it was also an instant API for your database, allowing for easy access with the Supabase npm package. Unsurprisingly, it also had Realtime support.

Only about 6 weeks later, they had overhauled the existing dashboard with a fresh UI and announced that they got into the **Y Combinator Accelerator Program** (`https://www.ycombinator.com/about`), which is one of the most famous startup founding accelerators, having boosted companies such as Heroku, AirBnB, and Dropbox. At that time, they also announced that the maintainer of PostgREST, Steve Chavez, had joined the Supabase team. Just a few weeks after that, he had helped in adding the authentication feature in Supabase.

The speed of development with the given quality shows the experience of those people. Back then, they were primarily active on Hacker News and Twitter (now X) to push this idea. Twitter was the place where I got hooked the first time. Fast forward to about one year later. Not only did a lot of developers realize that this is an awesome and fast-growing platform to build on but investors were also impressed, leading to Supabase gaining 30 million dollars of funding. Supabase, being completely open source, has 116 million dollars of funding in total at the time of writing this book. This is quite unusual for a non-closed source platform.

However, let's dive deeper into Supabase's capabilities today. I've led a lot of projects in my career of basically every scale, and all of them have ever-ongoing questions such as "Do we have the right tech stack for our needs?" in common. In many projects, developers recreate the same stacks repeatedly tailored to the *supposedly* specific needs of their software. However, at the end of the day, most software has more or less the same common day-one needs:

- Storing data in a database
- User authentication
- Saving files and serving them
- Permissions for handling data and files
- Running scheduled processes
- Connecting with APIs such as Stripe
- Logging data from services to be able to debug potential problems
- A web-based management UI

> **Note**
> Day-one needs are needs that projects will have from the start. Nowadays, Supabase also puts a lot of work into "day 100 needs" such as scalability, performance, observability, integrations, and so on.

Rethinking and recreating solutions for these needs for each product every time costs a lot of time. Often this recreation process is justified by saying that every product has specific needs. That is true. However, at the same time, it doesn't stay true when it comes to the aforementioned foundation of your application.

Platforms such as Supabase, which solve these common needs with low complexity, overcome a big part of developing a product's software architecture from scratch. So, you can therefore go from asking the question of "Where do I start?" to "Is there something I don't see solved with the BaaS and what do we need to add or change to achieve it?" The latter question not only costs less time but also allows all developers to come back to well-known standards (instead of having to learn new conventions and APIs all the time), helping them focus on their product.

Now some might argue as follows: "I can also build my own custom architectural setup and reuse that for my projects all the time. Then, I don't have the added costs for the base setup—which is the same as if I'd be using a platform such as Supabase." This is, sorry to disappoint, not true:

- You must maintain everything on your own
- Once you start the actual development of your application, your custom solution will already be outdated
- You won't be as fast at adding new features to your custom platform
- As your own setup is not public nor well-known, there's a bigger cost of developer onboarding and bigger costs of documentation

All of these points obviously mean more time consumption and more costs with a custom solution.

From my own experience, Supabase has significantly increased development speed, changing the way I work. An essential reason to choose Supabase is its open-source nature. Unlike closed-source alternatives such as Google Firebase, Supabase allows you to actively contribute ideas, code, or bug reports. This ensures that you're not locked in. The community's strong bond with Supabase further enhances its appeal.

Now, let's have a look at what Supabase consists of architecturally. The following diagram shows a visual representation of a Supabase instance. As a whole, the diagram represents the network that the Supabase services live in, namely Kong, GoTrue, PostgREST, Realtime, Storage, ImageProxy, Edge Functions, pg-meta, Studio, and the Postgres database.

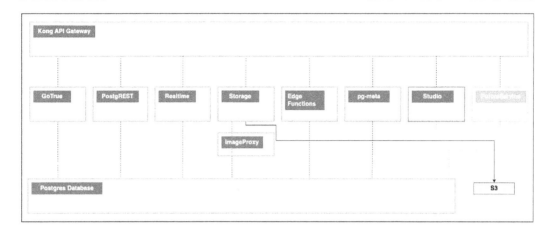

Figure 1.1: A visualization of a Supabase instance setup

We will now go through these well-separated services one by one to understand their function within Supabase. We will start with the Postgres database, as this is the common denominator and central point of it all.

> **Note**
>
> Before moving on I would like to clarify the term **services**. Whenever I talk about Supabase services in this book, I am referring to isolated **containers**, as this is what they are. Each Supabase service lives in its own container. The database is one service, living in its own Docker container. Realtime is another service living in its own container, and so on, but all those services live together in a shared network.

Demystifying the inner workings of Supabase with Postgres

Postgres is, and always has been, one of the go-to databases. It has proven to be a reliable and fast partner for projects of any size. Hence, it's no surprise that Supabase is not only using one but in fact *is* a Postgres database with a specific setup and specific extensions, enhanced with additional services.

Everything related to Supabase (excluding files) is stored within that exact Postgres database. That's why we will now jump into a mix of theoretical and practical thoughts about what differentiates Supabase, at its core, from just being a regular database.

One very interesting fact is that the database within Supabase can be controlled by you and is extensible. Often, when you work with any kind of platform that provides you with a database, your access will be limited in regard to safeguarding you from making fatal changes. Supabase, however, gives you full access to its underlying Postgres database.

> **Note**
>
> Supabase has restrictions in place to try to prevent you from destroying your own infrastructure. This means there are certain structures and users that you will not be able to delete or modify in a way that would allow you to do so. However, the `postgres` database user, which you have control of with Supabase, does have admin rights. This means that it has access to all resources and can also modify all resources to the extent of not destroying crucial systems.

The database can be extended with additional capabilities, which can be installed with the internal open-source package manager (`https://database.dev/`) or the click of a button (for example, extensions for testing the database, making HTTP requests, or executing cronjobs). It is worth noting that those extensions are just regular, but extremely powerful Postgres extensions—deliberately. However, that's just one huge thing of many.

One of the most exciting and clever concepts behind Supabase is that it gives you the possibility to give your users access to specific data in your database. For instance, you might grant access to a table but not to all of its rows.

However, what does that mean? Isn't that the case for every web app with authentication anyway? Don't they all integrate fine-grained access with their databases? Not really. Nowadays, most web apps use routes or a central service and combine actual backend logic to achieve access control. The database itself usually does not constrain that.

Enough talk though; let's take a closer look.

Access logic within a route

First, we want to have a look at when access logic is implemented as part of a specific route of your app.

Imagine that you have a text editor app called `text.myapp`, wherein users can write documents and share those documents to be edited by others. Now, imagine a user who wants to save a document with the `456` ID. This is done with a request to `text.myapp/document/456/save`. As an example, the following backend code logic shows how the app could make sure that only the owner or people with shared access can save the document:

```
// route = /document/[id]/save
if (isLoggedIn()) {
  canSave = false;
  userId = getUserIdFromLoginSession();
  db = getDbConnection();
  document = db
              .select('author').from('documents')
              .where({id: documentId});
  canSave = isEqual(document.author, userId);
```

```
   if (!canSave) {
    // check if shared, then also eligible to save
    canSave = db.exists('shared_docs')
      .whereEquals({ documentId, userId})
   }

   if (canSave) db.update(content).whereEquals({ documentId });
 }
```

So, when data (`content` in the code sample) is sent to the route for saving a specific document with a specific ID, the first check works out whether the user is logged in, while the second and third checks work out whether the user is either the owner of the document or if someone has shared this document with that user. In both cases, that user is allowed to update the document in the database.

These logic snippets can make the code unreadable pretty quickly. That's why abstraction into functions such as `canEditDocument(documentId)` is a must so that you only have something like this in the end:

```
 // route = /document/[id]/save
 if (isLoggedIn()) {
   canSave = canEditDocument(documentId);
   if (canSave) db.update(content).whereEquals({ documentId });
 }
```

With this, all of the checks would be inside the logic of `canEditDocument`. This makes it cleaner but definitely not less work.

We'll get back to this specific example in a moment, when we look at how Supabase can solve this specific sample. First, however, we'll have a look at how permissions can be solved with a central permission system.

Access logic as a central service

Controlling who can do what with an *explicit* central permission system is usually found in more complex apps. I'm talking about so-called **Role / Relationship Based Access Control** (**RBAC**) or **Fine Grained Authorization** (**FGA**) systems. Though they have different names, they are essentially the same, being gatekeepers to set and check permissions explicitly and declaratively.

In a more visual sense, let's imagine access control via FGA in a school setting. The school is fully digital and uses NFC cards for people to get access to certain rooms. Teachers are allowed to go to the teachers' room and the students' room, while students are only allowed to go to the students' room. However, some students are allowed to go to the teachers' room on demand (temporarily). You can see this example illustrated in the following figure:

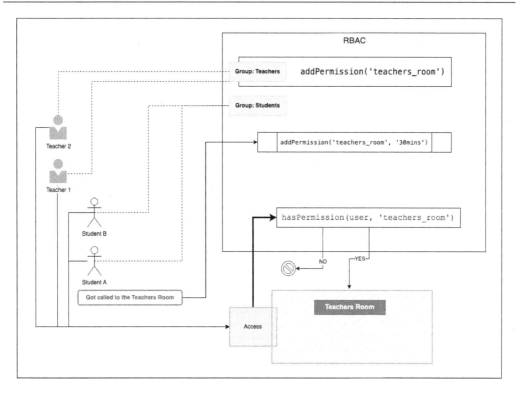

Figure 1.2: A visualization of an RBAC / FGA system accessing a teachers' room in a school

As every teacher is added to the Teachers group and the Teachers group has the addPermission('teachers_room') and addPermission('students_room') permissions, this means that all teachers have permission to use both rooms.

What's the exact difference between this and the text editor example from before? Well, the previous text editor example had implied permissions: the author, who created the document, had access no matter what. If the author wanted to give access to other people, it was necessary to also create a table where shared users could be mapped to a document and then check whether an entry exists.

That's different from the teachers' room example. Imagine that there's a person who created this room in the access system. That doesn't make that person a teacher, nor does it give them access. So, the *author* of the room does not have access. Also, just because you have access to that room doesn't say anything about whether you are a teacher or not. The only way to get access to the teachers' room is to have the teachers_room permission.

It's clear that the principal should also have access to the teacher's room, and in systems with implicit rights, there would probably be something such as if (isPrincipal()) THEN canAccessTeachersRoom(). With FGA, however, if the principal needs to have access, the principal needs to be given the appropriate permission, for example, by being added to the teachers' group or creating a school management group with that permission.

> **Note**
>
> The principle of managing rights explicitly, without making assumptions based on mental models, isn't new. However, there are new tools that are tailored to solve that for applications with the complexity of today. It doesn't come as a surprise that Google published a paper about it in 2019 (`https://research.google/pubs/pub48190/`) that quite a few of the existing access control services build upon, such as OpenFGA (`https://openfga.dev`) or Warrant (`https://warrant.dev/`).

However, what if we have many resources to manage, such as the items in the teachers' room (the microwave, the coffee machine, you name it)? Giving every single teacher access to each item individually would be quite cumbersome. Instead, you could give the teachers' room itself access to the resources. Then, the teachers are only assigned to the room, which implies that they will have access to its items. Pretty straightforward, right?

Be aware though; with that modeling, access to the teachers' room would mean access to its items. Hence, if a student was given temporary access to that room, they would also have temporary access to the items. So, it makes sense to split this up into two groups: **room access** and **room items access**. With this separation, you can give students access to the room without access to the items. Here's a visualization:

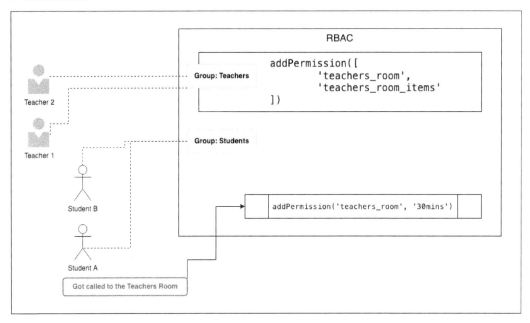

Figure 1.3: Splitting up the roles of room and room items with RBAC

Now, let's try to map the FGA solution to our previous text editor example so you can get a handle on the major difference.

When you create a new text document, the only reason you would have access to your own text document with an FGA-based permission system is because the code of the application has given you the **rights**, often called **relations**, explicitly. You can see an FGA sample here, where an author of a new document is given all rights (including saving, deleting, and sharing the document):

```
document = createNewTextDocument();
userId = getUserIdFromLoginSession();
getFgaClient()
  .subject({userId})
  .addPermissions({objectType: 'document', objectId: document.id})
  .relations(['save', 'delete', 'share'])
```

Now, there's no need to imply anything based on any kind of user role, such as being an author. Instead, we can ask the FGA system whether the current user has the right to save the document, like so:

```
userId = getUserIdFromLoginSession();
db = getDbConnection();
canSave = getFgaClient().check({
    object: {
     objectType: 'document',
     objectId: documentId
    },
    relation: 'save',
    subject: { userId }
  });
if (canSave) db.update(content).where({ documentId });
```

Here, we are not executing specific database comparisons; we are executing an explicit **permission check**. This allows us to move away from thinking about which commands to execute or re-rechecking mental access models, and instead moves us toward just asking whether permission is granted. The big benefit here is that you could also temporarily revoke rights from an owner without removing them as an owner (e.g., because of an audit process).

Now you understand why an FGA will allow you to think less and reduce complexity instead of trying to map mental models with the approach of "the person is an author, so the person must have access." Next, I want to show you that Supabase falls in a mix of both categories leading to massively reduced complexity within your application code.

How Supabase handles access control

Supabase does not use a typical generalized FGA System. It has built access control into the database. The access check does not need to happen in your application code; it happens under the hood, but how?

To answer this question, I'd like to show you the most minimal—but still functional—way of saving a text document using JavaScript, but only if the user has the proper rights:

```
const supabaseClient = createSupabaseClient();
supabaseClient.from('documents')
  .update({ content }).match({ documentId });
```

You might be thinking that some code is missing, but it isn't. Before continuing to read the explanation, I'd like to encourage you to think about it yourself first, visualizing in which way this could be possible, even though we are not checking anything.

So, the given code can run on the frontend or backend and will be equally safe, but how so? How can a database connection in the frontend even be safe at all? Where are the checks that verify whether the user is allowed to execute the action?

The reason why this works can be explained in multiple steps:

1. When you use the Supabase client, it will know whether you are authenticated.
2. The `.from(..).update(..)` code will send a request to the Database API, more specifically, in this case, to the `documents` table.
3. Inside the Postgres database are helper functions, which parse the authentication data from the Auth Service and know exactly which user is currently logged in (or whether the user is logged in at all).
4. By having permission policies defined in the database, the database will only allow access to the data access when the current user is authorized to do so, and will simply reject providing the data if not.

It's similar to RBAC, but here, the access service isn't an additional service. Instead, it is part of the database and the control mechanism also sits in the database, at the lowest level of data.

This is indeed innovative and a game changer because once your access control mechanisms are set inside of your database, you can use the database client—both on the frontend and backend—without having to think about permissions when executing commands (as given in the sample).

Now that you know how it works on the surface, let's look at it in more of a technical way.

How the access system works under the hood

Databases allow you to create multiple users and grant them specific rights to specific tables. So, in theory, you could create an actual database user for each user in your application. If your app had 20,000 users, you would have 20,000 database users (though there are pitfalls to this method, which you can see at `https://dba.stackexchange.com/questions/89275/feasible-to-have-thousands-of-users-in-postgres`).

Now Postgres allows you to set permissions for each user in the database. For example, if you wanted to give a database user, `thor`, read rights on the `documents` table, you'd do this:

```
GRANT SELECT ON TABLE documents TO thor;
```

However, this doesn't solve fine-grained access because `thor` could read all documents in that table. Also, with those read rights, `thor` could not create documents.

This is where **Row Level Security** (**RLS**) comes in, adding more fine-grained control. In Postgres, one can enable the RLS feature on a table; if it is enabled, all access to its data is denied by default.

In our case, `thor` could access the table but would always get no data returned from it. For them to have access to the table data, we would need to create an RLS policy. Assuming that `thor` has the `42` user ID, the following sample would give `thor` access to just their own rows in that table:

```
CREATE POLICY only_own_rows_thor
  FOR SELECT TO thor ON documents
  USING (owner = 42);
```

This policy-driven approach is what Supabase uses. We will see it in hands-on action—including all of its very interesting implications—starting in *Chapter 6*.

As indicated, having a database user for each user of your application isn't recommended. Supabase also bypasses this with a simple trick: it connects with a generic database user. You will create policies based on the built-in `auth` helper functions, which are connected to the authentication service. What might sound complicated at first becomes understandable with the following code snippet of a policy:

```
CREATE POLICY only_own_rows
  FOR SELECT TO authenticated ON documents
  USING (owner = auth.uid());
```

The authenticated role is a shared database user, which Supabase uses for all logged-in users. Here, `auth.uid()` will always return the specific ID of the **authentication service user** (e.g., you) that is currently logged in. Hence, this solution works universally.

> **Reminder**
>
> The code samples are here to improve the explanations. There is no need to remember them or stress yourself over not understanding what they mean in action—this will be taught throughout the book.

Let's strip down what happens, step-by-step, when a user within your web application tries to access data from the database with Supabase:

1. The user requests data from the database by sending an HTTP request to the Supabase API. The request, within Supabase, goes to a PostgREST Supabase service.

2. The service checks and processes the user's authentication.

3. The service tells the database which user is logged in (or if no user is logged in) and what data is requested.

The database uses this information to return data (or not), depending on its policies. The data is returned to the service and the service returns it to the user.

You can see a visual illustration of this in the following figure:

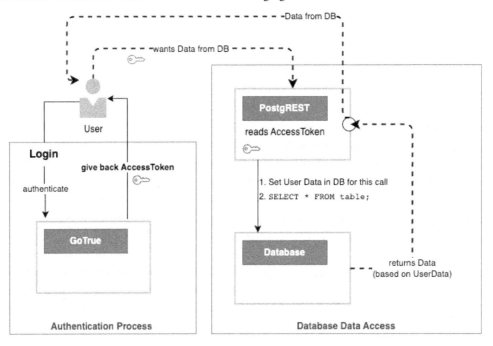

Figure 1.4: A user fetching data from a database

Even though RLS is a Postgres feature, Supabase has found the sweet spot, connecting it directly to user management (authentication service) that doesn't require actual database users. This very clear, though unique integration allows you to add or change data from your users in your database carefree.

Although this authentication system is probably one of my favorite things about Supabase, you'll see in the upcoming sections how Supabase integrates the other services and makes its Postgres database the integral part that connects everything. It essentially becomes more than just a database: it's a communication system and the cockpit of the services.

With this understanding of the importance of access management within Supabase, we'll move on to the explanation of the single Supabase services now.

Supabase Studio – the convenient web dashboard

At the heart of the Supabase stack, you'll find Supabase Studio, an intuitive dashboard designed to streamline the management of your Supabase instance. However, it's not a necessity—if you're a command line person and love to hack SQL commands and HTTP requests much more than using a convenient dashboard, you'd be free to ignore the dashboard. Hence, of all the Supabase services, Studio is the only one you could completely delete without negative impacts on your Supabase infrastructure as it isn't required by the other services.

Studio excels at simplifying complex tasks, from creating and managing table data to handling users and credentials, and even executing raw SQL commands for those who prefer that level of control. In *Figure 1.5*, you can catch a glimpse of the Studio start page for one of my projects, showcasing a quick statistical overview of its usage:

Figure 1.5: Supabase Studio

> **Note**
>
> In this fast-paced world, UIs are always bound to change quickly, so I will refrain from providing too many static explanatory screenshots of the dashboard in this book.

Throughout this book, Studio will be our go-to tool for creating our database structure, handling user management and permissions, and much more. By leveraging Studio, you'll not only increase your efficiency but also gain the capability to navigate through the complexities of Supabase with ease.

Enumerating every capability of Studio is a challenge, especially as Supabase consistently enriches its features, expanding the list further. Nonetheless, let's offer a brief look at some essential tasks that Studio effectively manages on your behalf:

- Managing database tables, including creating tables, inserting and updating data, and exporting data as `.csv`
- Managing the database beyond just tables, including creating access policies (RLS), adding custom functions, activating extensions, adding triggers, downloading or activating backups, and so on
- Executing SQL queries on the underlying Postgres database as the admin user
- Managing authentication settings, including login providers (email, phone, Google Login, Apple Login, GitHub Login, SSO, and so on)
- Managing file storage, including defining access policies (RLS), applying upload limits, browsing and manipulating existing files, and so on
- Creating reports, including getting usage and performance insights about your Supabase instance and building custom report dashboards for it inside of Studio
- Seeing all Supabase-related logs and building custom log queries
- Viewing general project settings, including API information or connection data
- Restarting your instance and changing the credentials

In essence, Supabase Studio is your partner for managing just about anything within your instance, as making Supabase usage as convenient as possible is one of the primary goals of the Supabase Team.

Next, we will have a look at the Supabase Auth service that is responsible for managing user authentication and how it differentiates its task from the access control happening in the database.

Supabase Auth (GoTrue) – the authentication handler

Supabase Auth, formerly known as **GoTrue**, is an auth API server written in Go. It's what I refer to as the *authentication service*. It was originally created by Netlify. However, Supabase forked it and now builds upon it (`https://github.com/supabase/auth`).

In the next chapter, you'll be setting up your first Supabase JavaScript client, which has a lot of convenient methods for the features of your Supabase instance. When you call the convenient `supabaseClient.signUp(..)` method, it requests the Auth server at its `/auth/signup` endpoint. Even though you can (and should) always use convenient functions such as `signUp(..)`, it is interesting to know that *everything* in Supabase is based on REST APIs and hence you are not dependent on any specific framework.

Auth itself does not care about *permissions*, it cares only about authentication, such as sign-ups, login, recovery, profile changes, and so on. The job of giving or denying access to a user is done inside of the database, as we've learned already. This access control that happens in the database is *based* on the authentication information that Auth communicates to the database.

Just to make it clear again: Auth tells the database, "Here, this is David" and the database knows whether David should have access to certain data. It provides endpoints that revolve around the possibility of users to sign in and manage the authentication sessions, for example, by verifying active logins and forwarding the information to the other services within the Supabase stack.

In the next section, you'll get to know the service that builds an automatic API on top of your database to allow fetching data from it.

PostgREST – a REST and GraphQL API for your database

The driver of any web application is selecting, updating, and inserting data from or into the Database. With Supabase, these actions are internal requests to the PostgREST service, which is reachable at the `/rest/` endpoint. That service is the PostgREST project, which essentially makes any Postgres database accessible via the REST API (`https://postgrest.org/en/stable/`).

However, how does it work? Relational databases consist of tables, just like an Excel spreadsheet. In one Excel file, there can be multiple tabs, each containing a table of columns and rows. Hence, the Excel file itself is like a container for different tables.

In relational databases, this container is called a **schema**. By default, any table within a Postgres database is created in the default schema, which is named `public`. A simple visualization of a `TodoItem` table with an `Upload` table resembling a `Todo` database is given in *Figure 1.6*:

public
TodoItem
id
title
owner
Uploads
todo_item_id
upload_path
uploaded_by

Figure 1.6: A public schema with two tables

Now, PostgREST can be told to inspect such a schema and create an easy-to-use abstraction layer in the form of an automatic REST API on top of it. This leads to the tables and their data being effectively accessible via REST API. As the default schema in Postgres is `public`, Supabase also uses `public` by default.

Then, PostgREST will expose endpoints that allow access to the database for that specific `public` schema. For example, let's say that you want to insert a `todo` item in a `TodoItem` table via command line with a raw PostgREST request:

```
curl "http://POSTGREST-URL/TodoItem" \
  -X POST -H "Content-Type: application/json" \
  -d '{ "title": "Learn Supabase" }'
```

In this code sample, we tell PostgREST that we want to execute something on the `TodoItem` table. By specifying a `POST` request with the JSON data, we can ensure that it treats that JSON data as a row to insert into that table.

> **Reminder**
>
> It's unlikely that you will touch these API endpoints directly. It's important to see that they exist but the recommended way is using a Supabase client, which we will use all the time in this book.

PostgREST within Supabase respects authentication, as seen in *Figure 1.4*. Hence, it doesn't matter whether PostgREST exposes the possibility of publicly manipulating any table. If you have a policy in the database that forbids adding a new `todo` item for the current, potentially unauthenticated user, then the request will fail at the database level, returning an error. Clever!

> **Note**
>
> Even though the Supabase client is very convenient, some people want to use GraphQL instead. For this reason, Supabase has also developed the open source `pg_graphql` GraphQL Postgres extension. The result is that GraphQL queries are supported at the database level (`https://supabase.com/blog/graphql-now-available`) and hence, Supabase exposes this specific endpoint for convenience at `/graphql`. Each sample in this book where we work with database data can hence be easily converted to a GraphQL request if you want that. As it doesn't add additional value to the explanation of Supabase, we will stick with the client-based samples though.

With the knowledge that PostgREST will always be the API responsible for interacting with data from the database, we will continue learning how the Realtime service adds value to the stack of Supabase services.

Realtime – elevating the user experience

Nearly any sophisticated application will come to a point where it makes sense to deliver live updates to a user without delay, whether it be gaming, a collaborative whiteboard, or simply new comments in your favorite social platform such as Signal. The **Supabase Realtime Server** solves exactly that.

It is an ultra-fast, open-source WebSocket communication implementation connecting to Postgres databases, written in Elixir Lang (`https://elixir-lang.org/`) by the CEO of Supabase, Paul Copplestone. It's worth mentioning that nowadays, it's driven by community contributors. It can be used for arbitrary real-time communication purposes and is able to handle millions of connections (`https://supabase.com/docs/guides/realtime/architecture`).

Even more interesting is the fact that it listens for changes within the database, so you can actually push live changes happening in the database to the user without any noticeable delay.

Of course, the same rules as for the other services apply. Realtime respects your authentication (GoTrue), and hence your access rights (RLS). One can only listen to changes in the database if one has the appropriate policies.

Now that you know about Realtime, imagine that you want to create a file hoster application such as Dropbox, Google Drive, or OneDrive and send users live notifications about changes to files. Sure, that's possible with Supabase, but you need a place for those files as well.

Storage – simple and scalable object storage

Supabase Storage is the service that allows you to store and manage files while respecting authentication. However, the service in itself isn't actually storing the files, it's only *managing* them.

As we learned, the PostgREST service isn't a database but only connects to it and provides an API on top of it to access the data from the Postgres database. The Storage service in the Supabase stack provides an API to manage the files which can be in arbitrary places or other servers. Let's get a deeper understanding of it.

A very common way of storing files is the **Simple Storage Service (S3)**. S3 is an Amazon-specific development that allows for failsafe, distributed, replicated object storage—in simple terms, it's a well-architected database for files.

Although Supabase uses Amazon's S3, nowadays, when talking about S3, one doesn't necessarily mean *Amazon S3*. There are alternative implementations that use the same API, so S3 can also just mean an S3-compatible object storage with the same API.

So, this Supabase Storage service will internally connect to the S3 and externally provide an API. This is conveniently used by the Supabase client methods to manage files within the S3 and saves the metadata of those (filename, ID, size, etc.) in the Supabase Postgres database.

Again, I want to emphasize that this Storage service, much like all the other services exposing REST APIs, communicates the authentication data from Auth to the database when accessing a file. Hence, the permissions, using RLS, are respected and the files are protected. This again confirms that the database is the central communication hub of Supabase. I recommend quickly having another look at *Figure 1.1* right now, as this can deepen the understanding of this connection.

With the knowledge about the Storage service, we'll now jump ahead to its partner in crime in helping to deliver transformed images: the Image Proxy.

Image Proxy – helping to transform images on the fly

The **Image Proxy service** is a standalone, but complementary, service to the Storage service. It is used internally by the aforementioned storage service to pre-render images; for example, if you only need a thumbnail of a saved image and not the full resolution, you can use the Image Proxy service.

In the following figure, you'll see a flow diagram illustrating what happens when you request a resized image from Supabase:

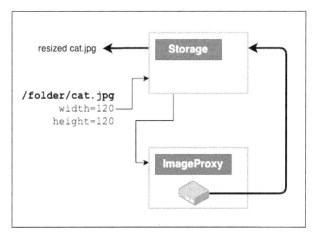

Figure 1.7: The service flow when requesting a resized image

In the figure, you requested a file called cat.jpg, which we assume is high-resolution, and you want it to be downscaled to a size of 120 pixels. In that case, the Storage service will download the file from the S3 and pass it on to the ImageProxy service, which returns it to the Storage service after the transformation. The Storage service will then pass the result on to the one who requested it.

With the JS Client of Supabase, such a request looks like this:

```
supabase.storage.from('bucket').download('/folder/cat.jpg', {
  transform: { width: 120, height: 120 }
})
```

In the next section, I'll explain the Edge Functions service.

Edge Functions – completing the optimization stack

Edge Functions are small and lightweight pieces of code logic that are separate from your application code. Their name comes from the fact that they best perform on the edge (`https://deno.com/blog/the-future-of-web-is-on-the-edge`). **Edge** is a rather new term and refers to a paradigm of delivering a fast result by being as close to the executor as possible. That usually means that the piece of code that is supposed to execute is distributed to different server locations in the world. Depending on where it's called from, the closest server takes the task.

Supabase's Edge Functions service allows you to add custom edge functions that you can then execute whenever you need them. However, that leaves us with a question: when do you use them, and if they are that fast, why isn't everything on the edge?

Edge functions can be considered to be *background workers*. They need to be small and have a limited processing time, execution time, and memory usage, which means that their field of use is constrained from the very beginning. Also, you cannot return HTML with a Supabase Edge Function, as `text/html` mime types are rewritten to `text/plain`, which would only show HTML source code in a browser.

> **Note**
>
> The mentioned limitation that you cannot serve actual web pages is due to the fact that Edge Functions in Supabase are meant to be logic-driven background workers, not page or app servers. So, this constraint is intentional, as your actual application is best hosted on platforms such as Netlify, render.com, Railway, Vercel, you name it.

So, from that perspective, we can derive the following: we cannot use it to host any kind of frontend activities and we cannot use it to do tasks that take long or where the CPU usage is high. This sounds pretty useless at first, but ultimately, this is part of the enormous optimizations that can only be achieved through certain limitations.

Now, let's say that for instance, you want to do regular, scheduled database cleanups of users that weren't active in your application in the last two years. That's a simple database execution task, but it's not necessarily a task that should be part of your application logic, as it isn't really application logic but rather housekeeping logic. This, being a small and isolated task, is a perfect task to hand over to a simple, minimal Edge Function logic, as they should be crafted to fulfill a task that is rather light.

Processing an image, for example, isn't something an edge function would ideally do, as it can be quite a resource-intensive task and would usually simply fail without a result.

In this book, you'll obviously experience edge functions in action. As edge functions are complementary in the Supabase stack, there is no necessity to use them other than optimizing your workflows.

It is also worth noting that edge functions are written in **Deno**, which is essentially TypeScript. Since the logic of such functions is usually very small, it doesn't necessarily add much complexity to implement one even if your application is written in a completely different language.

According to what we've now learned, having edge functions can be beneficial but is complementary and definitely not a must. In the next section, you'll get to know the pg-meta service, which adds management APIs for all things in the database that are not strictly related to data.

pg-meta – an internal helper service for the database

When you use the dashboard to create a new table or add some more columns to a table, you would think that this is done via PostgREST, right? Well, PostgREST is built for *data management*, not *database management*.

Instead, database management is the task of the pg-meta service. It is a REST API that allows all kinds of manipulations on the database that is used by the other Supabase services. For example, when you create a new table with Studio, it talks to the pg-meta service by making a POST Request to / pg-meta/tables. It is secured and only accessible for the internal services or admins (e.g., when you log in to Studio, you are an admin).

However, why is pg-meta needed when all the services in the Supabase stack could also access the database directly? The following are some key reasons:

- It abstracts away any underlying changes. In theory, you could even completely replace the database with a different database because the pg-meta API stays the same.

- Within the Supabase stack, it's a convention and best practice for services to communicate with REST APIs. Each service has a clear responsibility and should not directly manipulate data if it's not specifically the responsibility of that service to do so.

- If this service wasn't available, the Supabase team would need to build such an API for the dashboard (Studio) anyway, because it needs access to manage the Database from a frontend UI. Hence, a generic solution was built that can be used by all services equally.

That's all that you need to know at this point. In this concise section, you learned that the pg-meta service complements the PostgREST service to also add the possibility of managing the database beyond just managing its data.

In the upcoming section, you'll get to know Kong, which plays a big role in taking care of all of the services that you know about by now.

Kong – the overarching service orchestrator

The Supabase services are Docker containers, and those containers live in their own Docker network. So, whenever I say "Supabase service," I'm referring to a Docker container. The fact that all Supabase services live within the same Docker network means that if one service, such as Storage, has a server running with an API for managing files in its container at `localhost:3000`, all the other services can access the API of that service with `serviceName:3000` (so `storage:3000`). That also means that all the services can talk to each other via their REST APIs.

Usually, in production environments (such as `supabase.com` itself), none of the services are directly accessible to the outside, so from the outside, I cannot just use an API at `storage:3000`. Only containers that explicitly expose themselves to the public will be accessible to the outside, but none of the services in Supabase are—and this is best practice.

Now Kong, the API gateway, is also just a service within that network of Supabase services such as Studio, Storage, PostgREST, and so on. However, Kong exposes itself to the outside, and since it has access to the other services, it can expose specific parts of the services on specific routes. This is called **proxying**.

So, imagine that Kong itself is reachable at `https://your-supabase.domain/`. Then, it can map the internal `storage:3000` service (which Kong has access to) to `your-supabase.domain/storage`. Although the Storage service as a whole isn't accessible to the public, Kong can make parts of it accessible to the public, as it has access to the other services and exposes itself to the public.

Kong controls access to the service. It's basically the same as delivering packages. Any online seller won't bring you the package you ordered. They all use a proxy, such as DHL, UPS, FedEx, or similar services. The proxy delivers your package and guards and protects packages in general.

The advantage is that it doesn't matter where the underlying services provide and receive data—Kong can map it to a convenient path. So, Kong is basically the guardian of the services. It also ensures that certain services (such as pg-meta) are only accessible by logged-in Super Admins and it can ensure that certain services are not accessible to the outside when there is no need for them to be.

You now have an extensive overview of Supabase and its services behind the curtains. At this point, you most likely already stand out from other people who only scratch the surface of Supabase, which is awesome! You'll see that this deeper knowledge will help make decision-making even better in the future.

There's just one thing missing in this introduction at this point: the project that you'll build as you advance through this book. I'll unveil it in the next section.

Introducing the production-grade ticket system project

In this book, you will build a production-ready, AI-enhanced multi-tenant ticket system with Next.js and Supabase, which you can take, deploy, extend, sell, or do whatever you like with.

To illustrate how Supabase integrates into a real-world application, it was obvious that this book would have to build a real-world application. Hence, I had to choose a framework (Next.js) to build an application together with you rather than just pinpointing to Supabase features. This means that, in the beginning, it may feel a little bit like a Next.js tutorial, but rest assured that this feeling changes over the course of this book. Even if you use another set of technology, you should feel well-guided.

> **Note**
>
> The system that we will build will be functional, contain best practices, and be secure for production usage.

However, what do I mean by a ticket system? We're not talking about event tickets or bus tickets here; instead, we're talking about work or issue tickets such as **Improve the navigation in the app** or **Fix the missing payment button** with statuses such as **Not started**, **In progress**, or **Done**. The tickets will look similar to the one in *Figure 1.8*:

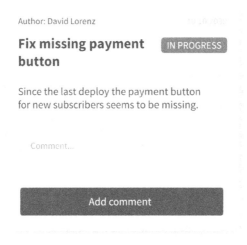

Figure 1.8: A sample ticket

But now that we've clarified what *tickets* refer to, what does *multi-tenant* refer to?

The project that we will create will be in the form of a SaaS product and should allow the creation and management of tickets bound to a specific client (tenant). That means that our final project will be able to determine whether you logged in on `company-cool.domain` or `company-foobar.domain`, and, depending on your permissions as a user, give you access to it and constrain the views to only show the respective tickets.

So, what do we need for that? Narrowed down, we will build the following:

- Pages that take care of the authentication flow (**Login**, **Registration**, **Password Recovery**)

- A way to create new tickets

- An overview of existing tickets with the possibility of searching and filtering those

- A way of viewing ticket details and the possibility of adding comments, as well as receiving real-time comment activity on the ticket

- A page that lists the existing users of the current tenant (**User's Overview**) for logged-in users to see who else can be working on tickets

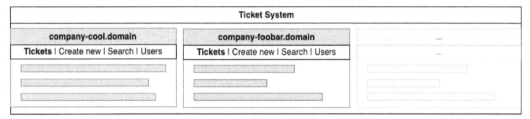

Figure 1.9: An ultra-simplified representation of the multi-tenant system

To be able to do so, there are a few things we will definitely want to consider:

- The signup process in our system needs to consider that users are only allowed to register with an email that has been approved by the tenant. In situations involving emails, especially during processes such as logging in via Magic Link, we will go beyond Supabase's default mailing system and send emails that are completely customized.

- For ticket management, we'll enhance the system to allow images, such as screenshots, to be attached to tickets, providing a visual context within the ticket details. Additionally, the overview page should clearly indicate whether a ticket has associated comments or attached images.

- From a technical standpoint, a single user account should seamlessly access multiple tenants without interfering with others. Introducing a bit of role management, only admin users of a tenant will have the privilege to delete any tickets, while other users can only delete their own tickets and comments.

- To enhance the ticket system's usability, we'll implement AI to identify similar existing tickets when creating a new one. Furthermore, a scheduler will be built to notify ticket creators via email if a ticket reaches its due date without being completed, adding a layer of convenience to the system.

Along this path, there are going to be exciting things to be discovered and unraveled that you wouldn't have guessed by reading the previous description, so get ready for a ride full of discovery and *aha*-moments—that's a promise!

> **Note**
>
> Since this book covers everything that you need to fully understand Supabase, including the management of permissions, we will not build an admin UI for the company to manage the ticket system. However, I want to encourage you to undertake this as a next challenge *after* completing the book, maybe extending our simple ticket system so that you can actually convince paying clients to use it.

Summary

Wow, what a deep dive into the Supabase ecosystem! We peeled back the layers and got our hands dirty with the inner workings of this powerful platform. From the clever RLS in the database to the seamless chat between services via REST APIs, we saw how Supabase isn't just smoke and mirrors—it's smart engineering at its finest.

We uncovered how Supabase blends traditional permission checks with fancy FGA, making data management a breeze for developers. Let's not forget the star of the show: those REST APIs that tie everything together, including the Supabase client that we'll be using. We learned about those in this chapter as well.

However, we're not just here to admire the architecture. All this knowledge is building toward something tangible: our multi-tenant ticket system. We've set the stage, and now it's time to put the theory into practice.

As we move on to the next chapter, get ready to roll up your sleeves. We'll be spinning up our very first Supabase instance and connecting it to a Next.js project. It's where the rubber meets the road, folks—a chance to see Supabase flex its muscles in a real-world scenario.

Are you ready to get your hands dirty? Let's go!

2

Setting Up Supabase with Next.js

This chapter will guide you through the technical setup for your future projects with Supabase, showing you not only what is required to get a proper connection to Supabase but also how to run your local instances. This will gear you up to work on our exciting multi-tenant ticket system project.

You'll learn what it takes to run local instances of Supabase and, in the context of multiple local instances, explore effective management of those to prevent conflicts.

You'll also learn how to seamlessly connect to Supabase both with a local instance or the **Software-as-a-Service (SaaS)** supabase.com version. As part of that, you'll prove that the setup is working properly by fetching the list of file storage buckets (don't worry; we haven't touched on Supabase Storage yet, but we expect it to always return an empty list).

Furthermore, you'll learn how easy it is to get proper TypeScript support with Supabase and how you can access the database directly – without a UI or a Supabase client.

Think of this chapter as our launchpad so that, by the end, you'll have a Supabase client in your Next.js application that you can use to do all of the wonderful things you'll want with Supabase: fetching data, calling remote functions, authenticating users, storing files, and more.

So, in this chapter, we will cover the following topics:

- Getting ready with Next.js
- Installing the Supabase CLI
- Running your first Supabase instance on your machine
- Managing multiple local Supabase instances
- Connecting to Supabase with the Supabase JavaScript client

- Connecting directly to the database

- Using Supabase with TypeScript

- Connecting Supabase to other frameworks

Technical requirements

Regarding prior knowledge, for the Next.js sections, you should have an understanding of how Next.js works (you can learn more about it or recap your understanding here: `https://nextjs.org/docs`).

As we will be working with Next.js at a technical level, this chapter requires you to have **Node.js** (`https://nodejs.org/`) installed. Usually, installing that also installs the **Node Package Manager** (npm and npx). However, on some Linux systems, it doesn't – so make sure you install it as well.

Since all containers in Supabase are Docker containers, you'll need to install Docker. At `https://docker.com`, you'll find a **Docker Desktop** installation executable for every operating system. It installs **Docker Engine** and a nice **graphical user interface (GUI)** for it.

Alternatively, you can install the convenient `https://orbstack.dev` application instead.

It is also expected that you have `git` installed.

Whenever you feel stuck or want to compare your progress, you can head to this book's GitHub repository, which contains the project: `https://github.com/PacktPublishing/Building-Production-Grade-Web-Applications-with-Supabase`. In each branch, you can call `npm run reset` to reset your local Supabase instance and get a fresh setup tailored for that branch.

In the upcoming sections and chapters, you will see references to which branches to look at. All the code in this chapter can be found in the `Ch02_AppRouter` branch. There's also an additional, optional branch named `Ch02_PagesRouter` for this chapter. It's only required for the optional Next.js Pages Router subsection within the *Connecting to Supabase with the Supabase JavaScript client* section.

> **Note**
>
> In this chapter, we will do the necessary things to finally get into the more practical part in the next chapter. A lot of explaining is done in this chapter to make you fully aware of how Supabase works and what you need to do with Next.js to get it running, depending on the context. If it feels a bit overwhelming in some places, then I recommend looking back at *Chapter 1* if you feel mentally disconnected from the inner workings of Supabase. Also, make yourself aware of the fact that you're only at *Chapter 2* – what we'll apply here will gradually become clearer with every upcoming chapter and with the more code you write.

Getting ready with Next.js

Since we will create a complete ticketing system software, you must first create a new Next.js project. For that, simply run the following command:

```
npx create-next-app ticket-system --app --src-dir --use-npm
--eslint  --js
```

This will create a Next.js project named `ticket-system` using App Router, as well as add sensible defaults such as `eslint`, the beloved `src` directory, and the activation of `npm` as the script runner (which is more widespread as opposed to `yarn`).

If you want to use it with TypeScript, simply exchange `--js` with `--ts`.

> **Note**
>
> This book does not use TypeScript for project creation. The reason is extremely simple: If you know TypeScript and want to use it, you can just write any of the upcoming code in this book with TypeScript instead. There's no added complexity. However, Supabase is technology-agnostic, so for those who don't know TypeScript, adding it would've been an additional layer of complexity.

Since the whole setup runs automatically, we can continue with setting up Supabase on our computer.

Installing the Supabase CLI

Before doing anything else, you'll want to install the **Supabase CLI** as it will – besides other great features – allow us to spin up a local Supabase instance with no effort.

In theory, you can install the Supabase CLI globally with package managers such as `brew` (`https://brew.sh`) or `scoop` (`https://scoop.sh`). However, from my experience, I advise against them as they usually confuse more than they help, particularly when you have multiple projects, as well as a different version of the Supabase CLI than other people in your team.

In short, install the Supabase CLI via `npm install supabase --save-dev` as a development dependency in *every* project where you want to use Supabase.

Do that now in the ticket system project you just created.

Afterward, you can test if the installation succeeded by running `npx supabase --help`. Running this command should *not* result in you being asked to install the CLI –since you've done that already, it should quickly output helpful information about the CLI.

Now that you've installed the CLI, we can manage our instance with it. We'll start by creating a new instance.

Running your first Supabase instance on your machine

Before we start a new Supabase instance, we need to clarify the term. In *Chapter 1*, you learned that Supabase is essentially a Postgres database with services surrounding it, all tied together via containers. One such setup is called *Supabase instance* (or *Supabase project*). So, for each of your web apps, you will have one instance.

When you go on `supabase.com`, sign in, and create a new project, that's your *Supabase instance*. However, we want to run an instance locally on our computer. Let's see how to do that.

Initializing a new local Supabase instance

Having a local installation is crucial in development as it allows us to test things before we deploy them. Hence, even if you plan to use the awesome `supabase.com` platform, you'll be better off *also* having a local installation.

Although you can have Supabase instances without a git repository (ultimately, creating a Supabase instance on `supabase.com` doesn't require any git repository), you still want to have your Supabase configuration inside your actual project. This allows you to manage a development instance from within your directory and link it to a live/production instance later.

Your previously created Next.js is a git repository by default. In that same project directory, run `npx supabase init`. This will create a new `/your-path/my-project/supabase` folder with a few files in it. This folder is used by the Supabase CLI and should be committed as part of your repository.

The command also creates a `config.toml` file in that new directory, which contains a complete, working configuration that the CLI uses to run your local setup. You can have a peek at it but don't expect to understand it just yet.

Starting your first Supabase instance

Now that you've initialized the required files for a local instance, you'll want to get it up and running so that you get the information from Supabase about the services it started, as well as the credentials it was deployed with. This will contain everything we need to make a connection to Supabase.

So, inside your `/your-path/my-project/` supabase project folder, call `npx supabase start`. This will automatically trigger the underlying Docker setup, pull all the required Docker images, and spin them up – in other words, the command creates and starts a fully running Supabase instance without you having to configure it.

> **Note**
> In case you've never worked with Docker before, imagine a Docker image like a template for something that runs in an isolated capsule (container).

The process can take a while, depending on your computer and internet speed, particularly the first time you do this. When the process has completed, you should see a confirmation message and something like this:

```
        API URL: http://localhost:54321
    GraphQL URL: http://localhost:54321/graphql/v1
         DB URL: postgresql://postgres...
     Studio URL: http://localhost:54323
   Inbucket URL: http://localhost:54324
     JWT secret: super-secret-jwt-token...
       anon key: eyJhbGciOiJI...
service_role key: eyJh...
```

This means your local instance is running. I recommend that you just copy all of the output you receive and save it somewhere for easy access later.

> **Note**
>
> Please be aware of the fact that this is your development version and is meant for your local development only. That's the only reason why all of these credentials are printed out in cleartext. The development version should always be complementary to a public-facing version from either supabase.com or run a secure self-hosted version of Supabase (which we will also discuss in this book).

The Terminal output provides a summary of the things you'll need to connect to Supabase (it is not a list of services started). Let's go through them one by one:

- **API URL**: The connection URL for our Supabase instance. It is the unified URL that takes all your API calls and, depending on the path, forwards them to the correct service to be processed.

- **GraphQL URL**: When explaining the PostgREST service in *Chapter 1*, I mentioned that Supabase allows you to work with GraphQL if that is your preferred way of working. This is the URL you'll need if you want to fetch data via GraphQL (you can find out more here: https://supabase.com/blog/graphql-now-available).

- **DB URL**: This is the connection URL of the database itself, bypassing the other services. It is already in the form of a PostgreSQL connection string and includes the PostgreSQL protocol, username, and password. For most use cases, you won't need this *DB URL* as you will be using the *API URL* – or more specifically, the Supabase client – for everything related to your project.

- **Studio URL**: This is the dashboard where we will conveniently manage and structure data within Supabase.

- **Inbucket URL:** This is a special service that's deployed to simplify your local development process. Every application that interacts with users needs to send emails – for example, when logging in, when changing an email address, or when resetting a password. To send emails, you need an email server. In development mode, you usually don't want to send out real emails to real people. Instead, you'll want to check and verify that the content in those emails is correct and functional.

  Inbucket is a dummy mail server. In your local installation, it will catch all emails that the services of your local Supabase might send, such as a magic login link or a password recovery link. Those caught emails can then be accessed through the Inbucket URL. So, even if you have five users with five different emails and you request a login link in your local setup, you will always "receive" those emails in the Inbucket web interface. Here, Inbucket will provide you with information about who was supposed to receive it. Very handy!

- **JWT secret:** JWT is the short form of **JSON Web Token** and is an encrypted JSON-formatted value that usually contains authentication data. The JWT secret is, hence, the secret key with which those encrypted token values are generated. This is done internally within Supabase so that you don't have to deal with this secret.

- **anon key:** This is the *anonymous key* that, complementary to the API URL, allows us to connect to our instance with the Supabase library – something we'll do in this chapter. This key provides no permissions by default, so it's considered safe in the frontend and backend. Even if someone "steals" that key, you have nothing to worry about.

- **service_role key:** The service role key sounds non-threatening but is the *superadmin key* and allows any kind of manipulation. You should never use it in the frontend as someone could steal it (something that can easily be done by sniffing the network requests) and access/change everything in your Supabase instance.

If you've watched your Terminal closely, you might have also spotted something similar to `Seeding data supabase/seed.sql`. Ignore it for now; we'll have a look at this file in *Chapter 5*.

At this point, I hope you're excited – your first local Supabase instance is running fine! Go ahead and have a peek at these URLs, especially Studio, and then come back so that you can continue your Supabase journey.

Now that this is done, a question remains: can you have multiple Supabase instances locally so that you can work on multiple projects? Yes. This is what we will see in the next section.

Managing multiple local Supabase instances

Developers often work on many different projects. Running multiple servers of any kind at the same time can be done by changing the port configuration (one on `localhost:3000`, the other on `localhost:3001`, and so on). But what if you have three projects, all using their own Supabase instances? Can we have three local Supabase instances run in parallel at the same time? Let's have a look at the options for how to handle this.

Option 1 – the start-stop technique

This is the recommended option when the projects are separated and you can spare a minute to switch between instances.

Imagine that you have two projects. If you currently run the instance from project 1, then go to the project directory and run `npx supabase stop`, it will automatically back up your data locally.

Now, if you go to project 2 and run `npx supabase start`, the project 2 instance will be running and consider the project 1 instance to be in sleep mode.

If you want to run the project 1 instance again, just do it in reverse: run `npx supabase stop` in project 2 and run `npx supabase start` in project 1. By doing so, everything gets fully recovered in no time.

This is a very good way to work because you have exactly one instance running at one time, so your computer saves resources.

Option 2 – change ports

Inside `config.toml`, you can change the ports that Supabase uses for its services. For example, for Studio, you'll see something like this:

```
[studio]
port = 54323
```

Change `port` to something unique, such as `port=9100`. Upon doing this, for this project, your Supabase Studio service will run on `localhost:9100`. Now, change the ports for the other services to something that your other projects don't use; you should be able to run it parallel to another project.

> **Note**
> Running multiple instances of Supabase on your computer simultaneously is rarely needed. In most use cases, you should go for option 1 as it is a pretty fast and efficient way of managing multiple projects.

With that, you know how to manage one or even multiple Supabase instances, one for each of your future projects. But how can you bring this into Next.js? We'll cover this in the next section.

Connecting to Supabase with the Supabase JavaScript client

As you saw in *Chapter 1*, you could theoretically just fire HTTP requests to the REST API of your Supabase instance for everything you want to do. But firing HTTP requests for everything you want to do is neither convenient nor safe against API changes in future versions of Supabase.

Instead, I'll introduce you to the Supabase client, which provides a convenient way to access all the features of Supabase so that you can successfully build an application.

In this section, we'll focus on getting the Supabase JavaScript client up and running within the Next.js framework. We'll also explore different integration methods, regardless of whether you're working with the Next.js App Router or the Next.js Pages Router, and whether you're on the frontend or the backend; we'll break down the details of each.

What's interesting is that at no point in this book will we use any kind of Next.js-specific Supabase package; we will work with universal JavaScript packages that can be used in any other JavaScript framework so that what you'll learn now doesn't only apply to Next.js.

> **Note**
>
> Even with languages other than JavaScript, the applied methods of using a Supabase client always follow the same design. Hence, moving from the JavaScript-based client to a Swift-based Supabase client, for example, is not a new mental model and doesn't require a new way of thinking.

Starting with the following section, I'll explain how to initialize a basic Supabase client with the base Supabase JavaScript (or TypeScript; both are supported) npm package called `@supabase/supabase-js`.

Initializing and testing the base Supabase JavaScript client within Next.js

In this section, you'll get a grip on the basics of the crucial, framework-independent Supabase JavaScript library. You'll initialize a Supabase JavaScript client with the credentials from your instance and fire your first request to test the Supabase connection.

Let's get started:

1. Install the Supabase library in your project:

    ```
    npm install @supabase/supabase-js
    ```

    You'll always need this library when using any JavaScript/TypeScript projects with Supabase. It provides a function called `createClient`, which allows us to initialize a client by providing it with the Supabase URL and a key.

    So, we need to define the URL and the key. We'll do this next.

2. The Supabase URL and key are configuration values. Configuration values are usually provided to applications via environment variables and should never be hardcoded. Next.js, just like many other frameworks, uses the **dotenv** (`https://www.npmjs.com/package/dotenv`) file format, where each line represents a key-value pair (`KEY_NAME=VALUE`) that can be used per its respective key in your application.

    In Next.js, this file is called `.env.local` (`https://nextjs.org/docs/pages/building-your-application/configuring/environment-variables#loading-environment-variables`).

    So, what you need to do now is create a file called `.env.local` and add the Supabase instance URL and the anonymous key to that file, like so:

    ```
    NEXT_PUBLIC_SUPABASE_URL=https://your-url
    NEXT_PUBLIC_SUPABASE_ANON_KEY=sYQMA369pT...
    ```

    The first line – the Supabase URL – is the *API URL*, while the second is the *anonymous key*. You'll find both of these values printed in the Terminal output of your local instance when you run the CLI with `npx supabase start`. You can also run `npx supabase status` to show those values again.

    If you want to configure the connection to an instance on `supabase.com` instead, you find these values via the project dashboard. Under **Project Settings**, click **API**:

    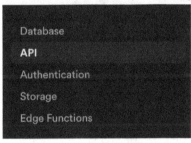

    Figure 2.1: Project settings in the dashboard (supabase.com)

    In the **API** settings, you should see **Project URL**, which is the equivalent of the *API URL* in your local Supabase instance:

    Figure 2.2: Project URL in the dashboard settings

In the same area, you should also see **Project API keys** details, namely anon (*anonymous key*) and `service_role`. You'll need the anon key to be defined for `NEXT_PUBLIC_SUPABASE_ANON_KEY`:

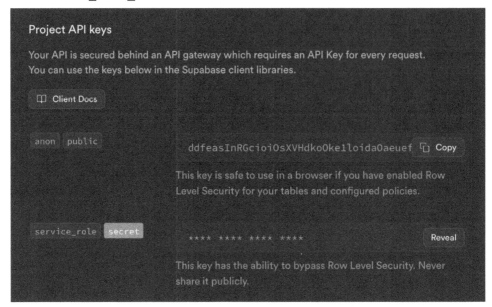

Figure 2.3: Project API keys

3. All functionality of Supabase (including data fetching, real-time updates, authentication, and more) is done with the Supabase client of the package you installed. Such a client is going to be *stateless*, meaning it doesn't check if your provided URL or key is correct (that's why we'll implement a little trick to test our connection in *Step 4*).

At this point, we want to initialize our first Supabase client. Since we want to have a clean separation of code logic, we will create a file in which we'll export a reusable function that creates the client. So, create a file called `your-project/src/supabase-utils/client.js` with the following code:

```
import { createClient } from "@supabase/supabase-js";
export const createSupabaseClient = () =>
  createClient(
    process.env.NEXT_PUBLIC_SUPABASE_URL,
    process.env.NEXT_PUBLIC_SUPABASE_ANON_KEY
  );
```

At this point, you have a function called `createSupabaseClient()` that initializes a new Supabase client with the given values. This function can now be imported from any other file. We'll make use of it in the next step.

4. As mentioned in the previous step, the client won't tell us if the provided connection values we passed to `createClient()` were valid. It's a wrapper for the APIs and you'll only notice if something doesn't work when you're using one of its services. But we can just fire any request to Supabase to see if it succeeds and confirm that the connection is okay.

   Since any app using Supabase will somehow communicate with Supabase in the very beginning, you will also notice when something's wrong. For instance, if you check if someone is logged in and the connection fails, it will throw an error.

   You can also use a trick to test the connection. Go to any page where you want to test the client (such as `app/page.js`) and import `createSupabaseClient` from your created client file:

   ```
   import { createSupabaseClient } from '@/supabase-client';
   ```

   Then, for testing purposes, on component startup, we'll want to list all existing storage buckets (don't worry about what they are for now; you'll learn about them in *Chapter 11*), like this:

   ```
   useEffect(() => {
       const supabase = createSupabaseClient();
       supabase.storage.listBuckets().then((result) =>
        console.log("Bucket List", result);
       });
   }, []);
   ```

   Now, if you start your Next.js server with `npm run dev`, open the page, and check the browser console, you should hopefully see something like this, getting an empty but successful result within `data`:

   ```
   Bucket List ▸ Object { data: [], error: null }
   ```

   However, you may also see something like this:

   ```
   Bucket List ▸ Object { data: null, error: StorageUnknownError
   ... }
   ```

   This means that you should check your configuration again because something's misconfigured (or your instance is offline); you'll need to check if you made a copy-paste mistake when adding the API URL or anonymous key or whether you forgot to start Supabase with `npx supabase start` after your last computer reboot.

With that, you have a correctly working Supabase client and have confirmed that it's working by requesting your first data from its API with it. In the next few sections, we'll dive deeper into that. Next, I'll give you a little bit of useful background knowledge regarding the Supabase client.

Understanding the base Supabase client

In the previous section, I showed you how to use the base Supabase client. Here, I want to clarify how the Supabase client works to help you understand why we have two different keys, and how the frontend and backend differentiate (or not).

The `@supabase/supabase-js` client is a versatile isomorphic package that can run anywhere that JavaScript runs – whether it's in the browser, on the server (Node.js), or even within a native iOS app (for example, with React Native). This flexibility allows developers to easily switch between environments without modifying their code.

However, you must know that how the client is instantiated does make a difference, either with the anonymous key or the `service_role` key. In *Chapter 1*, you learned that Supabase has robust authentication and authorization built into it. This means that, by default, Supabase is secure, and the anon key alone won't grant access to your instance without further configuration.

> **Heads up**
> Don't think too much about the differences between the Supabase keys as you'll see both of them being used very soon.

However, if you'd used the `service_role` key (don't ever expose it publicly!), you would have gotten unlimited access to everything – it's your admin access.

The anon key will always respect the access of the current user being signed in (or not signed in). But for the Supabase client to be able to know if a user is signed in so that it can authenticate a user, it needs to store some kind of authentication data. On the web, this is usually done via *cookies*.

Storing and managing cookies in a browser is always the same, no matter which browser you use (it's done via `document.cookie`; see `https://developer.mozilla.org/en-US/docs/Web/API/Document/cookie`). However, in every non-browser environment, getting these cookies is dependent on the framework you use. For example, you could be using the JavaScript client on the frontend but a Python Supabase client on the backend. Both need to be able to manage the same Supabase cookies to authenticate users and authorize their actions.

In our case, we're using Next.js for our project, so we need a way to manage cookies for Next.js when we use backend logic in Next.js. In the upcoming subsection, I'll show you a convenient, official package that solves the difficulties of using the Supabase client both on the frontend and backend with Next.js.

Using the Supabase client with Pages Router and App Router

You've just learned how to use the Supabase client with the base Supabase JavaScript client. In this section, you'll learn about the best way to use the Supabase client within Next.js, including the specific implications of using Supabase with Next.js' *App Router* versus *Pages Router* (or both).

Before Next.js version 13.4, there was only the so-called **Pages Router** for file-based routing. You had to create a file in the `pages/` directory (for example, `pages/hello.js`) that returned **Hello**, for example, and you had your page at the `your-site-domain.com/hello` route.

With Next.js 13.4+, **App Router** was added – it has similar file-based principles but for `/hello`, you would create a file called `app/hello/page.js` instead that comes packaged with a bunch of great new features.

Both can coexist in the *same* project, and there's a good reason for this. There are thousands of Next.js projects, from small businesses to corporate e-commerce sites, that were built before Next.js 13.4 with Pages Router. With the coexistence of both, new routes can be created with the new App Router approach while keeping the existing ones in the Pages Router structure. Hence, having the following folder structure in a Next.js project isn't unusual:

```
/
 /pages
  /old-hello.js

 /app
  /new-hello
   /page.js
```

Because Pages Router works differently than App Router, there are different setups, depending on which Router you'll want to use with the Supabase client.

> **Note**
>
> For our ticket system project, we will focus solely on *App Router* as this will be the predominant Router in Next.js in the future. However, you'll still be provided with the required information about using *Pages Router* later in this chapter.

We've already initialized a Supabase client with the plain, framework-independent Supabase JavaScript library (`@supabase/supabase-js`) and fetched an empty list of buckets from the storage service. So, everything seems to be working. Why would we need to do anything else?

Well, there are two problems to solve:

- If we need the Supabase client in multiple components on the *frontend*, we would need to call `createSupabaseClient()` multiple times, which would recreate a client multiple times. That is unnecessary. Also, the default Supabase client from `@supabase/supabase-js` will not store anything the server-side will be able to read. So, it will work for a solely frontend-based application.

- Although we can use `createSupabaseClient()` on the *backend* as well, it will not work as expected there and always assumes that no user is logged in, even if you are logged in (due to missing cookie management, as mentioned in the previous subsection).

Let me introduce you to the framework-independent `@supabase/ssr` package, which solves the process of instantiating a browser-based frontend client as well as a non-browser-based backend client in a streamlined way.

Utilizing createBrowserClient on the frontend

Let's start by solving the frontend problem first and then go back and re-evaluate the details of the backend problem.

Here, `@supabase/supabase-js` is always the base package that you install in any JavaScript-based environment where you use the Supabase JavaScript client. Then, there is a complementary package from Supabase called `@supabase/ssr`. It is framework-independent but wraps functionality to make cookie management easy on both the frontend and backend so that you can log in on the frontend and use the authentication on the backend and vice versa. We'll use the package to create a frontend Supabase client.

Although `ssr` stands for server-side rendering, it comes packed with a function called `createBrowserClient`, which solves our first problem, recreation: it is a wrapper on top of the `createClient` function from `@supabase/supabase-js` and uses a Singleton pattern (`https://en.wikipedia.org/wiki/Singleton_pattern`). So, you can call it as often as you want, and the client is still just created once.

To use it, first, install the package with `npm install @supabase/ssr`. Then, go to your `supabase-utils/client.js` file. Here, you will see the following code:

```
import { createClient } from "@supabase/supabase-js";
export const createSupabaseClient = () => createClient(...);
```

Now, replace `createClient` with the `createBrowserClient` import, as follows:

```
import { createBrowserClient } from "@supabase/ssr";
export const createSupabaseClient = () => createBrowserClient(...);
```

This would be sufficient, but I want to do two more things to add more clarity that this is about the frontend:

1. First, rename `createSupabaseClient` to `getSupabaseBrowserClient` so that it looks like this:

    ```
    import { createBrowserClient } from "@supabase/ssr";
    export const getSupabaseBrowserClient = () =>
    createBrowserClient(...);
    ```

    Just by changing the name, it's clearer that we aren't recreating the client all the time and that we're supposed to use this on the frontend.

2. Second, rename `client.js` to `browserClient.js` as this name provides more details about what it is – that is, a browser-targeted Supabase client. Don't forget to adapt your existing `createSupabaseClient` imports so that they match the name changes.

> **Tip**
>
> In VS Code, you can press *F2* to rename `createSupabaseClient`. It will automatically be renamed across your project. If you change a filename, VS Code should also pick up the references and refactor the path in other files accordingly.

With that, problem one has been solved. Now, let's get back to problem two – the backend.

Utilizing createServerClient on the backend

When the `@supabase/supabase-js` Supabase client is initialized on the backend, it doesn't run in the browser but on a server. In a *browser*, accessing a cookie is always the same. On the *server*, however, it depends on the framework you use. So, even though cookies are always set via HTTP headers, *how* to set those HTTP headers is different, depending on the framework you use.

So, what we need to do for the Supabase client to work properly on the backend is to provide it with instructions on how to read, write, and remove cookies in Next.js. This can be achieved with the `@supabase/ssr` package as it exports a `createServerClient` function, which allows us to provide additional cookie functions.

> **Note**
>
> There is a previous package from Supabase tailored specifically for Next.js with the name – that is, `@supabase/auth-helpers-nextjs`. However, it was marked as deprecated in October 2023 and replaced with the generic framework-independent solution called `@supabase/ssr`.

We want a proper Supabase client for the backend that can manage cookies. So, we'll begin by having a look at how to achieve this from a general, framework-independent perspective; we'll see how it's done with Next.js after.

So, first things first, import `createServerClient` from the `@supabase/ssr` package. Then, initialize a backend client – this is very similar to the frontend version with the difference that we pass a configuration object containing the `cookies` property with two functions Supabase can call. The functions are named `getAll` and `setAll`. Supabase calls these to retrieve or set cookies, respectively:

```
createServerClient(
  process.env.NEXT_PUBLIC_SUPABASE_URL,
  process.env.NEXT_PUBLIC_SUPABASE_ANON_KEY,
  {
    cookies: {
```

```
    getAll: () => /* code for getting all cookies */,
    setAll: (cookies) => /* a list of cookies to set
    }
  }
);
```

As you can see, we need to pass a `getAll` function that will retrieve all available cookies. The `setAll` function gets a list (an array) of cookies to set, with each cookie being an object of the form `{ name, value, options }`. You might be wondering why there is no `remove` or `removeAll` function. This is implicit because `setAll` with proper options (expiry date) will make the cookie disappear.

At this point, you might be thinking, "Now I have to know how to deal with cookies in HTTP headers and get and set them?" Not really. All you have to know is how to set cookies for Supabase, which I'll show you, and then you can forget about it afterward. First, let's see what options we have to set cookies in Next.js.

In Next.js, there are *exactly* two ways to get and set cookies, based on the context of your code:

- In some context, you manage cookies with Next.js based on the request (`https://nextjs.org/docs/app/api-reference/functions/next-request`) and response (`https://nextjs.org/docs/app/api-reference/functions/next-response`) objects, often named `req` and `res`

- In other contexts, only with App Router, you manage cookies in Next.js with the built-in `cookie()` utility

As there are two ways of managing cookies in Next.js, it leaves us with the questions of "How can we implement these methods?" and "When do we use which one?"

With App Router files, we have four places in which we can use the Supabase backend client – only one uses `req` and `res`; the others use the `cookies()` utility:

- In the `middleware.js` file: `req` and `res`

- In a server-rendered component (for example, `/app/foo/page.js`): `cookies()`

- In a Route Handler: `cookies()`

- In a server action: `cookies()`

Now, you might be wondering, "Why is `middleware.js` the only one where `cookies()` cannot be used?" Well, ask Vercel. There has been an implementation to retrieve cookies with the `cookies()` utility (`https://github.com/vercel/next.js/pull/53465`) but no way to set them. Hence, the plain answer is "Probably due to historical reasons."

> **Note**
>
> Even experienced developers don't necessarily have experience with Next.js' *middleware* (`https://nextjs.org/docs/app/building-your-application/routing/middleware`) function as many projects don't require it. That's why I want to slide in with a short remark. In Next.js, middleware is a possibility of intercepting *any* kind of request before it gets processed further. It also allows you to rewrite, manipulate, or redirect requests. In *Chapter 4*, we will use middleware as a protection layer.

Next, we'll create a Supabase client for the middleware (based on request/response objects) and then one for all the other three mentioned places with the `cookies()` utility.

Creating a Supabase client based on the request and response for the middleware

Create a new file called `supabase-utils/reqResClient.js` and fill it with the following code:

```
import { createServerClient } from "@supabase/ssr";

export const getSupabaseReqResClient = ({ request }) => {
  let response = {
    value: NextResponse.next({ request: request})
  };

  const supabase = createServerClient(
    process.env.NEXT_PUBLIC_SUPABASE_URL,
    process.env.NEXT_PUBLIC_SUPABASE_ANON_KEY,
    {
      cookies: {
        getAll() {
          return request.cookies.getAll();
        },

        setAll(cookiesToSet) {
          cookiesToSet.forEach(({ name, value, options }) => {
            request.cookies.set(name, value);
          });

          response.value = NextResponse.next({
            request,
          });

          cookiesToSet.forEach(({ name, value, options }) => {
            response.value.cookies.set(name, value, options);
          });
        },
```

```
      },
    }
  );

  return { supabase, response };
};
```

Here, we must realize that the getAll() part is very straightforward. From a given request object, we can access the cookies object, which itself follows the standard described at https://developer.mozilla.org/en-US/docs/Web/API/CookieStore/getAll. Hence, we can simply return request.cookies.getAll(). Great, that's easy.

But why is the second part of the code so big? Why do we define a response variable that has a value property? And couldn't we just call request.cookies.setAll and we're good? Coming back to what I said earlier, if Next.js had implemented the cookies() utility so that it could be read and written from every backend part (not excluding the middleware), none of this would be required. But let me elaborate to explain why we need all of that code.

In the following code, at the beginning of our getSupabaseReqResClient function, we create a response object that copies existing headers from the incoming request to retain existing cookies for whatever is rendered after the middleware:

```
let response = {
  value: NextResponse.next({ request: request})
};
```

At the end of getSupabaseReqResClient, we have return { supabase, response };. So, we return the Supabase client, but also the response to be used as a response object later.

So, why don't we simply use let response = NextResponse.next({ request: request})? If we were to do that, we'd return the response object as-is and cannot change it afterward. But the Supabase client has to change it. That's why we use the trick of returning the response = { value: ... } object instead – we can still manipulate the value reference inside it (again, this is an architectural problem of Next.js that we can work around, not a Supabase problem).

Ok, we resolved the beginning and the end. Now, let's look at what happens inside setAll(cookiesToSet). First, we do this:

```
cookiesToSet.forEach(({ name, value, options }) => {
  request.cookies.set(name, value);
});
```

What it does is clear – it sets the cookies that Supabase needs to set on the request object. But *why* do we set cookies on the request we received? We need to set it on the request to be able to forward it to any not-yet-processed files, such as a page, such that the page will have the cookies available.

Unfortunately, setting it with `request.cookies.set` isn't sufficient as every time a request object is provided in Next.js, it's a copy of the original request (`https://github.com/vercel/next.js/blob/f0abeb8e42f723e515f93e0062ebf0a78df52765/packages/next/src/server/next-server.ts#L1110`). That's why we need the following code block:

```
response.value = NextResponse.next({
   request,
});
```

All this does is tell Next.js, "Hey, here's my response. Please use this new request object for anything that you still have to process (like the actual page)." Since we modified the request object with the proper cookies, a page that's rendered hereafter will have those cookies available.

But why do we need the following code block?

```
cookiesToSet.forEach(({ name, value, options }) => {
   response.value.cookies.set(name, value, options);
});
```

Well, this sets the required cookies for the response being delivered to the user (for the browser to save the cookies).

If this was confusing, trust me, I can relate. Hopefully, Next.js will fix this architectural flaw soon. Take your time to understand it or simply create the file and forget about it forever – we will never have to look at it again once it's been created.

Using the request/response client in the middleware

With `getSupabaseReqResClient`, which we just created, I'll show you how to use this client inside middleware. Just add the following code to the `src/middleware.js` file:

```
import { getSupabaseReqResClient } from "@/supabase-utils/
reqResClient";
export async function middleware(request) {
   const { supabase, response }
      = getSupabaseReqResClient({ request });

   return response.value;
}
```

This does nothing yet and is still foundational knowledge. There's no need to create a middleware file right now but it will come in handy to have a boilerplate that can be enhanced later – we will create a proper middleware in *Chapter 4*.

Next, we'll dive into the specifics of initializing a Supabase client on the backend routes.

Creating Supabase backend clients with App Router

In App Router, we have the `cookies()` utility from the Next.js `next/headers` package, which you can use to read and write cookies. Just like what we did for `getSupabaseReqResClient`, we want to create a reusable function named `getSupabaseCookiesUtilClient`. Since we will use the built-in `cookies()` function, there's no need to pass anything from the outside.

Go ahead and create the `src/supabase-utils/cookiesUtilClient.js` file and add the following code to it:

```
import { createServerClient } from "@supabase/ssr";
import { cookies } from "next/headers";

export const getSupabaseCookiesUtilClient = () => {
  const cookieStore = cookies();

  return createServerClient(
    process.env.NEXT_PUBLIC_SUPABASE_URL,
    process.env.NEXT_PUBLIC_SUPABASE_ANON_KEY,
    {
      cookies: {
        getAll() {
          return cookieStore.getAll();
        },
        setAll(cookiesToSet) {
          try {
            cookiesToSet.forEach(({ name, value, options }) => {
              cookieStore.set(name, value, options);
            });
          } catch (err) {
            console.error("Failed to set cookies", err);
          }
        },
      },
    }
  );
};
```

At the beginning of the file, we get access to the cookie store so that we can read and write cookies by calling the `cookies()` function from Next.js. In `setAll`, we can do what we couldn't do in `getSupabaseReqResClient`: set the cookies with the `cookies()` utility by calling `set()` for each cookie that we need to set.

The only surprising thing might be the `try-catch` part. This is added for the rare case that Next.js throws an error for whatever reason. We want to know the error (that's why we use `console.error`), but we use the `try-catch` statement to prevent things from breaking. So, let's see how we can make use of that function.

With *App Router* in Next.js, there are three different backend scenarios in which we could initialize the `cookies()`-based Supbase-client – in server components, Route Handlers, and Server Actions. In those, we can simply import our new function and initialize a proper Supabase client. It's essentially the same for all of them, but let's look at some exemplary code for each:

- **Server components**: Rendering and serving data from the server, as part of the **server component**, instead of waiting for the frontend is a lovely way of making your app faster for your customers. With the Supabase backend client, we are now able to do so and initialize it on server-only components, as follows:

```
import { getSupabaseCookiesUtilClient } from "@/supabase-utils/
cookiesUtilClient";

export default async function ServerComponent() {
  const supabase = getSupabaseCookiesUtilClient();
  const buckets = await supabase.storage.listBuckets();
  return <div>{JSON.stringify(buckets, null, 2)}</div>;
}
```

- **Route Handlers**: The former API routes (Pages Router) of Next.js have found their successor in **Route Handlers**. Route Handlers can now be everywhere, not just in one specific folder as with the former API routes. Say you want to execute certain code at the good old `/api` path (for example, `/api/demo`); you can do so by creating a file called `app/api/demo/route.js` and exporting a function named after the type of request you want to respond to (GET, POST, PUT, or DELETE).

  When it comes to initializing the Supabase client there, it's the same thing as in the server component. So, you can use the following code for a GET Route Handler that will return the bucket list from Supabase in JSON form:

```
import { getSupabaseCookiesUtilClient } from "@/supabase-utils/
cookiesUtilClient";
import { NextResponse } from "next/server";

export async function GET(request) {
  const supabase = getSupabaseCookiesUtilClient();
  const buckets = await supabase.storage.listBuckets();
  return NextResponse.json(buckets);
}
```

- **Server actions**: Server actions in Next.js are server-defined code that can be triggered from the frontend without the need to call a URL or be defined as part of a piece of code that can be reached via a certain URL.

We won't make use of server actions in this book. However, if you do know about them and want to use them with Supabase, you simply need to follow the same approach that you would for the other two request types:

```
import { getSupabaseCookiesUtilClient } from "@/supabase-utils/
cookiesUtilClient";

export default function PageWithServerAction() {
  async function serverActionWithSupabase() {
    "use server";
    const supabase = getSupabaseCookiesUtilClient();
    const buckets = await supabase.storage.listBuckets();
    console.log("@server", buckets);
  }

  return (
   <form action={serverActionWithSupabase}>
     <button type="submit">Run Server Action</button>
   </form>
  );
}
```

Here, submitting the form will cause the *server action* to execute and log the buckets in your terminal running the Next.js app.

Note

You'll find all of these files in the GitHub repository to play around with.

At this point, you know how to initialize and use a Supabase client with App Router from Next.js – both on the frontend and backend. Next, you can either get some additional knowledge about usage with Pages Router or move on to the following section, where I'll show you how to bypass the Supabase client and make a raw connection to the underlying Postgres database.

Optional knowledge – using createServerClient with Pages Router

If you don't have any old projects or plan to migrate your existing Pages Router files to the new App Router, then this optional knowledge will be a waste of time, so skip it and jump to *Connecting directly to the database* if needed. But if that isn't the case, let me now show you how to use the Supabase client on the backend with Pages Router.

Creating Supabase backend clients with Pages Router

Starting with Pages Router, since we've only covered the middleware, we haven't discussed how to use the Supabase client for the two scenarios, so this is something that must be discussed:

- Using Supabase within the `getStaticProps` method of a page
- Using Supabase within the `getServerSideProps` method of a page
- Using Supabase within an API route (the files within the `/pages/api` folder)

We'll quickly look at how to use `getStaticProps` with Supabase first, after which we'll look at the `getServerSideProps` and API route options together.

Using Supabase within getStaticProps

The `getStaticProps` method is never called from a user; it's only called from the system that pre-generates static pages. This means that this method will never use cookies as there is no user to authenticate. Hence, for `getStaticProps`, you must instantiate a default `createClient` value from `@supabase/supabase-js`.

Using Supabase within getServerSideProps or API routes

In both the `getServerSideProps` method of Pages Router pages (for example, `pages/dashboard.js`) and API routes (for example, `pages/api/my-data.js`), Next.js passes a *request* object (`req`) and a *response* object (`res`) with which we must manage cookies for our Supabase instance.

Your first natural thought is probably, "That's great, so we can just pass the `req` object to our existing `getSupabaseReqResClient` and it will work." Great thought! But that doesn't work because the request objects work completely differently for App Router and Pages Router. App Router uses an extended version of the web standard request (`https://developer.mozilla.org/en-US/docs/Web/API/Request`) and response (`https://developer.mozilla.org/en-US/docs/Web/API/Response`). However, Pages Router uses custom-built implementations.

We cannot just pass the request object and return a new response object based on it. This approach isn't compatible with Pages Router. Instead, I've prepared a customized `getSupabaseReqRes` method where I've highlighted the changes:

```
import { createServerClient, serializeCookieHeader } from "@supabase/
ssr";
import { NextResponse } from "next/server";

export const getSupabaseReqResClient = ({
  request,
  response: responseInput,
}) => {
  let response = {
```

```javascript
    value: responseInput ?? NextResponse.next({ request: request }),
};

const supabase = createServerClient(
  process.env.NEXT_PUBLIC_SUPABASE_URL,
  process.env.NEXT_PUBLIC_SUPABASE_ANON_KEY,
  {
    cookies: {
      getAll() {
        if ("getAll" in request.cookies) {
          return request.cookies.getAll();
        } else {
          // pages router / not middleware
          return Object.keys(request.cookies).map((name) => ({
            name,
            value: request.cookies[name],
          }));
        }
      },

      setAll(cookiesToSet) {
        if ("getAll" in request.cookies) {
          cookiesToSet.forEach(({ name, value, options }) => {
            request.cookies.set(name, value);
          });

          response.value = NextResponse.next({
            request,
          });

          cookiesToSet.forEach(({ name, value, options }) => {
            response.value.cookies.set(name, value, options);
          });
        } else {
          // pages router / not middleware
          responseInput.setHeader(
            "Set-Cookie",
            cookiesToSet.map(({ name, value, options }) =>
              serializeCookieHeader(name, value, options)
            )
          );
        }
      },
```

```
        },
      }
    );

    return { supabase, response };
};
```

The modified code of `getSupabaseReqRes` can either take just a request object and will work with the middleware or it can take a request object and a response object from Pages Router. Let me explain this in detail.

The changed function determines whether it needs to fall back to the Pages Router cookie management based on whether `getAll` is available or not. Because `getAll` doesn't exist in Pages Router's implementation of `request.cookies`, we know it cannot be the Pages Router request when `getAll` hasn't been defined.

So, in the `else` case of getting all cookies within `getAll`, we read existing cookie keys with `Object.keys(request.cookies)` and then map them to the same format (an array of {`name, value`} entries) that the Supabase client needs.

Due to the different architecture in Pages Router, in the `setAll else` case, we only need to set the cookies on the `responseInput` object with the help of `setHeader('Set-Cookie', serializedCookies)`. Supabase provides a special function called `serializeCookieHeader` to make this as simple as possible.

> **Note**
>
> Since we execute different code logic in both `getAll` and `setAll`, you might be wondering why we adapted the existing file and not create a `pagesRouterClient.js` file. This would be a valid approach as well but since the general idea of handling cookies via request and response objects is the same, I preferred having them in one reusable function.

I want to emphasize that this modified `getSupabaseReqResClient` still works for the `middleware.js` file – it was only adapted to also work for the Pages Router files.

Now, using this client is really easy. Within *API routes*, you can use it like this:

```
import { getSupabaseReqResClient } from "@/supabase-utils/
reqResClient";

export default async function handler(request, response) {
  const { supabase }
    = getSupabaseReqResClient({ request, response });
  // ....
}
```

Within `getServerSideProps` of Pages Router pages, you can make use of the Supabase client as follows:

```
export const getServerSideProps = async ({ req, res }) => {
  const { supabase, response } = getSupabaseReqResClient({
    request: req,
    response: res
  });
  // ...
  return {};
}
```

With the additional Pages Router knowledge, you're all set to use the Supabase client within Next.js for any kind of backend. Next, we'll learn how to connect to the Supabase database directly.

Connecting directly to the database

> **Note**
> Building a raw database connection is helpful but complementary knowledge. In this book's project, we will use the Supabase client and not a direct database connection.

At the end of the day, Supabase comes down to just being a Postgres database with additional services surrounding it like a galaxy. Hence, you can also directly access the database. But why would you ever want to do this?

When you work with platforms such as Supabase that make your life easier by providing data storage, file storage, authentication, and more, you *often* don't get direct access to the underlying database or your access is extremely limited. The reason is that providers of such platforms often want to safeguard you and themselves from scrapping the project in a way that will break it irrevocably.

Having no or limited direct access to your database also means that you cannot extend it with additional features or use libraries of any kind that need direct access (such as `sequelize`, `drizzle`, or `pg_dump`). But with Supabase, you can. So, let's have a look at how we can connect directly.

On a `supabase.com` project, within the **Dashboard (Studio)** area, you'll find the database connection URI of the Postgres database in the **Project Settings | Database** section. In your local instance, the complete connection URI is shown in the Terminal after running `npx supabase start` or, for a running instance, when calling `npx supabase status`. It already contains the username and password, separated with a colon (on your local instance, this is usually `postgresql://postgres:postgres@localhost:54322/postgres`).

Then, you can connect to it with whichever tool you like – for example, via GUIs for databases such as DBeaver (`https://dbeaver.io/`).

To test if the connection to the database works, I prefer the `psql` command-line tool. For my local instance, I can simply use one of the following commands:

- The most minimal way to test a connection is by calling `plsql` with the connection string in `postgresql://username:password@host:port/postgres` format, like so:

  ```
  psql 'postgresql://postgres:postgres@localhost:54322/postgres'
  ```

- For real connections (not just local ones), you should prefer a more verbose form that doesn't keep the password in cleartext in the Terminal and prompts you for the password:

  ```
  psql -h localhost -p 54322 -d postgres -U postgres
  ```

  This is equal to the longest form where the parameter meanings become self-explanatory:

  ```
  psql --host=localhost --port=54322 --dbname=postgres
  --username=postgres
  ```

With that, you know how to connect to the database if needed. Please be aware that connecting to the database directly and changing data there can be dangerous if you don't know what you're doing as there's no protection layer in between.

Next, you'll learn what you need to do to get immediate TypeScript support with Supabase.

Using Supabase with TypeScript

Many projects nowadays use TypeScript instead of JavaScript. In this book, we'll focus on using Supabase with JavaScript instead of TypeScript. But still, I want to show you how easily it can be used in combination with the Supabase JavaScript clients, and which benefits it brings.

Supabase's npm library comes with TypeScript support out of the box. However, with TypeScript, Supabase can also tell you that the expected data from your database doesn't exist or help you find the correct table name for your database via autocompletion in your editor.

All you need for this is a specific TypeScript file that is generated specifically for your Supabase project. The following steps show how to trigger the Supabase CLI so that it creates such a `supabase.ts` file containing the needed types for TypeScript – depending on whether you want the types from a `supabase.com` project, a local instance, or an instance hosted somewhere else than `supabase.com`:

- If you want types for a project based on `supabase.com`, follow these steps to get a `supabase.ts` file:

  I. Go to `https://supabase.com/dashboard/account/tokens` and create an access token.

  II. Run `npx supabase login`. You'll be asked for the access token you just generated. After pressing *Enter*, it will tell you that the login process has succeeded.

III. Now, open your project via `supabase.com`; you'll see a link in your browser that looks like `https://supabase.com/dashboard/project/YOUR_PROJECT_ID/....` You'll also find the same project ID as part of your API URL in the **Settings** | **API** section. Copy this project ID.

IV. Generate your custom `supabase.ts` file by running `npx supabase gen types typescript --schema public --project-id **YOUR_PROJECT_ID** > supabase.ts`

- If you're running a local instance, which you should have by now, and want to grab the types from there, you don't need an access key. You only need to run the following command in your project folder (this is where we ran `npx supabase init` previously in this chapter):

```
npx supabase gen types typescript --schema public --local  >
supabase.ts
```

Note that if you run it *outside* of the project folder, it won't know which local instance you're referring to and fail.

- If you have an instance that's self-hosted on a remote server or running with a provider other than `supabase.com`, then the previous steps won't work and you'll need the generalized variant of fetching types with a direct database connection. To do that, you must generate the `supabase.ts` file, as follows:

I. Find your database URL (see the *Connecting directly to the database* section). For example, in your local instance, you'll find it in the Terminal output after starting Supabase with `npx supabase start`. It will be in the following format: `postgresql://**USER:PASSWORD**@**DB_HOST**:PORT/postgres`.

II. Run `npx supabase gen types typescript --schema public --db-url postgresql://**USER:PASSWORD** @**DB_HOST**:PORT/postgres > supabase.ts`. You'll receive the file.

With this `supabase.ts` file, it's easy to make your client type-safe and get proper type hints – simply import the `Database` type from `supabase.ts` and pass it to the client creation process. For example, if you want to make the `createReqResSupabase({req,res})` function type-safe, you just pass the `<Database>` type when creating the client:

```
import type { Database } from './supabase';
export const getSupabaseReqResClient = ({ req, res }) => {
  return createServerClient<Database>(...);
};
```

With that, your Supabase client is type-safe. But let's understand what that means and what it implies. Say, for example, you're fetching data from a specific table of your database: the Supabase client will exactly know which columns to fetch and provide proper type support for the returned data.

But what happens when I change anything in my instance? Won't it be outdated immediately as my `supabase.ts` file contains outdated types?

Let me try to answer this question with another question: How can you use a new feature on your smartphone if the new feature is only available in a newer software version? The simple answer is that you update the software version.

The same goes for the Supabase types. Anytime you change something in your Supabase project and it doesn't give you the proper TypeScript hints, run `npx supabase gen types typescript . . .` again and you'll be all set.

With this, you can use Supabase in a TypeScript-based project. Before finishing up this chapter, we'll have a look at some samples of how a Supabase client can be used with other frameworks so that you're familiar with Supabase's flexibility.

Connecting Supabase to other frameworks

Imagine that you've set up an awesome project with Next.js and Supabase. However, one day, you want to add another feature to your project – an extremely fast API that does complex calculations based on data from your Supabase instance. You notice that JavaScript won't be the best choice and decide to build a small Python server for this feature that can be called from your primary project.

This is what I did in one of my projects at Wahnsinn Design GmbH where the web application, with Supabase at its heart, was built with Next.js. However, a new feature was added using another project with Python. Since there is a Python library for Supabase, the connection was seamless.

Since Supabase is not framework-dependent, since it's just REST APIs, the options for integrations are endless, from *C#*, *Swift*, and *Kotlin*, to JavaScript-based frameworks such as *Nuxt* or *refine* (you'll find the most recent list at `https://supabase.com/docs`).

Although we will focus on JavaScript with Next.js in this book, you can use most samples, especially in the upcoming chapters, and translate them into other languages or frameworks with ease. This is because using the Supabase client for the different languages will have similar syntax (as far as the language allows).

Let's have a brief look at how to connect Supabase in Nuxt and Python.

Nuxt 3

Nuxt is the Vue-based full-stack competitor to Next.js. Connecting with Nuxt comes down to installing the `@nuxtjs/supabase` package – which, again, is just a convenient wrapper for the `@supabase/supabase-js` package.

Once installed with `npm install @nuxtjs/supabase`, add the module to your Nuxt configuration, like so:

```
export default defineNuxtConfig({
    modules: ['@nuxtjs/supabase'],
})
```

Similar to our Next.js application, add the `anon` key as SUPABASE_KEY and your API URL as SUPABASE_URL to the `.env` file of your Nuxt project.

Now, you can use the client in Vue composables, like so:

```
<script setup lang="ts">
    const supabase = useSupabaseClient();
</script>
```

Alternatively, you can use proper TypeScript types, as we've already learned, like so:

```
<script setup lang="ts">
    import type { Database } from '~/supabase';
    const client = useSupabaseClient<Database>();
</script>
```

You can find a detailed explanation of Nuxt 3 at `https://supabase.nuxtjs.org/get-started`.

Python

Python is fast and has become more popular than ever with many AI applications. This is because it is convenient to use for scientific calculations.

The Python Supabase package is one of the easiest to use:

1. Install the Supabase package and the `dotenv` package with `pip install supabase` and `pip install python-dotenv`, respectively.

2. Create a `.env` file with two lines, one being your SUPABASE_ANON_KEY=... value and the other being your SUPABASE_URL=... value.

3. Initialize the Supabase client in a file such as `supabase_client.py`, as follows:

   ```
   import os
   from dotenv import load_dotenv
   from supabase import create_client, Client
   load_dotenv()
   ```

```
supabase_url: str = os.getenv("SUPABASE_URL")
supabase_anon_key: str = os.getenv("SUPABASE_ANON_KEY")
my_supabase: Client = create_client(supabase_url, supabase_anon_
key)
```

4. Use it in any file via `import`:

```
from supabase_client import my_supabase
...
```

You can find the full Python documentation here: `https://supabase.com/docs/reference/python`.

I'd be lying if I said all frameworks and languages are equal concerning updates and support within the Supabase community. On the web, there is a general trend toward JavaScript-based environments (Vue, Next, React, Nuxt, Remix, Svelte, Deno, you name it) and at the time of writing this book, several client libraries exist, including JavaScript, Flutter, Python, C#, Swift, and Kotlin.

However, it is extremely important to keep in mind that Supabase can be used in any framework or language due to its REST-based nature and that Supabase is also very keen on contributions. Lastly, you can always just use the direct database connection – but with that, you'd be bypassing all authentication and permissions.

With this at hand, you are well-positioned to tackle any project with Supabase, no matter if you are using a framework-specific client, the RESTful API, or the direct database connection.

Summary

In this chapter, you started setting up the ticket system project by initializing not only a new Next.js project but also a local Supabase instance. At this point, you are capable of managing multiple Supabase instances locally.

By completing this chapter, you've seen what the Terminal output of your instance means and how it connects the dots back to *Chapter 1*, where you learned about the inner workings of Supabase.

On top of that, you installed various Supabase helpers, added connection credentials for your Supabase instance to your Next.js environment file, and proved that your Supabase client is connecting to it by fetching an empty list of storage buckets.

You also know where you have to look for the connection credentials if you're not using a local instance but one via `supabase.com`. You also learned how to directly access the database if needed – that is, by bypassing the Supabase REST APIs and their protection.

In the next chapter, we will build the basic structure of the ticket system project while using our Supabase client to protect our application, which will eventually allow users to log in with a password or magic link.

3

Creating the Ticket Management Pages, Layout, and Components

After successfully setting up your Next.js application and establishing an instantiable Supabase client in the previous chapter, it's time for a refreshing change. In this chapter, we will shift our focus to building the project's design interface – the second foundational layer. That means there is no Supabase-specific knowledge to be found in this chapter.

The objective here is to create a navigable design that mimics real content using mock data. Designing with mock data not only liberates your mind from implementation details but also offers a crucial psychological boost. Visualizing the project in action, even with mock data, enhances our understanding and excitement.

Not only that but having a nice design will give us a lovely sense of achievement in this chapter, as well as lay the ground work for future sucesses when revitalizing each portion of the app through code logic in upcoming chapters.

In *Chapter 1*, you were provided with a brief introduction to the ticket system you are creating throughout this book. Now, in this chapter, you'll build its most necessary design foundation using file-based routing, the correct elements and components, and a bit of CSS.

The following topics will be covered in this chapter:

- Setting up Pico.css with Next.js
- Building the login form
- Visualizing the Ticket Management UI
- Creating a shared layout with navigation elements
- Designing the Ticket List page

- Making the Ticket Details page

- Implementing the page to create a new ticket

- Implementing a user overview

- Enhancing the navigation component

By the end of this chapter, you'll have solid-looking pages and components created that you can test in your browser, and which will be brought to life and polished piece by piece in the upcoming chapters. Exciting times ahead!

Technical requirements

In this chapter, you'll face the same technical requirements as in the last one. If you want to further improve the style of the application, it can be very useful to have good CSS knowledge, but there's no given requirement for it.

Also, again, make yourself aware of the fact that, at any point in time, you can check the application code in the book's GitHub repository: `https://github.com/PacktPublishing/Building-Production-Grade-Web-Applications-with-Supabase`. You will find the final project code for this chapter in the `Ch03_TicketSystem_UI` branch.

Setting up Pico.css with Next.js

Since this isn't a CSS book, we don't want to be bothered by designing too much, but everybody loves a clean UI. That's why we will add Pico.css (`https://picocss.com/`) to our project now. It is a CSS file that will set default styles without the need to add CSS classes – everything gets useful default styles. On top of that, it also seamlessly supports dark and light themes out of the box.

The easiest way to install Pico.css is by using a public **content delivery network** (**CDN**). All you need to do is pop a `<link>`- tag into your Next.js project's root layout, `app/layout.js`, like so:

```
<head>
  <link
    rel="stylesheet"
    href="https://cdn.js
          delivr.net/npm/@picocss/pico@1/css/pico.min.css"
  />
</head>
```

In that same file, in the `<body>` tag, you should have something like this right now:

```
<body>
    <main>{pageProps.children}</main>
</body>
```

Pico has a useful class named `container` that will center an element at a certain desktop width, instead of 100% width on huge screens, so let's add it such that it looks like this:

```
<main className="container">{pageProps.children}</main>
```

That's all we need to do to have solid, good-looking base styles. We can still add more CSS manually if we want to. Now, let's start building pages and components for our app.

Building the login form

Let's start building the **Login** page for our ticket system. To keep pages clean and readable, let's create a component in a new file named `app/Login.js`.

> **Reminder**
> `app/Login.js` is not a page but a component! In the Pages Router, every file within the `pages/` directory was a page. In the App Router, pages are always named `page.js`, and their routed path is defined by the directory they are in. This allows you to create non-page files (such as component files) co-located with the actual page files.

In this component file, we will start by defining the `Login` component skeleton, like so:

```
export const Login = () => {
  return (
    <form>
      <article style={{ maxWidth: "420px", margin: "auto" }}>
        <header>Login</header>
        <strong>Hello Login!</strong>
      </article>
    </form>
  );
};
```

The `<article>` tag in Pico, as per the documentation, is used for slightly highlighted containers with gracious padding, and that's what we want here. We also give the card two inline styles to center it and give it a maximum width, such that it looks more like a well-known login screen.

> **Note**
> Even though `<article>` might seem odd semantically, it is the best element for such a use case according to `https://developer.mozilla.org/en-US/docs/Web/HTML/Element/article`.

Before adding more elements to the component, we want to make sure it will be shown when visiting the main page. So, open up `app/page.js` and change the `Page` component so that it only will return the `Login` component for now:

```
import { Login } from "./Login";

export default function LoginPage() {
  return <Login />;
}
```

Now, when you run the Next.js server, you should see a card with heading text that says **Login** and, below that, **Hello Login!**.

> **Note**
>
> Why don't we just put all of the code in the page itself without having to create another `Login.js` file? Well, you could, but there are two simple reasons why I split it. First, I wanted to clarify the difference between a page and a component again. Second, I usually tend to put components with a specific task within their own file. But it's equally okay to just put the code in the page file.

Later, we want to allow logging in with a password, as well as with a magic link, so we'll add the required HTML elements now. To do this, we want to have two different types of forms:

- One that allows you to sign in with a password and email address
- One that allows login via a magic link – you provide your email (no password) and get a login link sent to your email address

Both will need the email field, but for the password login, we need an additional password field.

For most frontend people, the first thing that comes to mind is a frontend-based toggle to switch the second input on or off, using something such as `const [isPasswordLogin, setIsPasswordLogin] = useState(true)`. That would indeed be handy, but it would only work on the frontend; with disabled or broken JavaScript, it wouldn't work anymore.

In the next chapter, I'll show you how to add progressively enhanced authentication that works on the backend and frontend. Hence, we want to prepare a UI that also doesn't just rely on frontend capabilities. We want to hide or show the password field based on a query parameter in the URL, as that works on the backend too.

That's why we toggle the field by checking if the `GET` parameter, `magicLink=yes`, is part of the URL (`http://localhost:3000/?magicLink=yes`). If it is set, then we hide the password field, as we want to use the magic link approach; otherwise, we show the password input field.

So, let's read this parameter in `LoginPage`. Within a Next.js page file, you get easy access to all `GET` parameters directly on the server via the passed property, `searchParams`. So, let's make sure we read the `magicLink` value and pass it on to the `Login` component:

```
export default function LoginPage({ searchParams }) {
  const wantsMagicLink = searchParams.magicLink === "yes";
  return <Login isPasswordLogin={!wantsMagicLink} />;
}
```

We'll then implement the switch within `app/Login.js`, as follows:

```
import Link from "next/link";
export const Login = ({ isPasswordLogin }) => {
 return (
  <form>
   <article style={{ maxWidth: "480px", margin: "auto" }}>
    <header>Login</header>

    <fieldset>
     <label htmlFor="email">
      Email
      <input
       type="email"
       id="email"
       name="email"
       required />
     </label>

     {isPasswordLogin && (
      <label htmlFor="password">
       Password
       <input
        type="password"
        id="password"
        name="password"
       />
      </label>
     )}
    </fieldset>

    <p>
     {isPasswordLogin ? (
      <Link
       href={{
```

```
      pathname: "/",
      query: { magicLink: "yes" },
    }}
  >
    Go to Magic Link Login
  </Link>
) : (
  <Link
    href={{
      pathname: "/",
      query: { magicLink: "no" },
    }}
  >
    Go to Password Login
  </Link>
)}
</p>

<button type="submit">
  Sign in with
  {isPasswordLogin ? " Password" : " Magic Link"}
</button>
</article>
</form>
);
};
```

As the page passes the value of the query parameter on to our Login component, we hide the password field, change the button text, and change the link for the password login or the magic login – all of this is based solely on one value, isPasswordLogin. Now, you can easily jump between the two modes if you click the link below the input fields.

Next, we want to store a reference to both the email and password input elements so that we can access the values the user entered when submitting them. Saving the input references will be done using useRef(), and we require the Login component to be a Client component. It should look like this:

```
"use client";
import { useRef } from "react";

export const Login = () => {
  const emailInputRef = useRef(null);
  const passwordInputRef = useRef(null);
  return (
```

```
        ...
    )
  }
```

Then, for both elements, apply `ref` accordingly. So, for the email input, type `<input ref={emailInputRef} type="email" name="email" id="email" required />`, and do the same for the password input field. Now, we can access the text-based input field values.

At this point, by default, your UI should look like this:

Figure 3.1: A default login UI

And when **Go to Magic Link Login** is clicked, it should look like this:

Figure 3.2: The login UI with the password option

Let's wrap this up by including a small piece of code to test the form submission. This will trigger an alert box, clearly indicating the expected action when the form is submitted:

```
<form
  onSubmit={(event) => {
    event.preventDefault();
    if (isPasswordLogin) {
      alert("User wants to login with password");
    } else {
      alert("User wants to login with magic link");
    }
  }}
>
```

By using `event.preventDefault()`, we prevent the page from reloading, and then, depending on the state, we know whether the user wants to log in via a password or a magic link.

> **Note**
>
> Instead of saving references, a good approach in React is using **controlled forms** – this involves not just reading what is inside the input but also setting it at every keystroke, effectively controlling the value and potentially filtering or limiting it. For our use case, reading the value through references is totally fine. Check out the following resource to learn more about controlled forms: `https://react.dev/learn/sharing-state-between-components#controlled-and-uncontrolled-components`.

Now, we have a proper login form, which we will bring to life in the upcoming chapter. Let's move on to design the part that comes after logging in – the actual ticket management system.

Visualizing the Ticket Management UI

We'll now build the pages and components we will need to manage tickets. Let's recall the key requirements of that area:

- Navigation and the possibility of logging out
- An overview page for the ticket list
- A possibility to view ticket details with a comments section
- A possibility to create a new ticket
- A tabular list of users for the tenant, which will allow other users to see who is available and who is an admin user with higher permissions

To help you, I have sketched a conceptual design for you to follow along:

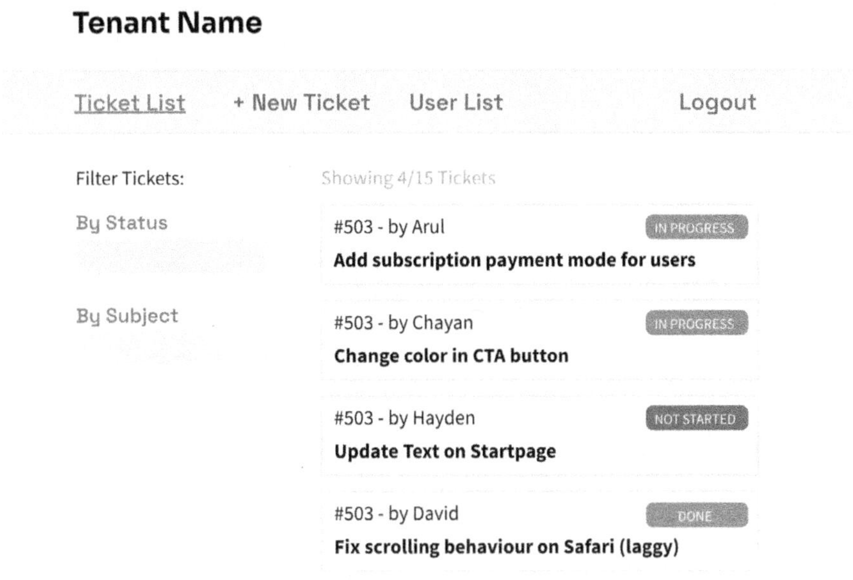

Tenant Name

Ticket List	+ New Ticket	User List		Logout

Filter Tickets:

Showing 4/15 Tickets

By Status

#503 - by Arul   IN PROGRESS
Add subscription payment mode for users

By Subject

#503 - by Chayan   IN PROGRESS
Change color in CTA button

#503 - by Hayden   NOT STARTED
Update Text on Startpage

#503 - by David   DONE
Fix scrolling behaviour on Safari (laggy)

Figure 3.3: A conceptual design for the ticket system

In our project, we don't want to have complex navigation. At the top of the screen, we want to have the company (tenant) name. Then, in the navigation bar, we will only need three primary navigation links – **Ticket List**, **Create New Ticket**, and **User List**. At the end of the navigation bar, we will add a **Logout** button. For each of the links in the navigation, we also want to make sure that the active route is highlighted (active) in the navigation component. So, let's start designing!

Creating a shared UI layout with navigation elements

Our **Login** page is the root page of our application, as shown when you visit `http://localhost:3000/` when running the development server with `npm run dev`. Later, when logged in, the entry page for the management of the tickets shall be reachable at `http://your-domain/tickets`.

Due to the file-based routing in Next.js, we will put all files related to ticket management below the `app/tickets/` directory, so all routes related to ticket management will start with `/tickets`. At `/tickets`, we will have the dashboard (Ticket List); at `/tickets/new`, it should be possible to create a new ticket, and so on. Everything related to ticket management will be at `http://your-domain/tickets/*`, explained in one wildcard URL. This not only makes sense semantically and structurally but also allows us to easily protect everything that sits within `/tickets` in the next chapter.

To get started, create the `app/tickets` directory. Then, we want to create the one thing that all of the different pages have in common – the layout. Across all of the pages, the tenant name and navigational structure will stay the same. We can do this by creating an `app/tickets/layout.js` file with a placeholder for the tenant name and navigation, as follows:

```
export default function TicketsLayout(pageProps) {
  return (
    <>
      <section style={{ borderBottom: "1px solid gray" }}>

  {/* tenant name component goes here */}

  {/* navigation component goes here */}
      </section>

      <section>{pageProps.children}</section>
    </>
  );
}
```

This layout file has two `<section>` elements – the first we will fill with the tenant and navigation component, while the second contains the current page's content by using what Next.js passes as a `children` prop.

Now, create two files, `app/tickets/TenantName.js` and `app/tickets/Nav.js`, which we will use to create the named components respectively.

For `TenantName.js`, we will use the following component code, which simply displays the word **Ticket System**, and then the name of the tenant, which can be passed as a `tenantName` property:

```
export default function TenantName(props) {
  return (
    <header style={{ marginBottom: "10px" }}>
      <div
        style={{
          borderLeft: "4px solid orange",
          display: "block",
          padding: "4px 10px",
          fontSize: "1.1em",
        }}
      >
        Ticket System
        <strong style={{ marginLeft: "1ex" }}>
          {props.tenantName}
        </strong>
```

```
        </div>
      </header>
   );
}
```

The added CSS in the `style` attribute adds a bit of white space and highlights the brand-like name (see *Figure 3.4*). Before adding this new component to the layout, let's also create a navigation component.

Switch to the empty `Nav.js` file you created and fill it with the navigation elements, like so:

```
import Link from "next/link";
export default function Nav() {
  return (
    <nav>
      <ul>
        <li>
          <Link role="button" href="/tickets">Ticket List</Link>
        </li>
        <li>
          <Link role="button" href="/tickets/new">Create new Ticket
          </Link>
        </li>
        <li>
          <Link role="button" href="/tickets/users">User List</Link>
        </li>
      </ul>
      <ul>
        <li>
          <Link
            role="button"
            href="/logout"
            className="secondary">
            Log out
          </Link>
        </li>
      </ul>
    </nav>
  );
}
```

In this code block, there are two `<ul>` elements within the `<nav>` element. In Pico.css, this is how to have the first one at the left edge and the second one at the right edge of the design.

In the first `<ul>`, there are three links, targeting pages that we will create one by one in the following sections of this chapter to view tickets, create tickets, and view the list of users.

In the second `<ul>`, there sits only the **Log out** button, for which we won't create a page in this chapter; we will deal with the intricacies of logging out in the next chapter. It has an additional class secondary, which gives it a more subtle, grayish look.

Now, prepare your base layout by importing and using the tenant and navigation components in `app/tickets/layout.js`, as follows:

```
import TenantName from "./TenantName";
import Nav from "./Nav";

export default function TicketsLayout(pageProps) {
  return (
    <>
      <section style={{ borderBottom: "1px solid gray" }}>
        <TenantName tenantName="Packt" />
        <Nav />
      </section>

      <section>{pageProps.children}</section>
    </>
  );
}
```

Your base layout is now ready to be used with a page. But a `layout.js` file in Next.js is always just a container for the pages. There's no page file yet at `/tickets` that Next.js could render. Hence, we will move on to creating the required pages now.

Designing the Ticket List page

The first button in our navigation component points to `/tickets`. Let's create a page for that by creating the `app/tickets/page.js` file. This will, later, be the first page you'll see after a successful login.

To get started and test whether our code works, we'll add some dummy code to `app/tickets/page.js` first:

```
export default function TicketListPage() {
  return <div>Ticket List should go here</div>;
}
```

Now, you have a page alongside your layout, which you can test by running the server (`npm run dev`). When you visit `localhost:3000/tickets`, you should see something like this:

Figure 3.4: The page at /tickets

Next, let's go one step further and create a list of tickets, with some test tickets on that page. First, define some dummy tickets in the page file:

```
const dummyTickets = [
  {
    id: 1,
    title: "Write Supabase Book",
    status: "Not started",
    author: "Chayan",
  },
  {
    id: 2,
    title: "Read more Packt Books",
    status: "In progress",
    author: "David",
  },
  {
    id: 3,
    title: "Make videos for the YouTube Channel",
    status: "Done",
    author: "David",
  },
];
```

Then, let's create a component file, `app/tickets/TicketList.js`, which takes an array of tickets like we just created and lists them neatly. We want to list them in a table, so let's start by defining a table structure in `app/tickets/TicketList.js`:

```
export function TicketList({ tickets }) {
  return (
    <table>
      <thead>
        <tr>
          <th>ID</th>
```

```
          <th></th>
          <th>Status</th>
        </tr>
      </thead>
      <tbody>{/* tickets go here */}</tbody>
    </table>
  );
}
```

With this structure, we have the ID in the first column, whatever we want to display as ticket data in the second column, and the status in the last column.

What's missing in this component is listing the table rows, showing the data of the passed `tickets`. By iterating through the tickets, our final component becomes this:

```
export function TicketList({ tickets }) {
  return (
    <table>
      <thead>
        <tr>
          <th>ID</th>
          <th></th>
          <th>Status</th>
        </tr>
      </thead>
      <tbody>
        {tickets.map((ticket) => (
          <tr key={ticket.id}>
            <td>{ticket.id}</td>
            <td>{ticket.title}</td>
            <td>{ticket.status}</td>
          </tr>
        ))}
      </tbody>
    </table>
  );
}
```

Now, in your `app/tickets/page.js`, import the `TicketList` component and use it as follows:

```
export default function TicketListPage() {
  return (
    <>
      <h2>Ticket List</h2>
      <TicketList tickets={dummyTickets} />
    </>
```

```
    );
}
```

In your browser, the page should now look like this:

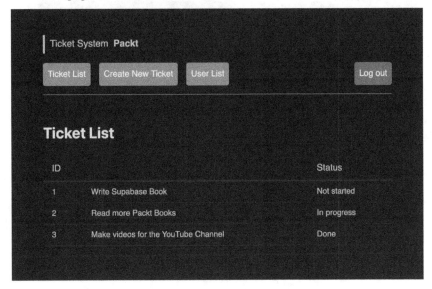

Figure 3.5: Dummy tickets showing on the page

From here, we want to enable viewing ticket details. Each ticket shall be clickable and lead to a Ticket Details page. That's what we will do in the next section.

Constructing the Ticket Details page

When a specific ticket is clicked, we need a page that shows the details of it. For example, the ticket details for a ticket with the ID 857 should be reachable at `http://your-domain/tickets/details/857`. This URL uses dynamic parameters in which we can simply create a suitable directory/file structure, using `tickets/details/[id]/page.js`. By doing so, Next.js will recognize that the part after `details/` needs to be passed to us as `id`.

To do this, create the `tickets/details/[id]/page.js` file and put the following code inside of it:

```
export default function TicketDetailsPage({ params }) {
    return (
        <div>
            Ticket Details page with <strong>ID={params.id}</strong></strong>
        </div>
    );
}
```

If you visit any URL using that scheme – for example, `/tickets/details/123` – you should now see this:

Figure 3.6: The parametrized Ticket Details page

The next step is to link the previously created ticket list with this details page, and then afterward, we will elevate the Ticket Details page to contain a visual ticket representation.

Go back to your `app/tickets/TicketList.js` component. Here, you want to make sure that each ticket title is a link that targets the Ticket Details page, so when the link is clicked, we are forwarded to the corresponding ticket details route. To do this, you can simply use the `Link` component from Next.js again, such that `<td>{ticket.title}</td>` becomes the following:

```
<Link href={`/tickets/details/${ticket.id}`}>{ticket.title}</Link>
```

That snippet links the title to the `/tickets/details/TICKET_ID` path.

Open up `/tickets` in the browser again, and you should see that all titles are clickable and lead to the details page, with the correct `id`.

Now, let's move forward and improve the Ticket Details page, making it show more than just a placeholder sentence with the ID. When a user of your ticket system clicks an existing ticket, the user should be able to get all the required information to process the ticket, such as a detailed description, the author, the creation date, its status, and comments. In *Figure 3.7*, you can see our target design without a comments section yet.

Figure 3.7: A minimal Ticket Details view without the comments section

As with the login form, we want to use the card style, splitting the core information into a header-like box and more information below that. To do this with Pico.css, we need the article tag, with a header element inside that gives us this style. In the header part, we'll have the ticket ID with a leading hashtag, #, and the ticket status, and then below that, we'll have both the name of the ticket author as well as the creation time. Then, below that, there will be the title of the ticket and the box with the ticket description text.

To create this, open up your app/tickets/details/[id]/page.js file and add the appropriate HTML elements, like so:

```
export default function TicketDetailsPage({ params }) {
  return (
    <article>
      <header>
        <strong>#{params.id}</strong> - <strong>Open</strong>
        <br />
        <small>
          Created by <strong>AuthorName</strong> at{" "}
          <time>December 10th 2025</time>
        </small>
        <h2>Ticket title should be here</h2>
      </header>

      <section>
        Some details about the ticket should be here.
      </section>
    </article>
  );
}
```

When you save the code and view the result in your browser, it won't look exactly like *Figure 3.7* but more like *Figure 3.8*. This is because it misses some custom styling for the separating borders, some color on the **Open** status, and the light/dark contrast in the line containing the author's name.

Figure 3.8: The Ticket Details page (without styling)

To add this custom styling and make the UI look like *Figure 3.7*, create the `app/tickets/details/[id]/TicketDetails.module.css` file, where we will add some custom CSS classes.

> **Note**
>
> Next.js, as with many other frameworks, supports CSS Modules (`https://github.com/css-modules/css-modules`). CSS Modules are usually CSS files that contain CSS classes, with the specialty that the file itself can be imported inside JavaScript, so the classes are then available as an object and can be automatically compiled and added to your application.

To apply a greenish color to the ticket status, add the following class to the CSS file:

```
.ticketStatusGreen { color: greenyellow }
```

Then, for the author/date section, the following class will make the text look lowlighted and highlight the strong tag inside of it (the CSS variables come from Pico.css):

```
.authorAndDate { color: var(--secondary) }
.authorAndDate strong { color: var(--contrast) }
```

Lastly, to add a border style to the card, we will create a class for the wrapper container, as follows:

```
.ticketDetails { border: 1px solid #3c5c60 }
.ticketDetails > header { border-bottom: 1px solid #244538}
```

Now, the only task left is to assign those classes inside our page code. On the Ticket Details page file, use the following statement to import the classes as an object:

```
import classes from "./TicketDetails.module.css";
```

Since we want the `.ticketDetails` class to style the `article` element of the page, we simply write `<article className={classes.ticketDetails}>`. For the status element, we add the greenish color with `<strong className={classes.ticketStatusGreen}>Open</strong>`. Finally, the `authorAndDate` class is added to the appropriate `small` element – `<small className={classes.authorAndDate}>`.

Now, your custom styles are in place, and everything should look like *Figure 3.7*.

Adding the comments section to the ticket details

To get a full picture of our ticket details, we want to have a `textarea` element so that any user can write a new comment. For existing comments, we want to list the time and date, the author, and, obviously, the comment itself.

As I like well-separated files, we'll create a `TicketComments` component within the `app/tickets/details[id]/TicketComments.js` file. In that file, I will create and export the following `TicketComments` component with some mocked `comments` data:

```
'use client';
import { useRef } from "react";
const comments = [
  {
    author: "Dave",
    date: "2027-01-01",
    content: "This is a comment from Dave",
  },
  {
    author: "Alice",
    date: "2027-01-02",
    content: "This is a comment from Alice",
  },
];

export function TicketComments() {
  const commentRef = useRef(null);
  return (
    <footer>
      <h3>Comments</h3>
      <form
        onSubmit={(event) => {
          event.preventDefault();
          alert("TODO: Add comment");
        }}
      >
        <textarea ref={commentRef} placeholder="Add a comment" />
        <button type="submit">Add comment</button>
      </form>
      <section>We have {comments.length} comments.</section>
    </footer>
  );
}
```

The `footer` element is used instead of `article` this time, since it will be the footer of the ticket card. And since we use frontend functionality such as `useRef` and `onSubmit`, we need to mark the component as the client component (`'use client';`).

Let's place it in the ticket card now by importing it and putting `<TicketComments />` at the end of the `<article className={classes.ticketDetails}>` element of your `TicketDetailsPage` component. The result in your browser should then look like *Figure 3.9*.

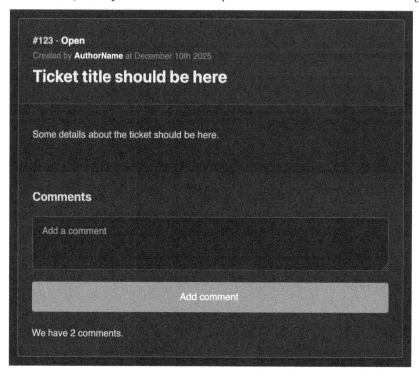

Figure 3.9: The Ticket Details page with the comments area

Now that already looks pretty! But let's enhance it such that we actually list the comments and not just state the number of tickets. The final ticket component should look like this:

```
<footer>
  <h4>Comments</h4>

  <textarea placeholder="Add a comment"></textarea>
  <button>Add comment</button>

  <section>
    {comments.map((comment) => (
      <article key="{comment.date}">
        <strong>{comment.author} </strong>
        <time>{comment.date}</time>
        <p>{comment.content}</p>
```

```
        </article>
    ))}
    </section>
</footer>
```

Without further changes, the result will look like *Figure 3.10*.

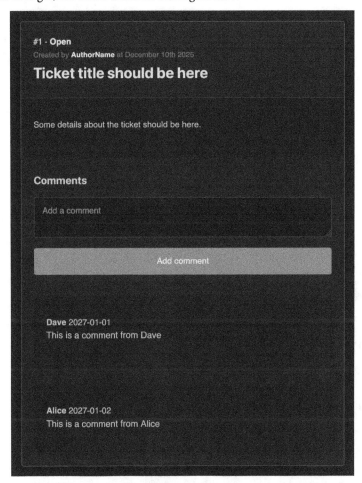

Figure 3.10: The Ticket Details page with comments added

However, I still wasn't quite happy with the previous result. As I didn't like the dark-on-dark shade and the big gap between the comments, as well as the color on the `time` tag, I added another CSS class named `.comment` to the existing CSS Module file. Then, I imported it into the `TicketComments.js` file and added it to each comment (`<article key={comment.date} className={classes.comment}>`).

You can do this too – here's the CSS to add to the `TicketDetails.module.css` file:

```css
.comment {
  --color-card-brighter: color-mix(
    in srgb,
    var(--card-background-color),
    var(--secondary) 20%
  );
  margin: 0;
  padding: 0.7em 1em 0.7em;
  background-color: var(--color-card-brighter);
}
.comment:not(:last-child) {
  margin-bottom: 5px;
}
.comment time {
  color: var(--secondary);
  padding-left: 0.5em;
}
```

Figure 3.11 shows my final result, with a working submit button triggering an alert:

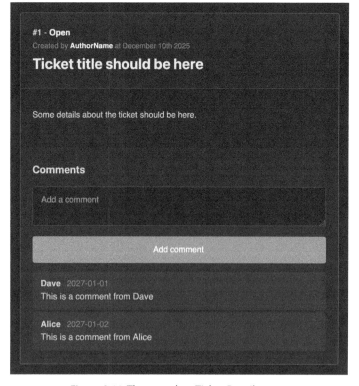

Figure 3.11: The complete Ticket Details page

Awesome! Hopefully, you agree that we have a good-looking Ticket Details page. We can see a ticket list, and we can click a specific ticket and get routed to the correctly parametrized Ticket Details page, where we will see the design of the ticket details.

However, we don't have a function to create a ticket yet. Let's do that in the next section.

Implementing a page to create a new ticket

We'll now create a page that is shown when the **Create New Ticket** navigation button is clicked.

Since, per our navigation component code, we wanted the ticket creation to be at `/tickets/new`, you now have to create the `app/tickets/new/page.js` file.

Then, we want to show a form that allows us to set the title and a description for a new ticket – that's sufficient at the moment, but you can extend the form to allow more input later (such as **Urgency**, **Reference Link**, or whatever comes to your mind with regards to ticket management).

We also want frontend functionality, such as onSubmit, in the ticket creation form, so we'll certainly need a client component (`'use client';`). This time, we will simply make the page itself the client component.

To do this, go to `app/tickets/new/page.js` and add the following code:

```
"use client";
import { useRef } from "react";

export default function CreateTicket() {
  const ticketTitleRef = useRef(null);
  const ticketDescriptionRef = useRef(null);

  return (
    <article>
      <h3>Create a new ticket</h3>

      <form
        onSubmit={(event) => {
          event.preventDefault();
          alert("TODO: Add a new ticket");
        }}
      >
        <input ref={ticketTitleRef} placeholder="Add a title" />
        <textarea
            ref={ticketDescriptionRef}
            placeholder="Add a comment" />
        <button type="submit">Create ticket now</button>
```

```
        </form>
      </article>
  );
}
```

Other than the fact that it's a component to add a new ticket, there's nothing in here you haven't seen already. We will use the Pico.css Card (`<article>`) within a form, and then store references of the input fields so that we can easily access the value.

Now, click on **Create New Ticket** in the navigation, and you should see a similar screen to *Figure 3.12*.

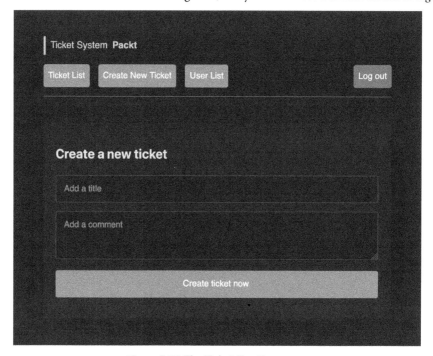

Figure 3.12: The Ticket Creation page

How wonderful – we're nearly done with the foundational app design! In the next section, we'll add the last page – the **User List** page.

Implementing a user overview

The idea of the **User List** page is that users of a tenant get an overview of all the users in the system for that specific company/tenant. This allows each user to see who is in the system, who is available (free or out of office), and who has what role in the company (e.g., manager, engineer, designer).

So, let's create a user list with mocked users. In the navigation bar (Nav.js), we link to the /users route, so create the app/users/page.js file. Then, in that file, we first add a variable with mocked user data:

```
const users = [
   {
     name: "Alice Ling",
     job: "Software Engineer",
     isAvailable: false,
   },
   // ... add as much users as you want
];
export default function UserList() {
   return <div>We have {users.length} users</div>;
}
```

Next, we replace the returned value with a user list table, as follows:

```
return (
 <table>
  <thead>
   <tr>
    <th>Name</th>
    <th>Job</th>    </tr>
  </thead>
  <tbody>
   {users.map((user) => (
    <tr key={user.name}>
     <td>
       {user.name}
       ({user.isAvailable ? "Available" : "Not available"})
     </td>
     <td>{user.job}</td>
    </tr>
   ))}
  </tbody>
 </table>
);
```

As I personally find it to be a better UX when statuses are accompanied by symbols in front of a name, I installed the `@tabler/icons-react` npm package and imported two icons from it – a check icon, indicating that the user is available, and a stikethrough icon, indicating that the user is not available. Then, I changed the first `<td>` to use the appropriate icon depending on the user's availability, as well as added a red color when they are unavailable. The code for this can be seen here:

```
<td style={{ color: !user.isAvailable ? "red" : undefined }}>
  {user.isAvailable ? <IconCheck /> : <IconUserOff />} {user.name}
</td>
```

In *Figure 3.13*, you can see the user list and the symbols:

Figure 3.13: The user list table

Now, we're done with all the pages, and the foundational design has been built! Next, we'll do some final optimization to conclude this chapter – making the active navigation item visually active.

Enhancing the navigation component

We are nearly done with the UI, which will allow us to add our first real functionality with Supabase in the next chapter. However, there's one thing that's annoying – the navigation items don't visually reflect the active route. They look slightly highlighted when you click them, but this is only because the item has focus after being clicked. When you click anywhere else on the page, the item loses focus and its visual highlight; we want the active navigation item to stay highlighted.

In Pico.css, you can make a link stand out by giving it the `contrast` class and `role="button"`. We can use this to mark an active route in a navigation item as active. Hence, you need to adapt your `app/tickets/Nav.js` file slightly.

To be able to check which link is active, we will import the usePathname() from next/navigation, which will tell us the active pathname of the URL. This only works in client components, so we need to add "use client"; to the top of the file as well:

```
"use client";
import Link from "next/link";
import { usePathname } from "next/navigation";

export default function Nav() {
  const pathname = usePathname();
  /** render return **/
}
```

Next, you want to add className="contrast" if the active pathname is the same as href in each link. Inside of your Nav Component function, define that as an object to be reused:

```
const activeProps = { className: "contrast" };
```

I will also add the complementary variable with an empty object for inactive links:

```
const inactiveProps = {};
```

Then, you can use this variable that contains the properties for the component, plus the spread operator (...), to add the properties to the link if – and only if – the active pathname matches:

```
<Link
  role="button"
  href="/tickets"
  {...(pathname === "/tickets" ? activeProps : inactiveProps )}>
  Ticket List
</Link>
```

Do that for all links, and you will have properly highlighted active states in your navigation.

Also, give the non-active links the secondary and outline CSS classes, as they give a nice, contrasted look. Just adapt your inactiveProps variable accordingly with the className property:

```
const inactiveProps = { className: "secondary outline" };
```

Here is the result in the browser:

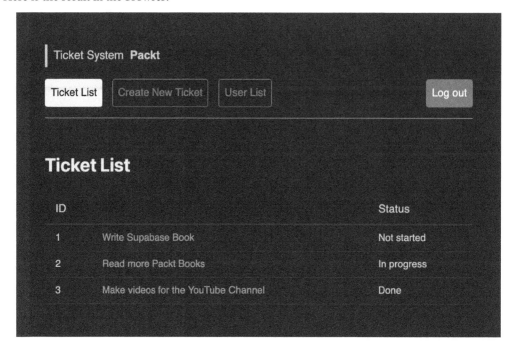

Figure 3.14: The /tickets page with dummy data and highlighted active navigation

> **Reminder**
> The final code can be found in the repository in Ch03_TicketSystem_Start.

Now, the navigation highlights according to the active route. Lovely!

Summary

Throughout this chapter, you've done the lion's share of the design work for the app. You implemented all the required pages, which we filled, one by one, with data from the connected Supabase instance. This will free your mind from mixing up logic and design parts and let you focus on the upcoming crucial Supabase topics. Whether the topic is login, ticket creation, or ticket details, we're prepared to dive right into it with Supabase without the need to implement design first (or afterward).

In the next chapter, you'll dive right into Supabase authentication, where we will implement the login logic with Supabase. Are you excited? Let's keep rockin'!

Part 2:
Adding Multi-Tenancy
and Learning RLS

In the second part of this book, we will transform your application into a multi-tenant powerhouse. You'll master the intricacies of **row-level security** (**RLS**) and implement tenant-specific permissions and domains. This part will equip you with the skills to handle complex multi-tenancy scenarios, ensuring your app is both secure and scalable. Let's unlock these powerful capabilities together!

This part includes the following chapters:

- *Chapter 4, Adding Authentication and Application Protection*
- *Chapter 5, Crafting Multi-Tenancy through Database and App Design*
- *Chapter 6, Enforcing Tenant Permissions with RLS and Handling Tenant Domains*
- *Chapter 7, Adding Tenant-Based Signups, Including Google Login*

4

Adding Authentication and Application Protection

Building upon the solid foundation laid in the previous chapters, your project is now equipped with essential routes and UI elements for our multi-tenant ticket system and allows you to instantiate a Supabase client everywhere easily.

At this point, envision the capability to log in with different user accounts across various tenants, enabling actions such as creating tickets, uploading files, and commenting. To unlock this functionality, proper user authentication is crucial.

In this chapter, we will dive into the integration of authentication with Supabase. This journey involves understanding the intricacies of adding users in Supabase, facilitating their sign-in and out processes, and dynamically redirecting based on their authentication state.

You'll also explore the realm of one-time passwords through magic links, gaining insights into fully customizing such emails. I'll guide you through the rationale behind avoiding built-in authentication emails in most projects and provide insights on adapting them if necessary.

Looking at the capabilities of Next.js in server-rendering dynamic applications, you'll discover how the aforementioned processes can seamlessly function even when JavaScript is disabled or encounters issues.

So, in this chapter, you'll embark on a journey that includes the following topics:

- Adding authentication protection with Supabase
- Adding a log out button
- Understanding server authentication
- Enhancing the password login
- Authenticating with magic links
- Adding password recovery

- Learning about the site URL and redirect URLs

- Optional knowledge – adapting built-in templates

It may seem like a lot to take in at first but rest assured, you'll find that it becomes much more manageable than you might anticipate. I'm excited about this chapter; I hope you are as well!

Technical requirements

In this chapter, you'll face the same technical requirements as in *Chapter 2*. You'll find the code related to this chapter in the Ch04_TicketSystem_Auth branch (https://github.com/PacktPublishing/Building-Production-Grade-Web-Applications-with-Supabase/tree/Ch04_TicketSystem_Auth).

In the package.json file of the repository, I've added a reset script that allows you to create a fresh, hassle-free Supabase instance in each branch. Just run it with npm run reset.

Adding authentication protection with Supabase

So far, we have organized our code so that the parts where we expect logged-in users to interact with our app are within the app/tickets/* folder structure. That also means that every route that is only expected to be accessible to logged-in users will start with your-application.domain/tickets/*. Since those routes all share the /tickets parent path, we want to protect that path and all of its subpaths. Here, protecting means only allowing authenticated users to have access and redirecting everyone else.

However, before doing so, we have to create users and allow them to log in. Here's what we will do now to achieve this:

1. We will create our first users. Without users, we cannot use and test our application and protection.

2. We will prepare the middeware.js file to make sure that authentication will work as expected.

3. With the login form we created earlier, we will allow the created users to log in by adding the required code logic.

4. We will adapt our middleware.js file such that only logged-in users can access the Ticket Management UI by protecting the /tickets/* routes.

Let's get into it.

Creating users

Obviously, you cannot log into the ticket service application if you don't have an account. Usually, this isn't a problem because in any typical web application, you'd simply register as a new user. However, we haven't done anything other than instantiating the Supabase Client and creating a bit of UI yet – which means that we cannot even register an account yet.

For now, we will create users, which will allow you to get familiar with Supabase Studio, and once this is done, we will implement the login functionality.

> **Note**
>
> In our multi-tenant project, typical user registration requires matching a user with a tenant (or else how would you know that a user can access a specific tenant?). You don't have to wrap your head around that in this chapter but that's the reason why a sign-up form will be part of the next chapter when we architect the multi-tenant code and database design.

To create a new user, go to your Supabase Studio (Dashboard) and, in the navigation menu, click **Authentication**. From there, click **Users**.

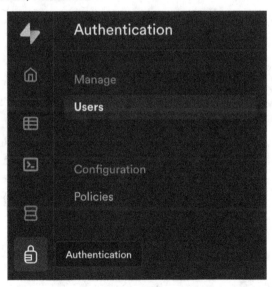

Figure 4.1: The Authentication button on the Dashboard

Now the user manager will open. Then, click on the **Add user** button, and in the drop-down menu, click **Create new user**:

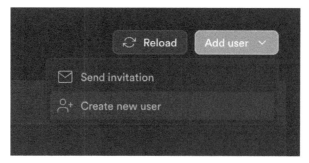

Figure 4.2: The Create new user button

A form should open where you can enter an email address and password for that user, as well as an option to autoconfirm the user:

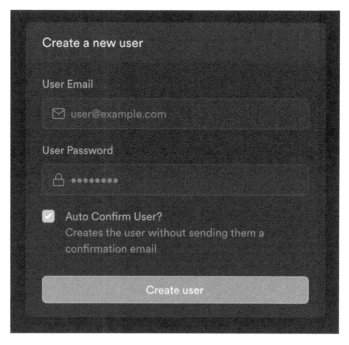

Figure 4.3: The user creation mask

Go ahead and enter an email address first. If you are using your local Supabase instance, as I am, then you can just enter any arbitrary email that you don't even own, such as icecream@cool.local. Locally, the emails won't get sent out to the public but instead stored for you to access within the Inbucket service, as you will see in the *Authenticating with magic links* section.

Then set a password of your choice and check the **Auto Confirm User?** box.

Usually, when you create an account on any app, you'll be asked to activate that account first. We don't want to do that here, which is why we must check the **Auto Confirm User?** box (however, you'll see how to handle confirmation as part of registration in *Chapter 7*). If you don't check **Auto Confirm User?**, then the user will be created as **inactive** and cannot log in.

Once the form has been filled, click **Create user** and your user will be created and activated!

> **Note**
>
> A few people asked me whether it's possible to simply create a user with a username (not an email) and a password. In web apps, the use cases of users without emails are super rare and this exact thing isn't planned to be part of Supabase itself (`https://github.com/supabase/gotrue/issues/903#issuecomment-1430708566`). However, it would be doable by using `username@your-custom-domain.com` email addresses where the domain is static and hence only the username needs to be entered. This book won't go into this specific use case though.

Feel free to create as many users as you want. However, for this chapter, one will be enough. Now, we can start preparing to authenticate our active users.

Preparing the middleware for authentication

No matter what kinds of features you are building for your users, or whether you need to protect certain areas of your application, the use cases for users within an application are vast. However, regardless of what you are planning, what I am showing you now is *always* necessary when you allow users to log in.

Let's first talk about what being logged in means and why that implies that we have to adapt our middleware file.

To do this, we'll use an analogy. When you have tickets for a concert, you usually show your ticket at the entrance – here, you can consider the ticket as your key and the process of getting it scanned as the authentication process. When everything is okay and your ticket is valid, you often get something in exchange, such as a stamp on your hand. The stamp has a specific color and a date below it. That night, as long as you have that stamped logo on your hand, you can walk in and out as you like without the need to show your ticket again and be granted access to the concert venue's facilities such as the toilet or the actual concert hall. You will have to show your stamped hand for this. If you visit a different concert the next day, your stamp will be outdated; hence, your authentication state will technically have expired.

In programming, we call this stamp the **session**, and this session is saved in cookies. So, when you log in, (for example, by entering a username and password), in exchange, you get a cookie with a certain expiry date and the values that allow the server to say, "Ah okay, it's you, David; come on in."

Now let's highlight the crucial technical part. For security reasons, the authentication session is only valid for a short amount of time – in Supabase, that's usually 60 minutes. If the user becomes inactive, the session will expire and the authentication cookie will not be valid anymore. However, as long as the session is valid, the user can refresh it such that the timer starts again at zero and is valid for another 60 minutes.

That leaves us with questions. Who does that? Is it done automatically? If yes, how?

In *Chapter 2*, I showed you how to initialize Supabase clients on the frontend and the backend. On the frontend, the Supabase client is set up by default with an additional `auth.autoRefreshToken=true` option. This has the effect that, if you initialize a client on the frontend, it will periodically whether there is an existing session that needs to be refreshed check in the browser and, if there is, it will refresh it by requesting a new session with the still-valid existing session.

This refresh process is visualized in *Figure 4.4*:

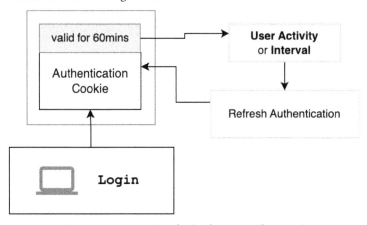

Figure 4.4: Automatic refresh of sessions (frontend)

As you can see, the Supabase frontend client automatically takes care of avoiding the expiry of your session as long as the user has the web application opened. That's handy!

Does that mean it's required to have a Supabase frontend client even if we want to work with server-side rendering? Definitely not! However, you must ensure that the session is refreshed in time (e.g., when the user navigates through your app, hence showing activity).

This can easily be achieved with the Next.js middleware as it runs before each request that hits the application, and requests hit the application when, you guessed it, there's user activity!

So, let's make sure that we equip the Next.js middleware with the capability to manage existing sessions just as it would in the frontend. The first thing you need to do is to make sure that you have a basic `src/middleware.js` file as follows:

```
import { NextResponse } from "next/server";
export async function middleware(req) {
  const res = NextResponse.next();
  return res;
}
```

This file does nothing yet as it only passes through the request. That's what we will change by initializing the Supabase client with our `getSupabaseReqResClient` function from *Chapter 2* and just adding another line waiting for `supabase.auth.getSession()` to finish, like so:

```
import { getSupabaseReqResClient } from "@/supabase-utils/
reqResClient";
import { NextResponse } from "next/server";

export async function middleware(request) {
  const { supabase, response }
    = getSupabaseReqResClient({ request });
  await supabase.auth.getSession();
  return response.value;
}
```

Here, we ask Supabase to retrieve the session. By doing that, it will also refresh the session if expired. That was simple, wasn't it?

> **Note**
>
> If you already have some experience with Supabase, you might wonder why we use `getSession()` over `getUser()`. In that case, quickly peek ahead at the upcoming Note box in the *Protecting access to the Ticket Management system* section. If not, just keep reading.

Earlier, I said that `middleware.js` is triggered before each request, which is true. This allows full control over each single request that hits the application. The problem is that there are requests we don't want to control because they would only slow down the server. For example, your browser requests static images when a website is opened, such as the typical `/favicon.ico` icon or images of the layout. There is no benefit in passing them through the middleware. So, you want to ignore certain files in your middleware that are publicly accessible static files.

> **Note**
>
> You can find detailed information about `matcher` at `https://nextjs.org/docs/`
> `app/building-your-application/routing/middleware#matcher`.

You can do so by exporting an object, named `config`, inside the `middleware.js` file. That `config` object allows for a `matcher` property, which Next.js will read and then only execute the middleware for the matched files. If you export the following, it will only execute for the two given paths:

```
export const config = {
  matcher: [
    "/foo",
    "/bar",
  ],
};
```

We want to run the middleware for all requests that aren't static files such as a favicon. So, instead of matching the files that we want to match, we can also tell the matcher which files we don't want to match instead. You can do so with a regular expression and a **negative lookup** with `?!`. The following matcher uses a pattern that matches requests that match a file naming in the `somefile.` `extension` form. By using `?!`, we can ensure that it will only execute the middleware when those patterns are not matched:

```
export const config = {
  matcher: ["/((?!.*\\.).*)"],
};
```

Let me explain the matcher: there is a `/` in the beginning and then, inside the inner parentheses, it searches for `.*\\.`, which means any character (`.*`) following a dot (`\\.`). So, put together, this would match `/favicon.` but since we used `?!`, we inverted the meaning and it specifically does not match. However, `/(?!.*\\.)` alone wouldn't work because then it would only dismiss `/favicon.` but not the actual `/favicon.ico` icon. That's why we use the additional `.*` to express that there can be any character following after the previous pattern.

Now, this matcher will serve us well and skip the unnecessary matches of mostly static files.

Your preparation of the middleware for authentication is now done. I bet that you had guessed that it would be more complicated. That allows us to implement the login functionality in the next section.

Implementing the login functionality in our app

Let's enable the app/Login.js component to log us in with our freshly created user. To do so, open the component and go to your form tag where we have the following onSubmit listener and code waiting for us to bring it to life:

```
onSubmit={ (event) => {
  event.preventDefault();
  if (isPasswordLogin) {
    alert("User wants to login with password");
  } else {
    alert("User wants to login with magic link");
  }
}}
```

As seen in the code logic, in our existing UI, we have prepared two login options: **Sign in with Password** and a **Sign in with Magic Link**. Here, we will look at the password-based login first, and in the *Authenticating with magic links* section, we will look at the magic link variant.

First, make sure that you initialize a Supabase client in the Login component. Just add one right below your input references:

```
...
const passwordInputRef = useRef(null);
const supabase = getSupabaseBrowserClient();
...
```

As you've seen in the first chapter, the Supabase services are neatly separated by concern but seamlessly connected; this is the same way the Supabase client exposes functionality to those services. If you want to do anything related to authentication, you'll find it within supabase.auth. Within that, we get access to a method named signInWithPassword(..), which takes an email and password to authenticate with. That's what we will use now.

The only thing we need to do is to call that method, provide it with the values from the form, and then wait, within then(), for the result. So, on submission of the form within the isPasswordLogin case, we simply read the email and password via the form references (emailInputRef and passwordInputRef) that we have set up in *Chapter 3*.

That process becomes very clear when looking at the final code:

```
onSubmit = { (event) => {
  event.preventDefault();

  if (isPasswordLogin) {
    supabase.auth.signInWithPassword(
      {
```

```
      email: emailInputRef.current.value,
      password: passwordInputRef.current.value,
    })
    .then((result) => {
      if (result.data?.user) {
        alert("Signed in");
      } else {
        alert("Could not sign in");
      }
    });
  }
}}
```

What we're doing here is simply accessing the Auth Service through `supabase.auth.`
`signInWithPassword()` and passing it an object with the email and password that we have
from the input fields. After calling it, we wait for a response with `.then()`. If the result contains not
just the `data` property but also a non-empty `data.user` property, then the login was successful,
as it will return our user information.

> **Key learning**
>
> Supabase-related functions often work by returning an object containing a `data` key. The sole
> existence of the data property doesn't imply any success of whatever you call. You always need
> to check what's inside `data` to verify that it didn't just return `data: null`.

What now? Well, the most logical choice after a successful login is to redirect the user to our Ticket
Management UI at `/tickets`. To be able to redirect within a `Client` component (which our
`Login` component is one of), you need to add the `useRouter` import at the top of your file, like so:

```
import { useRouter } from "next/navigation";
```

Make sure to import from `next/navigation`, not `next/router` – the latter is for usage with
the Pages Router, whereas the one we need is for the App Router.

Next, assign a variable to an instance of that router with `const router = useRouter();`.

Then finally, replace `alert("Signed In");` with the following:

```
router.push("/tickets");
```

Now, your login is ready! Test it by going to the form and entering the email and password of the
user you created earlier. In the case of providing the correct credentials, you should be redirected to
the main page (the ticket management layout) of the application. If not, you should see an alert box
popping up stating **Could not sign in**.

Believe it or not, the process of getting authenticated is now done. However, there's something else to consider: no guarding layer exists yet. If someone knew the path, that person could also just enter / `tickets` in the browser and the system wouldn't care whether they were logged in or not. That is actually less of a security concern* and much more of an annoying factor to users who regularly visit our application where the browser history would lead them to this path. Going back to our concert analogy, it would be like getting your ticket validity checked, but knowing that you can also use the door at the back of the concert hall, which is wide open. However, using that backdoor would really just be boring because you wouldn't be able to get into the concert facilities such as the concert hall without your stamp.

What we have to do is to make sure that at no point in time is there an easily accessible backdoor. In the next section, we will learn to implement restrictions that handle this based on the authentication status.

---

*An important clarification

As explained with the concert analogy where only the stamp would provide you with immediate access to the concert facilities, you'd have a boring time inside of the concert building even if you used the backdoor. The way that Supabase works with RLS (which we will discuss in the next chapter) is that a user has access to no data from the database (except their user data). So, if someone breaches the middleware with a fake session, that person will see a layout with zero data. You'll experience such missing data on your own in *Chapter 6*.

Protecting access to the Ticket Management system

Let's make sure that a user who is properly logged in will not see the login form again, but instead is redirected to the Ticket Management UI at /`tickets` when hitting the login route. Then, vice versa, we want to show unauthenticated people the way to the login form at / if they hit the route of the Ticket Management UI.

As part of this chapter, you have already created an `app/middleware.js` file that contains the Supabase client executing `supabase.auth.getSession()`. This `getSession` function returns a session if one exists – and one should exist if you are logged in. A session will also contain the user data within `returnedSession.data`.

Since the `middleware.js` file is executed before anything else, we can also use it as a guardian to check whether someone is authenticated and what resource is accessed. If there is an active session, we want to allow access to all routes starting with /`tickets`. You can easily achieve this with the following code:

```
export async function middleware(req) {
  const res = NextResponse.next();

  const { supabase, response }
```

```
= getSupabaseReqResClient({ request: req });
  const session = await supabase.auth.getSession();

  const requestedPath = req.nextUrl.pathname;
  const sessionUser = session.data?.session?.user;

  if (requestedPath.startsWith("/tickets")) {
    if (!sessionUser) {
      return NextResponse.redirect(new URL("/", req.url));
    }
  }

  return response.value;
}
```

In this adapted `middleware.js` code, we first get the path of the URL. If it starts with `/tickets`, then we check whether the user data is *not* available; if that's the case, then it's someone trying to access a protected area who is *not* authenticated. In that case, we would redirect back to the page where the login sits (at `your.domain/`); we do so by using the `NextResponse.redirect()` function and passing it a newly constructed URL.

Why don't we simply use `NextResponse.redirect("/")` though? The reason is that Next.js expects an absolute URL (`https://nextjs.org/docs/messages/middleware-relative-urls`) – this can be constructed with `new URL()` using the given URL (`req.url`) and changing its path (`https://developer.mozilla.org/en-US/docs/Web/API/URL/URL`).

> **Security note**
>
> `getSession()` from the Supabase client is a helpful function that loads – and, if expired, refreshes – a session. It also returns the data from the existing session, but that data is not verified; it parses that data from your cookie, which isn't encrypted and can easily be manipulated.
>
> For our application, in `middleware.js`, that's completely fine because the only thing we will do is make a redirect decision. So, if a user manipulates session data in the cookie and hence is redirected to `/tickets`, the user still won't be able to do anything or load any data there. This is because, for everything else, in future chapters, we will use sophisticated trustworthy methods to protect the data as stated already in the previous note about the backdoor.
>
> Still, never make decisions that have an actual security implication with the data from `getSession()`. Trustworthy user data can instead be retrieved with `supabase.auth.getUser()`. The reason why we use `getSession()` in `middleware.js` is because it's much faster than `getUser()`, as `getUser()` involves an additional API request. In other words: if this makes you feel unsafe, simply replace `getSession()` with `getUser()` in your middleware file.

However, we're not done yet. There's one thing missing – when we are logged in and visit our domain at the / root path, we still face the login form even though we're signed in. That's not convenient. Instead, it should immediately redirect to the /tickets section for authenticated users. Let's add that in the middleware code as well:

```
if (requestedPath.startsWith("/tickets")) {
  if (!sessionUser) {
    return NextResponse.redirect(new URL("/", req.url));
  }
} else if (requestedPath === "/") {
  if (sessionUser) {
    return NextResponse.redirect(new URL("/tickets", req.url));
  }
}
```

Now, when you run the server and visit localhost:3000/, the following should be true:

- If you previously logged in (and hence have a valid session cookie), it should redirect you to /tickets

- If you're not logged in but try to open localhost:3000/tickets, you should be forwarded to the login at localhost:3000/ (if you want to test this, delete your cookies in the browser to remove an existing session)

Your application now has guarded routes – that's a major milestone!

At this point, if you are thinking forward, keeping in mind that we are developing a **Software as a Service (SaaS)** system that allows users to log in to multiple tenants, you might be asking yourself now: "How can I be logged in when there are going to be many tenants or companies within the same system with their own tickets to manage?"

Good question, but it's just a little bit too early to answer that at this point, as we don't even have any tenants set up yet. We will get into that **tenant problem** very soon, in the next chapter.

Right now, our application acts like a single-tenant service. That's fine for the moment, as I need you to get equipped with more of the foundational authentication knowledge first, such as the mechanisms and implications of logging out which are coming up next.

Adding a log out button

Now that you can log in, you want to provide an intuitive way to log out. Deleting a cookie manually isn't a convenient way of signing out, so we want the possibility of logging out through our UI. That's why we will now build a log out button as part of our Nav.js navigation component.

Logging out can be either done on the frontend or the backend. They don't replace each other but complement one another: if the frontend JavaScript fails to load, the backend solution kicks in.

Let's start with the frontend version and complement the backend variant afterward.

Logging out using the frontend

We will start by adding the logout functionality to our existing **Log out** button.

Open the app/tickets/Nav.js component file. As we will need Supabase, you must make sure to import getSupabaseBrowserClient() and assign it to a variable as follows: const supabase = getSupabaseBrowserClient();.

In our existing code, we already added a Link component to be our log out button in our Nav.js component. We'll use that and add an onClick listener to it as follows:

```
<Link
  role="button"
  href="/logout"
  prefetch={false}
  className="secondary"
  onClick={(event) => {
    event.preventDefault();
    supabase.auth.signOut();
  }}
>
  Log out
</Link>
```

As you can see, all we are doing is preventing the link (/logout) from opening and executing supabase.auth.signOut() instead. That way, your existing session gets cleared and you're logged out. We also added the prefetch attribute and set it to false to prevent Next.js from automatically loading this page and hence unwantedly triggering a logout.

When you test it, you will notice that nothing happens when you click the **Log out** button – well, nothing happens *visually*. Technically, your session is gone but you're still staring at the page that was rendered before you clicked the button. When you refresh the page, it will bring you back to the login page after logging out. This is obviously weird because you want to be redirected to the login page immediately after logging out and not have a stale state that looks like the logout button didn't work.

We will have a look at two different ways to solve this problem now.

Explicitly redirecting after signOut()

Just like we did in `Login.js`, where we redirected the router to the **Ticket Management** dashboard after successfully logging in, the first logout option is to await the response from the `signOut()` method and then redirect the user back to the login page.

For that, you need to import `useRouter` and assign it to a variable accordingly. The upper part of your `Nav.js` code will resemble this structure:

```
// ... imports
import { useRouter } from "next/navigation";

export default function Nav() {
  //... other variables
  const router = useRouter();
```

Then your logout button's `onClick` event should become this:

```
onClick={(event) => {
  event.preventDefault();
  supabase.auth.signOut().then(() => {
    router.push("/");
  });
}}
```

So, when the **Log out** button is clicked and everything goes well, the callback within `then()` is called to trigger a redirect and you are brought back to the login page at `/`.

However, there's an even cleaner solution for this.

Event-driven redirection after logging out

When you log out with the `supabase.auth.signOut()` method, an internal JavaScript event is triggered that you're not logged in anymore. The previous solution was just fine, but there are caveats:

- What if, for example, due to your design, you have different logout buttons so you'd have to always add the redirection logic in multiple places?

- Or what if the logout doesn't happen explicitly but implicitly, for example, because you had your browser open, idling for a week, then closed the laptop and your user session expired?

However, what if we could simply say, "Hey, Supabase, please tell me whenever the user is logged out – no matter the reason"? Well, we can.

First, remove the following highlighted code in your **Log out** button:

```
onClick={(event) => {
  supabase.auth.signOut().then(() => {
    router.push("/");
  });
}}
```

Hence, the code becomes this again:

```
onClick={(event) => {
  event.preventDefault();
  supabase.auth.signOut();
}}
```

Now you need a `Client` component that is available on all authenticated pages, such that it can grab the logout event, no matter on which page we are. The `app/tickets/Nav.js` component is on every page, so we can simply use that one.

We want that component to subscribe to an event from Supabase telling us that a logout has happened. With `supabase.auth.onAuthStateChange(/** listener **/)`, you have a method that allows you to listen to all events related to authentication. You can pass it a function and the function will provide the `event` and `session` values whenever the authentication state changes. The method returns a subscription that can be unsubscribed from when it is not needed anymore.

Since we only want to have one subscription, we call this with the help of the `useEffect` React Hook (`https://react.dev/reference/react/useEffect`). The result looks like this:

```
export default function Nav() {
  // variables ...
  const supabase = getSupabaseBrowserClient();
  const router = useRouter();

  useEffect(() => {
    const {
      data: { subscription },
    } = supabase.auth.onAuthStateChange((event, session) => {
      if (event === "SIGNED_OUT") {
        router.push("/");
      }
    });

    return () => subscription.unsubscribe();
  }, []);
  // .....
```

As indicated, we execute the `onAuthStateChange()` method and pass it a function that executes every time an authentication event is triggered. As we are only interested in the event for logging out, we will check whether the fired event is `SIGNED_OUT` – if so, the user is logged out. In that case, we want to go back to the login page with `router.push("/")`.

Lastly, with React's `useEffect`, you should always return a cleanup function if there are things that can be cleaned up. Cleaning up a subscription is done by unsubscribing; hence, we return a function that does this. This avoids having multiple listeners even if the component re-renders.

Sure, this is a bit more code, but it's just in one place and provides a clean, universal approach to checking for a logout.

At this point, you can log in with a password, and you can log out. Regardless of whether you're signing in or out, you will be forwarded to the proper path and the route guard works.

Next, we want to adapt this **Log out** button to log out even with JavaScript disabled; this also enables you to get rid of the frontend code in case you are developing a fully server-rendered application. In our project, we will just allow both, which is a good practice.

Logging out using the backend

Logging out on the server should happen when the frontend code does not kick in. Right now, you have a **Log out** button that calls `supabase.auth.signOut()` in the frontend. That button also has an `href="/logout"` property defined but opening that link is suppressed in the frontend by `event.preventDefault()`.

If JavaScript was disabled or failed to load, you'd be redirected to that page – which does not exist, so you'd face a 404 message. That's what we want to change.

We want to create a Next.js Route Handler that takes care of logging out on the backend. We could create a `app/logout/route.js` Route Handler file such that it would match the currently set `href="/logout"` route of the button. However, following the *Understanding server authentication* section, we will add more server-related authentication files and, semantically, it makes sense to group them neatly in one shared folder named `auth`.

So, create the `app/auth/logout/route.js` file now. Obviously, this route is reachable at `your.domain/auth/logout` but we will use a trick to map it back to just `/logout` – so don't be too hasty in changing the `href` attribute of the button to `/auth/logout`.

Let's bring that Route Handler file to life by adding code that signs out the current user:

```
import { getSupabaseCookiesUtilClient } from "@/supabase-utils/
cookiesUtilClient";
import { NextResponse } from "next/server";

export async function GET(request) {
```

```
  const supabase = getSupabaseCookiesUtilClient();
  await supabase.auth.signOut();
  return NextResponse.redirect(new URL("/", request.url));
}
```

As we mentioned in our discussion on the usage of Supabase clients in *Chapter 2*, we need `getSupabaseCookiesUtilClient` in a Route Handler. Other than that, the given code is executing the exact same function as our frontend-based implementation by calling `supabase. auth.signOut()`. When that function has finished doing its job of clearing users' sessions, we redirect back to the / base path, where the login page is.

Now we need to make sure that opening the /`logout` route will trigger the actual /`auth`/`logout` route. There is a file at the root of your project named `next.config.js` that exports an object named `nextConfig`. One of its possible configuration options is adding a `redirects` property (`https://nextjs.org/docs/pages/api-reference/next-config-js/redirects`) to the object within the file, like so:

```
const nextConfig = {
  redirects: async () => [
    {
      source: "/logout",
      destination: "/auth/logout",
      permanent: true
    },
  ],
};
```

With that configuration in place, when visiting /`logout`, you'll always be forwarded to the existing route at /`auth`/`logout`. This way, we can maintain a clean file structure but still get the super-short route.

Go ahead and try it: make sure you're signed in, disable JavaScript, and click the **Log out** button.

> **Note**
>
> To disable JavaScript in your browser, I recommend that you search for browser plugins that can switch it on or off with a simple click, such as NoScript.

You'll see that everything works properly, the session is cleared, and you're moved to the login form. Logging out now works on the backend as well. Although we mentioned logging out on the backend on the fly, we haven't covered logging in on the backend yet.

In the upcoming section, you'll be guided through the world of server authentication, where you will first get to know what has to be done to make the password-based login work without the frontend, as well as how to send magic links for logging in.

Understanding server authentication

Back in the golden age of server-side languages such as PHP (`https://php.net`), the only thing JavaScript was used for was to add a little bit of interaction or animation to a website. Most user-based processes, including authentication, happened by submitting a form, and a new page request was made that processed it – there was no background request or asynchronous authentication. Plus, internet connections were slow and we wanted as few frontend assets as possible.

In the end, authentication itself always happens on a server (where the user credentials are checked against the database). However, nowadays, the frontend often calls the server in the background instead of doing a submit that causes a new page request.

However, I want to show you exactly how to do that. The question is: why would we want to go back to authentication, which means loading a new page, instead of doing it in the background with the help of the frontend?

With a frontend-only solution, you face the following challenges:

- When you're using a web application with your mobile phone, you are sometimes involuntarily brought into a state of extremely slow or broken connections (e.g., in a tunnel during a train trip). A broken connection can lead to fully or temporarily missing JavaScript, even though the page itself is already loaded. If your application works with frontend capabilities only, the application becomes unusable.

- Similarly , your JavaScript can also simply fail unexpectedly due to a bug you have somewhere in your code.

- In some environments, there are hard restrictions for JavaScript, which could lead to a disablement of such. Hence, the page becomes unusable again.

However, it's not just about frontend failure. There are more reasons to add server-based authentication:

- Server-based authentication doesn't need to replace background or async authentication via frontend but can instead complement it.

- We live in an era where fully server-rendered applications are seeing a revival because they can outperform applications that are either fully client-rendered or built using a hybrid approach.

- A common, recommended, and safe approach for logins is the magic link. This link is valid once and only for a short time. Even though it can be used on the frontend, it's usually faster not having to load the backend and frontend first to verify such a link but instead to let it check on the backend only without additional overhead. Also, constructing the magic link on the backend will give you much more flexibility in terms of templating the email, as you'll see in the related section of this chapter.

In the next sections, we'll tackle server-based authentication with Next.js and Supabase, complementing the frontend functionality.

Our current app allows users to log in via email and password on the frontend. We can check whether a user is logged in using Supabase, both on the frontend and backend (e.g., in the `middleware.js` file with `getSession()`). While our UI has a login form for using an email and password or requesting a Magic Login Link, the latter has not been implemented yet. Additionally, the sign-in process is limited to the frontend, and we lack a password recovery feature.

Moving forward, we'll implement backend-based login, the Magic Login Link functionality, and a password recovery feature.

> **Note**
>
> I hope it has become clear that server authentication isn't a necessity for your project to be fully functional; it's just that, because of the aforementioned reasons, it can come in handy to implement it with little additional effort.

Enhancing the password login

Imagine the following scenario with JavaScript disabled in your browser: you open the ticket application login page, enter your credentials, and submit the form. What would happen? It would submit the form to the page itself because if no `action` attribute is set in a `<form>` element, it defaults to `action=""`, which is the current page. With the submission, it would append your login credentials to the URL (`host?email=...&password=...`), as the default submission method of a form is GET. Other than that, nothing at all would happen as we don't have anything that handles form data on the server.

Instead, we want to create a Route Handler to process such a submission on the backend. So, let's create a Route Handler for a server-based password login process: `app/auth/pw-login/route.js`.

> **Note**
>
> Although a page in Next.js is rendered on the server and can fetch on the backend, it is not architecturally built for processing sent body data such as processing a form submission on the backend. That's what Route Handlers (or Server Actions) in Next.js are tailored for!
>
> You can read a general discussion on processing body/POST data in a page at: `https://github.com/vercel/next.js/discussions/44428`. Also, you can see the proposed solution for a similar discussion at: `https://github.com/vercel/next.js/issues/54957#issuecomment-1704982473`.

Before we add code logic to it, we need to make sure that the submission of the form will hit that Route Handler. To do so, go to the `app/Login.js` file and add `action="/auth/pw-login"` to the form. Then make sure to explicitly set the `method` attribute on the `<form>` tag. By default, it's set to GET, causing the form data to be included in the URL (e.g., `your-route-handler-url?email=yourmail&password=yourpassword`). To avoid exposing sensitive credentials in the URL, change the method attribute to POST.

Now your form should start like this:

```
<form action="/auth/pw-login" method="POST" ...>
```

The Route Handler file now needs to be told to process such POST requests. This is easy to do with Next.js; just add the following code to it:

```
import { NextResponse } from "next/server";

export async function POST(request) {
    return NextResponse.json({ message: "Hello from Route Handler" });
}
```

In *Figure 4.5*, you can see the flow of the current state that we have up to this point.

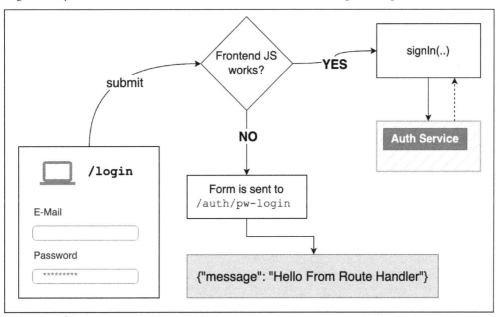

Figure 4.5: The submission flow with a dummy Route Handler

You can test this flow now. If you disable JavaScript in your browser and submit the login form with the **Sign in with Password** button, you should see the { message: "Hello from Route Handler" } JSON response.

However, the solution we strive for is shown in *Figure 4.6*. The Route Handler needs to, just like the frontend version, communicate with the Auth Service to check the credentials and, if valid, create an actual user session.

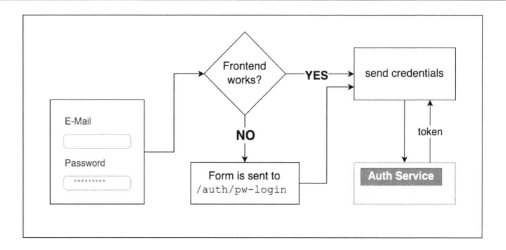

Figure 4.6: Authentication on both the frontend and backend

To achieve this, the next step is to change the Route Handler to process the actual login data. Within the app/auth/pw-login/route.js handler, we need to do the following:

1. Read the form data.

2. Instantiate a Supabase client.

3. Sign in with the Supabase client, passing the values from *step 1*.

4. Redirect the user to an error page if the login was not successful, or to the /tickets page if it was successful.

Here's the full code, including all steps:

```
import { NextResponse } from "next/server";
import { getSupabaseCookiesUtilClient} from "@/supabase-utils/
cookiesUtilClient";

export async function POST(request) {
 // Step 1:
 const formData = await request.formData();
 const email = formData.get("email");
 const password = formData.get("password");
 // Step 2:
 const supabase = getSupabaseCookiesUtilClient();
 // Step 3:
 const { data, error } = await supabase.auth.signInWithPassword({
  email,
  password,
```

```
});

// Step 4:
const userData = data?.user;
if (error || !userData) {
 return NextResponse.redirect(
  new URL("/error?type=login-failed", request.url),
  { status: 302 }
 );
}

 return NextResponse.redirect(new URL("/tickets", request.url), {
  status: 302,
 });
}
```

In the code, we first execute the `formData()` method from the `request` object to get access to the form values that were sent. Then, the Supabase client is initialized and we wait for the `signInWithPassword()` call to be done. Since we want to provide feedback for a failed login, we also grab the returned `data` and `error` values.

If an error happens or the data doesn't contain a user, the login won't be successful. That's what we check for with `if (error || !userData)`. In that case, we redirect to an error page (which we are about to create soon) passing the type of error that happened as a query parameter.

So, when that error-checking `if` statement does not hold true, the login must be successful. In that case, we forward users to `/tickets`.

Go ahead and disable JavaScript, then make sure that you are on the password-based login page. You should be able to log in with your valid user credentials and be forwarded to the Ticket List. Cool, right?

However, what about the error page? The error page has the sole task of giving information to the user when an error happens. With the possibility of reacting to a given `type` parameter, you can create an `app/error/page.js` file like this:

```
import Link from "next/link";

export default function ErrorPage({ searchParams }) {
  const { type } = searchParams;
  const knownErrors = ["login-failed"];

  return (
    <div style={{ textAlign: "center" }}>
      <h1>Ooops!</h1>
      {type === "login-failed" && (
```

```
      <strong>Login was not successfull, sorry.</strong>
    )}
    {!knownErrors.includes(type) && (
      <strong>
        Something went wrong. Please try again or contact support.
      </strong>
    )}

    <br /><br />

    <Link role="button" href="/">
      Go back.
    </Link>
  </div>
  );
}
```

With this adaptable error page, if a specific type is met, then the fitting error message is displayed. If the type isn't in the knownErrors array, it will display a universal error message instead. Finally, there's a button for the user to easily move back to the start page (we will extend this error page in this chapter with more types).

That's it about progressive enhancement and logging in with email and password on the server, as well as providing feedback with an error page. If you're like many people, who are still not using a password manager and keep forgetting their passwords, then the next section will come in handy as you'll see how to implement magic link functionality with Supabase.

Authenticating with magic links

When a user of our application switches to the magic link form, enters an account email and presses **Sign in with Magic Link**, the server should create a secret link that is then sent via email to the user who wants to log in.

Now, I will show you how to generate such magic links, allowing us to authenticate users with it. We will start with a frontend-based solution, which will use the built-in email templates, then gradually move to a server-based solution including fully customizing those emails.

Sending magic links with signInWithOtp() on the frontend

Similar to signInWithPassword(), in the supabase.auth file, there is a signInWithOtp({email}) function that takes an email address and sends a magic link to it.

> **Note**
>
> A magic link is a link that contains a **One Time Password (OTP)**, which can be used exactly one time to log in.

`signInWithOtp` works the same way on frontend and backend – calling it will make a request to the GoTrue Auth Service of Supabase, telling it that the user wants to login via magic link and asking it to please send one. The Auth Service will then construct a one-time link and send it via email immediately. When the user clicks it, we have to make sure to have the proper code in place to process it.

For now, it wouldn't matter whether we called it on the frontend or the backend. However, as we will dive deeper into server-specific code with the magic link later, we want to start right away by sending the form to a Route Handler file instead, which then calls this `signInWithOtp()` method.

So first, create a new, empty `app/auth/magic-link/route.js` Route Handler file. Then, jump to `Login.js` and make yourself aware that it has statically set `action="/auth/pw-login"`. Here, we need to set the form action based on the `isPasswordLogin` value. In case `isPasswordLogin` is `false`, we want the form to be sent to `/auth/magic-link` instead to hit the Route Handler we just created. We can do this like so:

```
<form
  method="POST"
  action={isPasswordLogin ? "/auth/pw-login" : "/auth/magic-link"}
  ...
```

As mentioned, at this very moment, it wouldn't make a difference whether we call `signInWithOtp()` on frontend or the backend. However, as opposed to password-based login, where you can immediately be forwarded from the frontend and hence it might feel a bit quicker, there's no benefit in requesting a magic link on the frontend when we have a solution on the backend. In both cases, the user has to wait to receive a link. Also, when we send our custom email in the *Implementing a server-only magic link flow with custom email content* section, there will be no option to achieve the same in the frontend, so we will save ourselves that work.

Hence, we want to skip the frontend entirely for requesting the magic link and always submit the form to the backend Route Handler. To make sure that the form is always sent to the backend for magic links, we need to make `event.preventDefault()` conditional in our form. Find the following code within the `Login.js` form:

```
onSubmit={ (event) => {
  event.preventDefault();
```

Then, replace the code with this:

```
onSubmit={ (event) => {
  isPasswordLogin && event.preventDefault();
```

This way, the latter part of the statement only executes if `isPasswordLogin` is `true`. With this approach, when the **Sign in With Magic Link** button is clicked, the standard form behavior is used and sends the form data to the backend at `/auth/magic-link`.

Now, within the Route Handler of the magic link, you can reuse a lot of code from `app/auth/pw-login/route.js`. The difference is that we will now call `signInWithOtp()` and also only need to read the `email` field, not the `password` field. Also, in addition to the error page we created, we want to have an info page that tells users that a magic link has been sent. The resulting Route Handler function looks like this:

```
export async function POST(request) {
  const formData = await request.formData();
  const email = formData.get("email");
  const supabase = getSupabaseCookiesUtilClient();
  const { data, error } = await supabase.auth.signInWithOtp({
   email,
   options: { shouldCreateUser: false }
  });

  if (error) {
    return NextResponse.redirect(
     new URL("/error?type=magiclink", request.url),
     302
     );
  }
  const thanksUrl = new URL("/magic-thanks", request.url);
  return NextResponse.redirect(thanksUrl, 302);
}
```

We call `signInWithOtp({ email })`, passing the email to trigger sending a magic link, but we also pass the `shouldCreateUser` option as `false`. This is required because the standard behavior of `signInWithOtp` is to create a user if it does not exist and hence send the magic link no matter what. I don't like this behavior as it can be very confusing. In particular, when there are typos in an email, it would create a user with that wrong email address instead of throwing an error, resulting in confusion and inactive ghost users.

With the disabled `shouldCreateUser` behavior, we also get an `error` value when a magic link is requested for a user that doesn't exist in our Supabase instance.

In the error case, we want to forward to our error page, which we previously created, and pass `type=magiclink` in the URL. I also added this `magiclink` type in the error page itself to the known errors with `const knownErrors = ["login-failed", "magiclink"];` and adapted the render template of the `error/page.js` file accordingly with the following conditional rendering code:

```
{type === "magiclink" && (
  <strong>
    Could not send a magic link. Maybe you had a typo in your E-Mail?
  </strong>
)}
```

So, our error case will be wonderfully informative.

As we immediately stop the execution in our magic link Route Handler when an error happens, we know that sending the magic link was successful if the error case deos not happen. In that case, we construct a redirection to `/magic-thanks`, which is a page that tells the user to check the inbox. We can create that page now at `app/magic-thanks/page.js` and fill it with some friendly content:

```
import Link from "next/link";
export default function MagicLinkSuccessPage() {
  return (
    <div style={{ textAlign: "center" }}>
      <h1>Magic on its way!</h1>
      Thanks! You should get a link to login in a few seconds.
      <br /><br />
      <Link role="button" href="/">
        Go back.
      </Link>
    </div>
  );
}
```

Now, enter the email of your created account on the **Magic Link** page and press the **Sign in with Magic Link** button. If the email is correct, you should be forwarded to the **Thank You** page we just created.

However, you will not receive an email in your usual inbox if you are triggering authentication emails with your local Supabase instance. Your local Supabase instance instead sends all emails directly to the Inbucket service, which is both an inbox and an email server where you are able to see all emails that were sent within a web-based interface.

When you started your instance with `npx supabase start`, it printed the Inbucket URL as well – usually, this is `http://localhost:54324` (for a running local instance, you can check this with `npx supabase status`). Now, when you open the Inbucket URL, click on **Monitor**. There, you'll see all emails that were recently sent by your local instance.

Figure 4.7: The Inbucket monitor

An email with a **Your Magic Link** title should be listed here. Before you click on it, make sure that you are logged out of your ticket application. Then come back to Inbucket, open the email with the magic link, and click on the link. This will then forward you to your page at localhost:3000.

You'll notice that it seems like the login didn't work at first. You clicked on it but you still see the login page. Why is that? Shouldn't it automatically forward to the Ticket Management UI? Did the magic link not work?

It did work, and you are logged in, but you have to understand the flow to understand why you weren't forwarded. signInWithOtp() will send a link containing a login token that is automatically parsed from the Supabase frontend client. That means that the following process happens:

1. You click on the magic link and, when it opens, your middleware.js file says you're not logged in (so far).

2. You land on the login page where the frontend JavaScript loads.

3. The Supabase frontend client, which we have initialized in our Login.js component, starts up and confirms the magic link automatically without you noticing it.

4. Then the session is immediately verified and stored. However, at that point we're already past the backend execution and past the middleware.js file, so we're not forwarded.

5. If you then refresh the page, you are forwarded because the middleware.js file then detects a valid session.

However, that obviously isn't a clean solution, so let's fix that.

Remember when we used onAuthStateChange in the frontend to listen for the event when a user has logged out? There's a complementary event to the SIGNED_OUT event: SIGNED_IN. All you need to do is to check for the SIGNED_IN event.

Just as we did in app/Nav.js to listen for being logged out in the authenticated part of your application, we want to listen in the login page for being logged in. You can essentially copy the useEffect code block containing onAuthStateChange from your app/tickets/Nav.js file and paste it to your app/Login.js file. Then, just change the event to SIGNED_IN instead of SIGNED_OUT and forward to the Ticket Management UI when the event happens. Here's the code part that changes; everything else will be the same:

```
if (event === "SIGNED_IN") {
  router.push("/tickets");
}
```

Now make sure that you are logged out and request a new magic link to see what happens. You should be forwarded to the login page and then, right after the authentication is processed by the Supabase client (which happens automagically), the SIGNED_IN event is triggered and your router.push code kicks in.

Awesome, the magic link functionality works, and you're able to be forwarded properly!

Since you have added the listener for being logged in, you can now clean up a bit of code for the password-based login, as it will also trigger the same event on the frontend. In your onSubmit function of the form, you will have the following code:

```
supabase.auth.signInWithPassword(...)
  .then((result) => {
    if (result.data?.user) {
      router.push("/tickets");
    } else {
      alert("Could not sign in");
    }
  });
```

In a success case, onAuthStateChange will fire and reroute, so you can clean this up by only handling the error case:

```
supabase.auth.signInWithPassword(...)
  .then((result) => {
    !result.data?.user && alert("Could not sign in");
  });
```

Back to the magic link topic: before we jump into the final backend-based solution, which works without calling signInWithOtp(), I want to elaborate on the reasons why I typically recommend skipping the built-in authentication mailing system that is triggered with functions such as signInWithOtp() in the next section

Why I usually don't use signInWithOtp()

signInWithOtp() uses the built-in mailing system to generate an email and send it. The problem, however, is that this system is too limited for many mature production applications. The following points are just a few limiting factors:

- Although the built-in email templates can be customized (https://supabase.com/docs/guides/auth/auth-email-templates), they are indeed limited and cannot even have different languages.

- It is impossible to customize emails depending on which user they are sent to or in which context they are sent. For example, imagine that your login was already branded per tenant – you wouldn't be able to include the appropriate logo in the email and you'd have to use a generic one.

- You don't get custom email functionality or statistics as you do with literally any other mailing service (Brevo, Mailjet, SendGrid, rapidmail, turboSMTP, Resend, Mailtrap, etc.)

- The generated magic links tend to look awkward, especially if the link itself is shown as text, such as how many enterprise inboxes configured with Outlook look. Hence, in one of my projects at Wahnsinn Design GmbH, we decided to make custom short links, hiding away the long magic link. With the built-in system, using short links is not possible; you have to take the long link that Supabase generates.

- For a tenant-based application, it adds trustworthiness to generate a link with the tenant's domain. This is not possible with the built-in mailing system.

- When self-hosting Supabase, you don't have a UI to manage the internal templates as opposed to using `supabase.com`. You need to use the configuration to change the email templates.

I could certainly list more points, but I just wanted to mention a few reasons why I generally avoid the built-in mailing system and instead choose to gain full control over the way in which I send emails. I would even go as far as saying that Supabase shouldn't provide built-in emailing, but on the other hand, I also fully understand the benefits of it:

- It's easy to focus on app functionality first and then, when the product is functioning, switch to customized emails

- It allows you to pull up projects without having any kind of backend at all and still be able to send all required emails

- It can be sufficient for some projects with simple needs, no tenant-based branding, and just one single language

> **Note**
> I'm not saying that the built-in mechanisms are bad. I'm just saying that they are too limited for many use cases.

The `signInWithOtp()` method is definitely a feature worth knowing in your Supabase repertoire. It comes in handy if you cannot use a custom solution, for example because you are building your app frontend-based and are truly backend-less. In that case, you'll simply have to live with certain limitations.

As promised, I will show you a fully backend-based solution for our magic link flow in which we distance ourselves from the non-customizable `signInWithOtp()` and evolve to a solution that gives us total freedom.

Understanding a server-only magic link flow

I know, you're on fire. You want to implement the server-only magic link now. If we were to jump into the implementation right now, you'd be missing out on so much additional background knowledge such as the `emailRedirectTo` option or the way in which the magic link internally works. Soak that in, and the implementation part will feel like a fresh breeze.

Aside from the fact that we *want* a backend solution, our current magic link solution has a flaw: you click the link, the login page opens, and then you are immediately forwarded to the Ticket Management UI. The visit to the login page hence feels unnecessary (it's a so-called **flash of content** because you only see the login page for a small amount of time until the frontend verifies your session and redirects you again).

In backend-less applications, a potential solution is to pass an `emailRedirectTo` property to the `signInWithOtp` method to tell it where to go after having clicked the magic link. So, you could say "Okay, then I am directly going to `/tickets` instead, and it will not have the flash of content because there is no need to redirect again."

It would look like this:

```
supabase.auth.signInWithOtp({
  email,
  options: {
    emailRedirectTo: "http://localhost:3000/tickets",
  },
});
```

However, that doesn't work with our application – our `/tickets` route is guarded by our backend (`middleware.js`), and before the frontend Supabase client will even have to chance to authenticate the magic link, you're redirected back to the login page. Dang!

What we need is a route that is not affected by the middleware and that doesn't load the frontend first, so we'll be using a Route Handler for it. To implement such a Route Handler, you first need to better understand the authentication flow that we have right now. This will help you comprehend what we will need to implement.

When you request a magic link, right now, the link in your inbox will look somewhat like this:

```
http://localhost:54321/auth/v1/
verify?token=pkce_...&type=magiclink&redirect_to=http://
localhost:3000
```

This is interesting because, as you can see, the link does not lead to your Next.js application. It goes to `localhost:54321/auth/....`. This is the Auth Server from your (local) Supabase instance.

When the link is clicked, that Auth Server will authenticate you with the provided token, then forward you to something such as `http://localhost:3000/?code=95bcb17f-0eb7-4e3e-b899-1a72bfd6db45`. This code is called the **exchange code** and usually, you won't even see the `?code` part in the URL because your Supabase frontend client immediately and automatically reads it when the page is loaded, saves it, and removes it from the URL. It uses that code to once again verify that the authentication session, initiated by the token, is valid. Only then does it save it as a cookie and change your status to `SIGNED_IN`.

That's the default flow. However, the Supabase client allows us to also parse such a token, generated by the Auth Service, on our own and initiate a user session on the server. However, to do that, we somehow need to pass that token (not an exchange code) to our backend. So, how can we manipulate or intercept that whole process?

To understand how to get the token on a route of our choice, let's dig one level deeper. The reason why you get a link to the Auth Service URL in the first place is because of the default email template that comes with any Supabase instance. It contains the following code:

```
<a href="{{ .ConfirmationURL }}">
```

The highlighted code is a placeholder that, when the email is generated, gets replaced with the Auth Service URL link containing the token.

When we trigger `signInWithOtp(..)`, before the email is sent, that template is loaded and the placeholder is replaced with a generated link from the Auth Server like so:

```
emailTemplate.replace(
  '{{ .ConfirmationURL }}',
  'http://localhost:54321/auth/v1/verify?token=pkce_...'
)
```

That works for what we have right now. However, what we want is something that goes directly to our backend like this:

```
emailTemplate.replace(
  '{{ .ConfirmationURL }}',
  'http://localhost:3000/verify?token=...'
)
```

That would allow us to read and verify the token ourselves. However, you cannot change the value of those placeholders. `{{ .ConfirmationURL }}` will always contain that link pointing back to the Auth Service; what you can do is change the type of placeholder that is used in the template! Instead of using `{{ .ConfirmationURL }}`, the Supabase email templates can use `{{ .TokenHash }}` to construct your own URL containing this `hashed_token` value. We can use this to do exactly what we want to do: verify an OTP (essentially the `hashed_token` value is our OTP) and initiate a user session.

We will definitely use that `hashed_token` value in our implementation, but *not* by changing the email template (for the reasons I have explained regarding why I generally avoid the built-in mailing system).

> **Note**
> There's an optional section at the end of this chapter if you want to know how to change email templates. In this book, you won't need them at all.

If you found this to be a lot of information to digest, don't worry. This knowledge, combined with the step-by-step implementation instructions in the next part, will soon feel like driving downhill: exciting but not too hard.

Implementing a server-only magic link flow with custom email content

As explained in the previous section, we want to send a magic link via email that contains the OTP and points to a route of our choice. Now, you'll get a straightforward explanation of how you can request OTP tokens from the Auth Service, send them via a self-constructed magic link, and verify it on the Next.js backend. As a nice side effect, you'll learn how to send custom-crafted emails.

The process will consist of these essential steps:

1. Prepare a Superadmin client.
2. Request an OTP from the Auth Service.
3. Craft a link containing the OTP.
4. Send the crafted link with a custom-made email.
5. Add a magic link verification route with a Route Handler that authenticates the user.

Let's start with the OTP generation step.

Preparing a Superadmin client

In *Chapter 2*, you learned about two keys: the **Anonymous Key** (`anon_key`) and the **Superadmin Key** (`service_role_key`). While the Superadmin Key grants unrestricted access to your Supabase instance, the Anonymous Key is as restrictive as possible and, by default, gives no permissions to an authenticated user.

Generating a token for a user is an administrative task and, hence, a security concern. It requires admin powers, which are exclusive to the Service Role Key. However, it's important to shield the Service Role Key from exposure to the frontend. Thus, never configure it as an environment variable with the `NEXT_PUBLIC_` prefix.

Let's configure it now by using the unprefixed SUPABASE_SERVICE_ROLE_KEY name. So, add another line inside .env.local that looks like this:

```
SUPABASE_SERVICE_ROLE_KEY=eyJhbGciOiOiJ...
```

Right now, we have two Supabase functions for the backend that allow us to get access to a Supabase client: getSupabaseCookiesUtilClient() and getSupabaseReqResClient(). Both use the Anon Key.

As I just mentioned, we need admin access with the Service Role Key. You might wonder whether that means that we need to create another two client functions such as getSupabase CookiesUtilAdminClient() and getSupabaseReqResAdminClient(). In theory, we do. Practically, we definitely do not.

Let me elaborate. We have the two existing different backend clients as there are two different ways of handling cookies. Now, keep in mind that we needed cookie management for handling user sessions. An admin client, however, shouldn't handle user sessions – you'd be mixing up specific user data with admin rights. That's not a good idea. That'd be like giving someone not just the keys to their apartment but also a master key for all apartments.

Therefore, since we won't use the admin client for features that require cookie handling, we can create a generic admin client that does not need cookie handling. Create a file at supabase-utils/ adminClient.js. Simply use createClient from the base package and pass it the two credentials, the latter one being the Service Role Key instead of the Anon Key:

```
import { createClient } from "@supabase/supabase-js";
export const getSupabaseAdminClient = () => {
  return createClient(
    process.env.NEXT_PUBLIC_SUPABASE_URL,
    process.env.SUPABASE_SERVICE_ROLE_KEY
  );
};
```

Your admin client is now ready to be imported where you need it (only in backend files)!

> **Key learning**
>
> Never use a Supabase client with a Service Role Key for calling user-specific features such as getSession() or getUser(). Use the Superadmin client only for tasks that really require admin powers. If you have a Superadmin client and you need to do user-specific tasks, your first priority should always be to instantiate another client with low permissions (the one we have used so far) for that.

Requesting an OTP from the Auth Service

Let's adapt our existing `app/auth/magic-link/route.js` file to generate an OTP. We want to keep most of the code , but we want to replace the non-admin client with the admin client and get rid of the `signInWithOtp` part. For now, that leaves us with the following:

```
export async function POST(request) {
  const formData = await request.formData();
  const email = formData.get("email");
  const supabaseAdmin = getSupabaseAdminClient();

  if (error) {
    return NextResponse.redirect(
      new URL("/error?type=magiclink", request.url),
      302
    );
  }
  const thanksUrl = new URL("/magic-thanks", request.url);
  return NextResponse.redirect(thanksUrl, 302);
}
```

We now want to generate an OTP (token). Unfortunately, there is no method called `supabase.auth.createOtpToken()`, but there is one that generates the token as part of generating a magic link.

Put the following code just below `const supabaseAdmin = ...` to request the link generation:

```
const { data: linkData, error } =
  await supabaseAdmin.auth.admin.generateLink({
    email,
    type: "magiclink"
  });
```

This will also define the `error` variable, which was defined earlier using `signInWithOtp()`, that we need for the existing `if (error)` check.

Then, after `if (error)`, knowing that no error happened, you can extract the `hashed_token` value from the link properties as follows:

```
if (error) { return ... }
const { hashed_token } = linkData.properties;
```

That's the one-time token with a limited time of validity and it's bound to the user (`email`) we requested it for. Now we need to put this token into a link.

Crafting a link containing the OTP

As a next step, you will want to construct a link that will point to a custom route of your application and that contains this `hashed_token` value as a query parameter so that, when a user receives and clicks that link, we can process it on our backend. Constructing a custom magic link is super easy:

```
const constructedLink = new URL(
  `/auth/verify?hashed_token=${hashed_token}`,
  request.url
);
```

So, when the form is submitted at `your.domain/auth/magic-link`, it takes that `request.url` value as a basis for `constructedLink` and sets its path name to where we want the user to be directed to – in this case, that's `/auth/verify`, containing the `hashed_token` value.

Make yourself aware of the fact that we don't have that `/auth/verify` route yet; we will create it in the last step. Next, though, we will make sure to send this link via email.

Sending the crafted link with a custom-made email

You might be thinking: don't we need an email server for delivering a link via email? With your local Supabase instance, you have one already: Inbucket. The reason why Inbucket catches emails is because Inbucket exposes an SMTP server with which you can connect to send emails – and those are listed in its web interface.

The Supabase services can directly connect to it because they live in the same internal container network. To be able to connect to the same SMTP server from your computer, you need to change just one tiny bit in your local configuration.

Go to the `supabase/config.toml` file and at the `[inbucket]` section, remove the comment (#) at the `#smtp_port = 54325` line so that it just says the following:

```
smtp_port = 54325
```

Save it and run `npx supabase stop` and `npx supabase start` to restart your instance. The Inbucket container will restart with the exposed `54325` SMTP port, which means that you're able to use that and send emails with it now. Don't worry, I'll now guide you through that process.

To send emails via SMTP, I recommend that you install the famous `nodemailer` package via `npm install nodemailer` in your project. In your `app/auth/magic-link/route.js` file, import it via `import nodemailer from "nodemailer";`.

After your `constructedLink` code, you can initialize a mailing transporter, which builds a connection to the Inbucket SMTP:

```
const transporter = nodemailer.createTransport({
  host: "localhost",
  port: 54325,
});
```

Then, you can use this `transporter` object to send any email – in our case, the magic link email:

```
await transporter.sendMail({
  from: "Your Company <your@mail.whatever>",
  to: email,
  subject: "Magic Link",
  html: `
  <h1>Hi there, this is a custom magic link email!</h1>
  <p>Click <a href="${constructedLink.toString()}">here</a> to log
    in.</p>
  `,
});
```

With this, you are obviously sending a completely customized email now. When you request a magic link with the login form, you should now see the customized email in your Inbucket inbox. The link doesn't work yet; we will fix that in the last part, *Adding the magic link verification route.*

Your final `app/auth/magic-link/route.js` structure should resemble the following structure:

```
export async function POST(request) {
  const formData = await request.formData();
  const email = formData.get("email");
  const supabaseAdmin = ...;
  const { data: linkData, error } = await
    supabaseAdmin.auth.admin.generateLink(...);

  if (error) { return ... }
  const { hashed_token } = linkData.properties;
  const constructedLink = new URL(...);
  const transporter = nodemailer.createTransport(...);
  await transporter.sendMail(...);
  return NextResponse.redirect(
      new URL("/magic-thanks", request.url), 302);
}
```

Let's quickly summarize the code – when someone enters the email in the form and presses the **Sign in with Magic Link** button, the form will be submitted as a POST request to the /auth/magic-link URL path, executing the POST function within app/auth/magic-link/route.js. Via generateLink, it will create a hashed_token value. We use this token and add it as a parameter to a custom-constructed link. Then, we send the email containing that link.

> **Note**
>
> Unfortunately, as of August 2024, the generateLink() function does not allow us to pass the shouldCreateUser=false option yet. There's an open issue, so it will hopefully be fixed in the near future:
>
> https://github.com/supabase/gotrue-js/issues/358
>
> Until this is fixed, you can try one of the workarounds which I (**@activenode**) added to the linked issue. However, I'd rather have a shouldCreateUser option soon.

We're so close. There's really just one thing missing now: creating the verification route for the link we constructed such that a click will actually authenticate the user. Let's do that now.

Adding the magic link verification route

In the current state, when a user requests and clicks a magic link, they are sent to a URL such as your.domain/auth/verify?hashed_token=HASHED_TOKEN_VALUE. Now we need to create the Route Handler that responds to this URL. When the user clicks on the verification link, it will simply open that URL with a GET request, so our Route Handler definitely needs to be the GET handler this time.

Go ahead and create the app/auth/verify/route.js file for it. With the request object that Next.js passes, we can grab the full URL and get the query parameters containing the token value as follows:

```
export async function GET(request) {
  const { searchParams } = new URL(request.url);
  const hashed_token = searchParams.get("hashed_token");
}
```

In that code, we extract the query parameters by accessing searchParams, which is automatically parsed when you pass the whole URL to the new URL() web API standard. This allows access to the query parameters via the get() method – so we can grab the hashed_token value from there.

Next, we need to use the Supabase client to verify the token. As the verification will, in a successful case, lead to session handling and therefore cookie handling, remember what one of the key learnings was: we don't need an admin client for that. The verification code comes down to this:

```
const supabase = getSupabaseCookiesUtilClient();
const { error } = await supabase.auth.verifyOtp({
  type: "magiclink",
  token_hash: hashed_token,
});
```

You have to understand that verifyOtp() now confirms whether the hashed_token value, which we generated and provided earlier, is indeed a valid token. If it is not, it will return an error. This means that we can simply check whether the error is falsy and if it is not, the session was successfully established and we can forward to /tickets.

The final code of our app/auth/verify/route.js file looks like this:

```
import { getSupabaseCookiesUtilClient } from "@/supabase-utils/
cookiesUtilClient";
import { NextResponse } from "next/server";

export async function GET(request) {
  const { searchParams } = new URL(request.url);
  const hashed_token = searchParams.get("hashed_token");

  const supabase = getSupabaseCookiesUtilClient();
  const { error } = await supabase.auth.verifyOtp({
    type: "magiclink",
    token_hash: hashed_token,
  });

  if (error) {
    return NextResponse.redirect(
      new URL("/error?type=invalid_magiclink", request.url)
    );
  } else {
    return NextResponse.redirect(new URL("/tickets", request.url));
  }
}
```

In case the token could not be verified, such as because it is expired, an error is returned by `verifyOtp` and we redirect to our existing error page. For a better user experience, you can add yet another error type, `invalid_magiclink`, to the `knownErrors` error pages array and then render it like so:

```
{type === "invalid_magiclink" && (
  <strong>The magic link was invalid. Maybe it expired? Please request
    a new one.</strong>
)}
```

That's all. Your project now allows you to log in with a password or with a magic link, even when the frontend isn't functioning, and you've also generated and sent your fully customized email. That's a reason to celebrate!

After the next section, you'll also know how you can add password recovery to the project.

Adding password recovery

There are two things missing for a complete picture of authentication: the ability to register a new user account and the typical password recovery feature. Registration is covered in the next chapter, but we will look at the password recovery feature now.

Password recovery works in the exact same way as magic links, except for two differences:

- After successfully clicking the Magic (Recovery) Link, you should forward the user to a page where the password can be changed (otherwise, it would be just the same as the magic link).

- To be able to tell the different scenarios apart, we use `type: "recovery"` instead of `type: "magiclink"` for generating the token, though it's effectively the same. For the built-in mailing system (which we don't use), this ensures that a `PASSWORD_RECOVERY` event is fired when used on the frontend (`onAuthStateChange`), such that you could redirect to a page where the user can set the password. However, since we constructed our link on our own anyway, we can just pass this information as a query parameter.

Thanks to the differences, it totally makes sense to reuse the code from `auth/magic-link/route.js` and `auth/verify/route.js`. Hence, in the final code, in the `Ch04_TicketSystem_Auth` project repository branch, you'll find a solution that just adds a query parameter to differentiate between the usual magic link and password recovery. Thanks to this similarity, I will not explain that process in detail but share the essential code differences now.

For requesting the password recovery, you add a `type=recovery` parameter to the constructed link such that we can differentiate the magic link from a (magic) recovery link later:

```
const constructedLink = new URL(
  `/auth/verify?hashed_token=${hashed_token}&type=recovery`,
  request.url
);
```

Within the `verify/route.js` verification route, you'd then read that type query parameter and pass `'recovery'` instead of `'magiclink'` to `verifyOtp` when the recovery type is set and, in the recovery case, redirect to a page where the password can be changed:

```
if (searchParams.get("type") === "recovery") {
  return NextResponse.redirect(
    new URL("/tickets/change-password", request.url)
  );
} else {
  return NextResponse.redirect(new URL("/tickets", request.url));
}
```

Since the user will be logged in at that point, you can have a `app/tickets/change-password/page.js` page where you have a field for setting the new password directly in the frontend with the Supabase client with `supabase.auth.updateUser({ password: newPassword })`.

Here's a view of the final result, which you can find in the `Ch04_TicketSystem_Auth` repository branch:

Figure 4.8: A simple password reset page

Now you know that the password recovery mechanism is just a magic link with a twist. Before finishing this chapter, I'll give you information on what redirect URLs are and why they are important in the context of all of the authentication matters.

Learning about the Site URL and redirect URLs

I'll finish this chapter by explaining to you the very important, but very easy-to-digest, information about the URL configurations in Supabase. I'll cover what they are, why they are important, and where to change them.

Previously, in our short trip to `signInWithOtp`, I showed the following code snippet:

```
supabase.auth.signInWithOtp({
  email,
  options: {
    emailRedirectTo: "http://localhost:3000/tickets",
  },
});
```

The `emailRedirectTo` value contains the URL that you want the Auth Service to redirect the user to after validating a token from the magic link. Although we didn't use any `emailRedirectTo` value in our first implementation with `signInWithOtp`, it still redirected us to `http://localhost:3000/`.

However, that wasn't because the Supabase client knew where our application runs. It was a happy coincidence that this is the default value. It would've even redirected to `localhost:3000` if our app ran on `localhost:8080` (or even `whatever-cool.domain`). In such a case, changing that `emailRedirectTo` value (e.g., to `http://localhost:8080`) wouldn't work out of the box. That's why it's crucial to understand site URLs and redirect URLs, as well as how to configure them.

In a Supabase instance, there exists exactly one site URL. I also like to call the site URL the **default redirect URL** because that's what it is. The site URL is mandatory. By default, it is set to `http://localhost:3000`. For a production instance, you can set it to whatever primary domain you are using. It's the URL that represents your page and the one that the Auth Service takes by default to redirect to when dealing with authentication and built-in mechanisms. It's also the URL that is exposed as a placeholder named `{{ .SiteURL }}` in the built-in email templates. If you have one definite primary domain for your project, feel free to set it to that (for example, `https://my-super-cool-project.domain`).

The project that we are building is tenant-based. Each tenant will also be able to have a custom domain – which essentially renders setting a primary site URL useless. That's why, for our project, we will just leave it at `http://localhost:3000` forever; just consider it to be one of many URLs because it is.

As opposed to the site URL, there can be as many redirect URLs as you want. Say that you run Next.js on `localhost:3000` but your colleague runs it on `localhost:4040`. To allow you to redirect to both without having to change the site URL all the time, you can just add `http://localhost:8080` as a redirect URL, which will then allow you to redirect the Auth Service to this URL as well.

For those of you that really dug deep into this chapter, you might be wondering why we even set any redirect URLs anyway when we don't use the built-in systems that redirect; on our backend, we can redirect to whatever we want. This is super easy to answer: there are way more authentication types in Supabase, such as GitHub, Google, or Apple. Those happen outside of your control, and for those, you need to set valid redirect URLs.

Let's have a look at how we can configure them.

How to configure site and redirect URLs

For supabase.com, it's super convenient to change the URLs. Within your project, you just click on **Authentication**, then **URL Configuration**. This will lead you to an intuitive UI allowing you to change the site URL and add as many redirect URLs as you want.

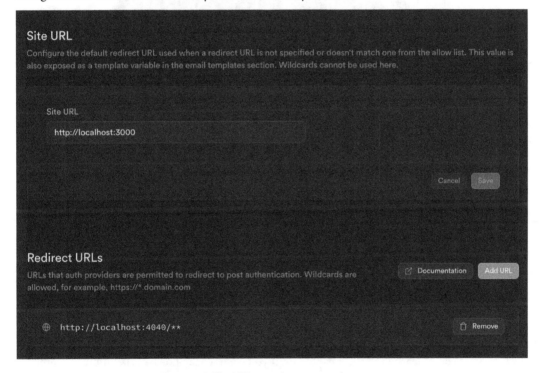

Figure 4.9: The URL overview on supabase.com

Let's say that we want to allow redirection to any path on localhost:4040. Adding http://localhost:4040 will not be enough. It would only allow this exact URL. Luckily, you can add a **wildcard**.

If I wanted to allow any path on that domain (which I usually want, as I only care about the domain, not the path), I would just add http://localhost:4040/** as shown in *Figure 4.10*. Here, ** is the wildcard modifier that matches any sequence of characters to the URL.

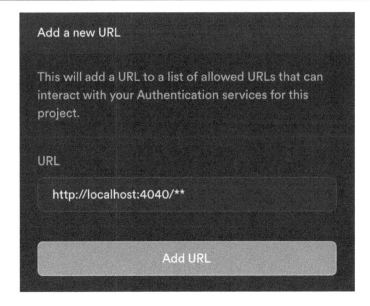

Figure 4.10: Adding a redirect URL on supabase.com

> **Note**
> There are more wildcard options available at `https://supabase.com/docs/guides/auth/concepts/redirect-urls#use-wildcards-in-redirect-urls`.

While the UI-based variant only works on `supabase.com` right now, in your local instance, adding redirect URLs or changing the site URL is done with `supabase/config.toml`. There, go to the `[auth]` section and set the `site_url` parameter and as many additional redirect URLs as you want:

```
[auth]
site_url = "http://localhost:3000"
additional_redirect_urls = ["https://localhost:4040/**"]
```

Your understanding of the URL configuration has been stored in your brain. Now, let's move on and reflect on what you've learned in this chapter, or, if you want, check how you'd edit built-in email templates by not skipping the next section.

Optional knowledge: adapting built-in templates

If you ever need to work solely with the built-in mailing system, changing an email template within the cloud-based `supabase.com` can be done easily via the UI. In your `supabase.com` project, click on **Authentication** and then on **Email Templates**.

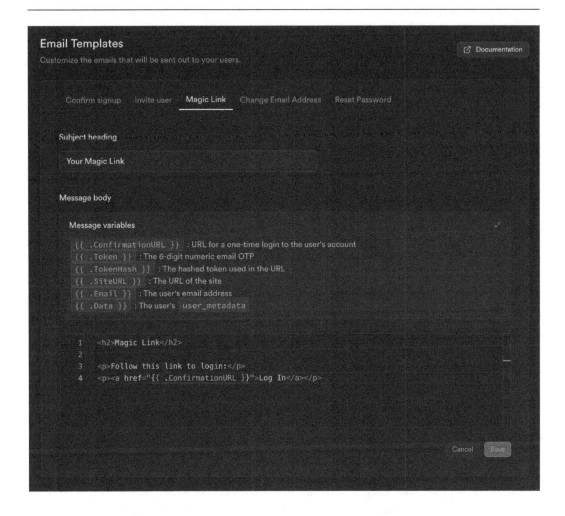

Figure 4.11: Changing built-in email templates on supabase.com

In your local instance, this UI isn't available. To adapt an email template for local development, you have to change the default template by creating a new template as a file and setting it in `config.toml` (in that config file, you'll find the related configurations as comments that are ready to be used).

For example, if you want to change the magic link template of the built-in system, you can create a new file that contains your template at `your_project_folder/supabase/templates/magic.html`. Then you open the `your_project_folder/supabase/config.toml` file and search for the `[auth]` section where, further down, you'll find the `[auth.email.template.magic_link]` section. Here, you can switch to your custom template as follows:

```
[auth.email.template.magic_link]
subject = "Your Magic Link"
content_path = "./supabase/templates/magic.html"
```

Then, just let Supabase refresh it by running `npx supabase stop && npx supabase start` and it will load the custom templates into your instance. It'll now send the customized emails. Here, you'll find an overview of all available templates to adapt: `https://supabase.com/docs/guides/cli/customizing-email-templates`.

Summary

This chapter took a deep dive into user authentication with Supabase in our project. Starting with user creation in Supabase Studio, we seamlessly integrated authentication. The strategic use of `middleware.js` became apparent, handling session refreshes and introducing route guards for authentication state-based redirection.

A key takeaway was the limitation of the session data from `getSession()`, which originated from the cookie and was deemed inadequate for security-related actions. We implemented two login options – password-based and magic link – designed to work seamlessly on the backend. Notably, Supabase's password recovery method was demystified as essentially a magic link with a parameter.

Exploring authentication emails, we bypassed Supabase's default mailing system. We learned to send customized authentication emails using `generateLink` with `nodemailer`, were introduced to Inbucket, and configured Inbucket via `config.toml` to expose its SMTP port for effective communication with `nodemailer`.

In summary, this chapter provided a solid foundation in authentication, equipping us with practical insights beyond the average Supabase developer – good for us!

With all these exciting new things in place, we will take a leap in the next chapter by architecting a database design that symbiotically integrates with our application and permissions such that we can match different tenants with specific users, effectively making it a multi-tenant application. I can't wait!

5

Crafting Multi-Tenancy through Database and App Design

The previous chapter equipped you with strong authentication knowledge and the ability to create as many users as you want. Those users can now log in and see the Ticket Management UI, where a mocked tenant name and some mocked data can be found. But there is no sense of multi-tenancy yet. This is going to change now.

After clarifying the meaning of multi-tenancy, you will lay the database foundations for it by creating tables, a few tenants, and permissions to define which user should be able to access which tenant. With an excursion to backing up your database, you'll ensure that none of your wonderful changes will be lost and see how that enables benefits teamwork as a nice side effect.

You'll also overhaul your Next.js application so that it becomes tenant-aware and you can enable multi-tenancy with ease. We'll finish up by learning how to combine everything we've learned to make sure that the login process can only be reached by existing tenants.

So, buckle up to soak up some knowledge as we cover the following topics:

- What kind of multi-tenancy do we need?
- Designing the database for multi-tenancy
- Committing your database state (if you don't need it, you lose it)
- Making our Next.js application tenant-aware

Technical requirements

Among the usual topics of Next.js and Supabase, this chapter will touch on the topic of database design (creating table structures and data). Hence, you must have basic database knowledge.

The related code for this chapter can be found within the `Ch05_TicketSystem_MultiTenancy` branch (`https://github.com/PacktPublishing/Building-Production-Grade-Web-Applications-with-Supabase/tree/Ch05_TicketSystem_MultiTenancy`).

You can use the `reset` script (`npm run reset`) to get a fresh, hassle-free Supabase instance that matches the progress of the branch.

What kind of multi-tenancy do we need?

In the world of SaaS, a **multi-tenant application** is one where tenants – usually companies/organizations – pay for usage of the app and use it as if the app is run and owned by them. They control their users, the settings, and whatever the application is capable of doing, but the code owner is the SaaS provider.

If you sell software that can be deployed inside of different companies (on-premises), then it is not considered a multi-tenant application, even if you sell it to multiple different companies. That's because every company will essentially run a copy of the software, which counts as multi-single-tenancy or pseudo-multi-tenancy.

Two typical kinds of SaaS multi-tenancy solutions exist:

1. **Same code base, different database/data storage**: This removes the complexity of managing permissions because when everyone in an organization (tenant) can only access their own database – which doesn't hold data from other tenants – this makes security better. However, there is an added complexity of having to manage multiple data sources.

2. **Same code base, same database/data storage**: This means there is only one code base and one source of truth, making scaling simpler as you oversee the usage of the application as a whole. However, there is added complexity in security to make sure that the users from one tenant will not be able to access data from another tenant, even though they are operating on the same system:

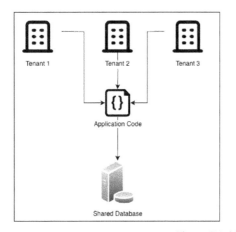 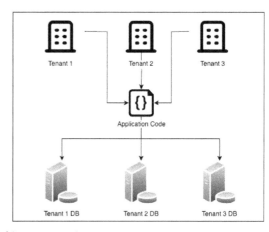

Figure 5.1: Multi-tenancy variants

We want to use Supabase and solve everything in one code base that's deployed at one central point, so we will be using the "same code base, same database/data storage" solution. In our case, Next.js will be the code base, and our Supabase instance will take care of database and data storage.

In the next few sections, we will implement this step by step, ultimately knowing which user belongs to which tenant as well as being able to differentiate between the tenants. Our first step in achieving this is by architecting the database accordingly.

Designing the database for multi-tenancy

Our database needs to be set up for multi-tenancy. But to do that, it's crucial to have a clear view of what our target vision is and what our data must look like. You know the project and you know that there are going to be tenants and users. But what does that mean database-wise?

Planning our database

Here are the semantical, architectural questions that will help us plan our database:

1. What is a **tenant** within the application?

2. What is a **user** in the context of our application?

3. What information must the database contain for us to know if a user has access to a tenant or not?

Let's go ahead and answer these questions.

Defining the tenant

A tenant, at the absolute minimum, is defined by its actual real name – for example, *Packt Publishing* – and a technical but human-readable, unique identifier – for example, *packt*, *packt-books*, or *super_cool_books* (a typical alphanumerical string).

With the knowledge we have right now, we could create a path-based system such as `your.application/TENANT_ID/tickets/...`, where the identifier in the path tells us which tenant our code logic applies to. However, we want to go one step further and match them by their own domain – for example, `tickets.packtbooks.com`.

This leaves us with the following minimum definition: a tenant has a *name*, a readable, unique, *technical identifier (an ID)*, and a *custom domain* (such as `packt.tickets.supercool`).

Defining the user

What will be the minimum definition for a user? I'd say a user is defined by a name and an email. There can be many users named *David* or *Hayden*, so the name is not necessarily distinct. But there can only be one user with the same email address. This means the email could be the value that unambiguously points toward a user while considering questions such as "Which user should get permissions XZY" or "Which user should I delete?" Since the email of a user uniquely identifies a user, answering these questions with an email address would provide some clarity.

However, emails are not commonly used as IDs. One good reason for this is that whenever we need to point to a user in our code, we would need to carry the email string as well, even if that is *super+long-and.weird-email@super-long-12345.domain*. Long story short, we will need a more simple, unique *technical identifier* as well. That means our user is defined by having an *ID*, a *name*, and an *email*.

Defining the tenant access mechanism

To be able to know which user has access to which tenant, we need something that points a user toward a tenant. Let's call this the *tenant permission*. In mathematical terms, a tenant permission would be defined as a tuple of [user_identifier, tenant_identifier] as this clearly states a relationship that a specific user has access to a specific tenant.

Given the previous statements, we don't necessarily need an extra ID for a tenant permission as it's already identified by its unique combination of user and tenant. However, later, to identify a tenant permission row for deleting a permission, for example, you'd need to provide two values. But it's even easier if we can identify the row with just one value, so we'll also add a technical identifier for the permissions.

Now, let's summarize the answers to the three semantical architecture questions:

- A tenant has an arbitrary *name*, a unique *technical identifier*, and a *custom domain*.
- A user has an arbitrary *name*, a unique *email*, and a *technical identifier*.
- A tenant permission is the combination of a *tenant identifier* with a *user identifier* clearly stating the permission of that user on that tenant. For the sake of convenience, a permission tuple will also get its own *technical identifier*.

Next, we'll translate these thoughts into our first database tables.

> **Heads-up**
> The database structure we are about to create will not enforce any kind of access management or permissions in our project. We will make use of the created structure to enforce permissions in the next chapter.

Creating the tenants table

We will start by creating the `tenants` table. To create a new table in your database, go to your Supabase dashboard and click on the **Table Editor** icon. You'll see a **schema** dropdown, space for a list of existing tables in that schema (this should be empty), and a **Create a new table** button:

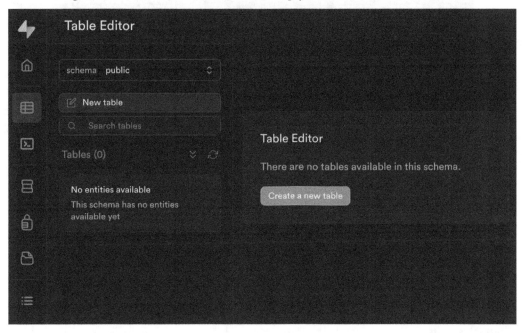

Figure 5.2: Table Editor

Click on **Create a new table**. A table creation UI opens, as shown in *Figure 5.3*:

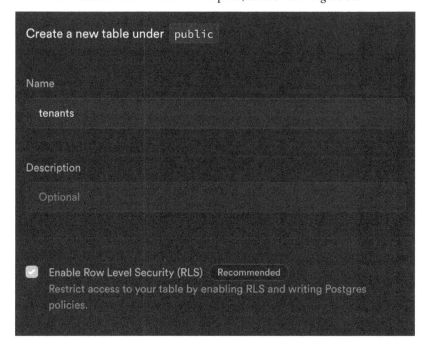

Figure 5.3: The Name and Enable Row Level Security (RLS) options for a new table

For the table's **Name**, enter tenants. The **Description** value is optional, so we'll leave that blank.

You'll also see the **Enable Row Level Security (RLS)** option, which is turned on by default. We will have a lot of touchpoints with RLS throughout this book but for now, I want to explain what it would mean if you unchecked that checkbox. Unchecking that box opens the table to the public in all aspects – everyone, authenticated or not, could read, write, and delete data. If you do that, you'll also face a big warning message telling you exactly that. So, don't!

> **Note**
> At the time of writing, I was trying to come up with a use case where it made sense to disable RLS. However, even with RLS, you'll be able to set up a policy that allows access for everyone. So, no matter how I twisted it, I don't see any benefit in disabling it. If you ever find a good reason to disable RLS on any table, though, I'm curious to know.

Scrolling down in the same table creation modal, you will see two pre-generated default columns: id (*Figure 5.4*) and created_at. Those are good defaults – I usually keep created_at in every table as it automatically adds a timestamp when a new table entry is created and it can never be a disadvantage to know when a row was created.

For this specific table, we want to change the default id value. Currently, it's set to the int8 type, but we want it to be a human-readable identifier as this will make the data's readability much easier. Instead, switch **Type** to text and make sure the **Primary** checkbox – indicating the column is the primary key of the row – stays checked:

Figure 5.4: The id column with the text type

> **Note**
>
> In relational database systems, the primary key is something that is not just unique across the table but also defines the key that unquestionably identifies one specific row. You can find out more here: https://stackoverflow.com/a/13349176/1236627.

Then, add two more columns to make our tenant definition complete:

- Add a column called **name** and set **Type** to **text**. This is the tenant's normal name. Then, click the gear icon next to the column to open the options and uncheck **Is Nullable** as we never want that name to be empty. The name doesn't have to be unique because it is possible to have companies in different countries with the same names, so we don't need to check the **Is Unique** checkbox.

- Add a column called **domain** and set **Type** to **text**. This value also cannot be null, so uncheck **Is Nullable**. However, it must be unique as we need to be able to map one domain to one tenant, so check the **Is Unique** checkbox:

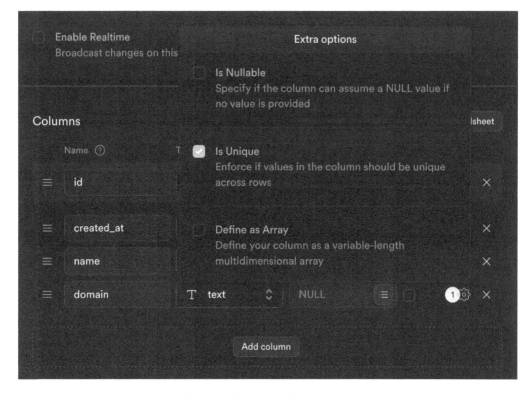

Figure 5.5: Opening column options

Now, click **Save** – your new `tenants` table will appear and be listed in the Table Editor. Awesome!

Your `tenants` table is ready. At the end of this chapter, a user should be able to log in to different tenants. So, let's add at least three tenants to our new table. In the table UI, click on the **Insert** button and select **Insert row**:

Figure 5.6: Inserting new data into a table

This will open a mask where you can add a new tenant:

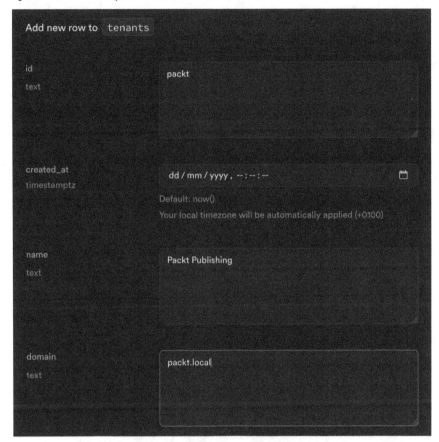

Figure 5.7: Adding the packt tenant row

Now, create three rows with the following tenant data:

id	name	domain
packt	**Packt Publishing**	**packt.local**
activenode	**activenode Education**	**activenode.learn**
oddmonkey	**OddMonkey Inc**	**oddmonkey.inc**

Figure 5.8: Tenant data

For all tenants you create, leave the `created_at` field untouched as it will be filled automatically.

Feel free to add more tenants but make sure that the given three are created so that you can easily follow along with this book.

At this point, the first part of the multi-tenancy database design is done – you have a `tenants` table and you have filled it with data. Let's go ahead and talk about how to design a `users` table (and if we even need one).

Designing the users table

As an architect, my first thought when talking about users in Supabase was, "Why would I need to create a `users` table when it somehow already exists?" I mean, Supabase already stores users – how else would we have been able to create a user and sign in with it? There's a `users` table already; it just sits in a different schema of the database called the `auth` schema.

You can view the `auth.users` table by switching to the `auth` schema in the drop-down menu of the Table Editor:

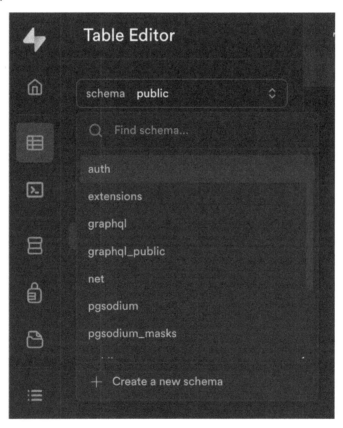

Figure 5.9: The schema dropdown

Clicking on the `auth.users` table will show you the data of the Supabase users that you've created so far:

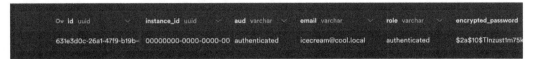

Figure 5.10: A section of the auth.users table view

Each Supabase user has an `id` value in the format of a UUID, which is just a unique, 128-bit string (`https://de.wikipedia.org/wiki/Universally_Unique_Identifier`). On top of that, a user has a `role` value with the value of `authenticated` being the default database group the user belongs to when signed in, an email column, a phone number column for phone-based logins, and more.

You can find a list of the most important columns of that table here: `https://supabase.com/docs/guides/auth/auth-user-management`. However, mostly, you'll be dealing with only a few in this book.

> **Tip**
>
> In the Table Editor, you can do anything you need related to tables. When you click on a table in the Table Editor, it shows the column headings at the top and the entries below, just like in a spreadsheet. Sometimes, you might be not interested in the data (the rows) at all and you'd rather have a nice overview of the table structure, especially when the table has lots of columns. You can do this by clicking on the **Database** icon on the left sidebar. From there, in the **Tables** section, all of your tables are listed. Then, by clicking on their respective columns button, you'll be directed to a vertical structural view with all the columns of the table.

Let's get back to the important question: Can we use this `auth.users` table instead of creating our own `public.users` table? Let's evaluate!

The `auth.users` table contains an identifier (`id`) as well as an `email` column – those are already two of the three things we defined in our requirement for users. The only thing missing is the `name` value.

Your first thought might be to add an additional `name` column to that `auth.users` table, but let me stop you right there! The `auth` schema and its tables are protected and must not be altered. It's the basis for Supabase to function properly. With admin rights and the correct SQL commands, you could alter it, but that's kind of like jailbreaking an iPhone: it works but once you want to upgrade, you might have a hard time, and the warranty is lost, so you'd be on your own for any future problems.

Does that mean we cannot add custom data such as a simple name to the Supabase user in `auth.users`? Well, you can, but not by adding an additional column. The `auth.users` table has a column named `raw_user_meta_data`, which is a JSON-type column. There, you can store as much additional data for a user as you want, so long as it's in JSON format – for example, `raw_user_meta_data` = `{ "name": "Jenna Hill" }`.

Be aware, though, that a signed-in user owns its own user metadata and can freely change that JSON object itself when signed in (`https://github.com/orgs/supabase/discussions/13091#discussioncomment-5360158`). That's okay for a name value but you shouldn't put security-relevant values in `raw_user_meta_data`.

Programmatically, with a Supabase client, the data of a signed-in user can be changed by calling `supabase.auth.updateUser()` (`https://supabase.com/docs/reference/javascript/auth-updateuser?example=update-the-users-metadata`), but that's just a side note.

If you want to put custom, arbitrary data alongside a Supabase user's data that cannot be manipulated by the user, then you should add it to the JSON object in `raw_app_meta_data` column instead. This cannot be changed by the user itself with `supabase.auth.updateUser()`, only by an admin role with `supabase.auth.admin.updateUserById` (`https://supabase.com/docs/reference/javascript/auth-admin-updateuserbyid`).

This is good knowledge to have but also brings us back to the question, "Should we use the `auth.users` table and store an additional name value inside the user's metadata?"

My best practice recommendation (and also that of one of the Supabase maintainers: `https://github.com/orgs/supabase/discussions/6363#discussioncomment-2527443`) is to create your own `users` table reflecting user data. Don't try to architect your database structure within the `auth.users` table.

Now, before you start running in circles about its potential redundancy, rest assured that we aren't replacing or duplicating the existing `auth.users` table by any means. Instead, we will create a new table that complements the existing table. I love naming this additional table `service_users` as it describes the fact that those are the Supabase users using our service (our application). Others call it the `profiles` or `accounts` table but use whatever floats your boat.

The idea is simple – the `service_users` table must contain columns to extend the user data beyond what `auth.users` has. But each row in that table must also reference a row in `auth.users` to determine which auth user belongs to which service user. This is something that becomes easier to understand once we do it.

Click on the **Create a new table** button and enter `service_users` as the name. The two default columns, `id` and `created_at`, can be left as-is – that way, every service user will get a unique number (`id`) automatically. Then, for the user's name, add a column named `full_name` with the **text** type, and make it **Is Nullable** to avoid forcing users to provide their name.

With the next and final column to add, we establish the aforementioned reference to `auth.users` – a relationship between `service_users` and `auth.users`. In relational databases such as Postgres, we call such references relations.

Add a column named `supabase_user`. You don't have to choose the type this time; you can simply click on the **Add foreign key relation** button afterward:

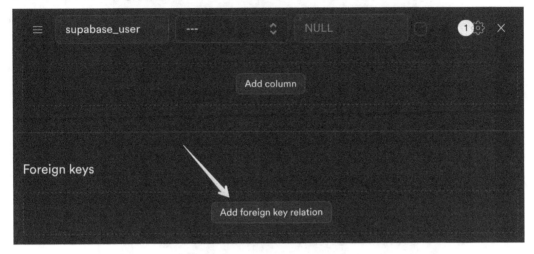

Figure 5.11: Add foreign key relation

When you click it, a new window will appear in which you can choose which foreign table column to link to the current column. The **public** schema is preselected. However, we want to build a relation to the **auth** schema, so select **auth** instead. Proceed by selecting the `users` table and choosing the `supabase_user` column to point to the `id` column of `auth.users`:

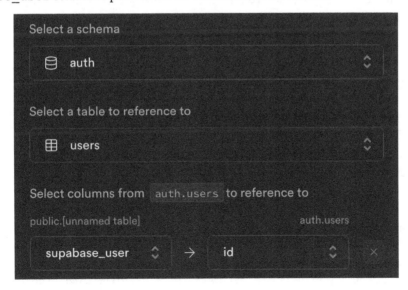

Figure 5.12: Selecting the related table

A few additional options will appear after you select the table you wish to reference:

Action if referenced row is updated

Cascade

Cascade: Updating a record from `auth.users` will also
update any records that reference it in this table

Action if referenced row is removed   ☐ Documentation

Cascade

Cascade: Deleting a record from `auth.users` will also delete
any records that reference it in this table

Cancel   Save

Figure 5.13: Cascade options

For both the **Action if referenced row is updated** and **Action if referenced row is removed** dropdowns, simply select **Cascade**. Cascading ensures a hard relation to the `auth.users` table. This means that when an entry in `auth.users` is deleted (in this case, a Supabase user), any row that references it will be deleted as well (our service user). However, deleting a `service_users` row will *not* trigger the deletion of the referenced Supabase user as the `auth.users` table doesn't reference the `service_users` table (see *Figure 5.14*):

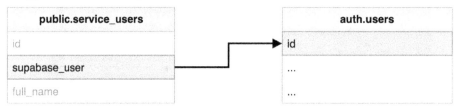

Figure 5.14: Visualizing the created relation

As we can see, a service user is bound to an existing Supabase user. Hence, in the settings of the `supabase_user` column, uncheck **Is Nullable** to make sure that each `service_users` entry will reference an entry in the `auth.users` table. Also, it wouldn't make sense to insert two `service_users` entries with the same `auth.users` reference, so check the **Is Unique** checkbox (same as in *Figure 5.5*).

Next, we want to create a service user row and connect to the *Supabase user* we created in *Chapter 4*. To do that, click the **Insert** button (the same as in *Figure 5.6*) in the `service_users` table to open the following mask:

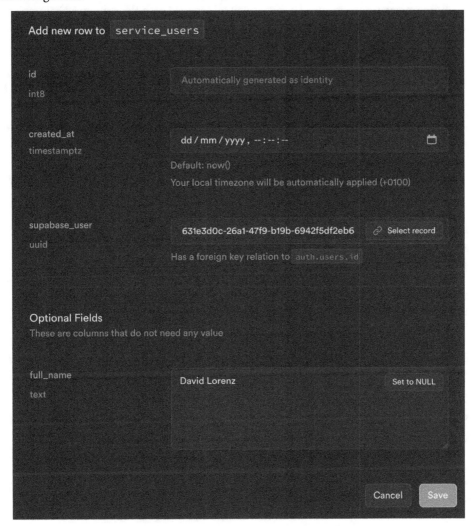

Figure 5.15: The table row before saving

When the mask opens, don't enter anything in the `id` or `created_at` field. Those values get generated automatically.

In the `supabase_user` field, click the **Select record** button to select the user from the `auth.users` table you want to link it to. When clicked, it shows all existing Supabase users as a row-based list, with all of its column values from `auth.users`. Despite all the column values being shown, it doesn't matter where you click inside of a row; it will fill the field with the referenced column value of the clicked row. Click a user; you'll see the `auth.users` table's `id` value (a *UUID*) automatically being filled in our `supabase_user` field.

In the `full_name` column, just fill in the name you'd like – I used `David Lorenz`.

Finally, click **Save**; you'll see the row show up in your table.

With that, you've built your first relationship. The created service user is now pointing (with `supabase_user`) to a previously created Supabase user in the `auth.users` table. Amazing!

Next, we'll bring the tables together to form a permission structure.

Designing the permission structure

Now, give yourself a little challenge and think about how you can structure the database so that it tells us which user has access to which tenant. Sit back and think about it for a minute before continuing. This will help you internalize the matter.

At the moment, we have two tables: `tenants` and `service_users`. This implies that our potential relationships between those are limited:

- We can add a `service_user` column to our `tenants` table that references a service user. But the semantic meaning would be that one row (tenant) has one user. It would be a so-called *1:1* relationship. That doesn't make sense as we couldn't add another user to that tenant.

- We can add a `tenant` column to our `service_users` table that points to a tenant. The semantic meaning here is that one user has one tenant. That's good as it allows multiple users to be assigned to a tenant (multi-tenancy), meaning it's a so-called *n:1* relationship. However, it doesn't allow one user to be assigned to multiple tenants.

If you've used business tools such as ticketing or time-tracking software a lot, you will have logged in to those with the company's email address (`your-name@company.name`). That's essentially the second option – it's usually super-sufficient if you want one user to be assigned to one tenant only.

But in this project, both options are not sufficient as we want to be technically able to give multiple users access to multiple tenants. We want a *n:n* relationship. As mentioned previously, a tenant permission is clearly defined by the `[user_id, tenant_id]` tuple. We can represent that in the form of a permission table, with each row representing a tuple.

First, let's clarify which `user_id` we are talking about in the tuple – the one from `auth.users` or the one from `public.service_users`? Both are IDs from *users* so to say. Regardless of whether we decided to create our permissions table to contain the `[auth_user_id, tenant_id]` or `[service_user_id, tenant_id]` combination, both would indicate the same permission information, semantically. It comes down to which one makes the most sense. Let's elaborate.

If we considered our `service_users` table to be only additional profile data, we probably would've named it `profile_data` instead and then used the ID from the `auth.users` table within our permissions table. However, I consider our `service_users` table to be the actual source of truth of our users and `auth.users` to be the related authentication data. Hence, I always reference the `service_users` table instead of `auth.users` when I reference a *user*. As a nice side effect, we don't have to deal with long UUID values from `auth.users.id` and can use simple numbers from `service_users.id`.

So, what we want is visualized in *Figure 5.16*:

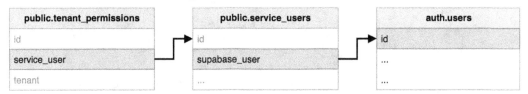

Figure 5.16: The relation chain

Now, let's create the permissions table. Click on the **Create a new table** button and enter `tenant_permissions` as the table name. Leave the two pre-defined columns as-is (`id` and `created_at`) and add two more:

- The first column represents the user id linking back to the `service_users` table. Hence, name this column `service_user`. Then, create a relationship to the `service_users` table. Again, use the **Cascade** options for both the **Action if…** options and unset the **Is Nullable** option; it wouldn't make sense to have a permission where a user would be NULL.

- The second column represents the tenant in our permission tuple. Follow the same approach as with the `service_user` column. However, for this column, we'll build a relationship to the `tenants` table, referencing a tenant ID:

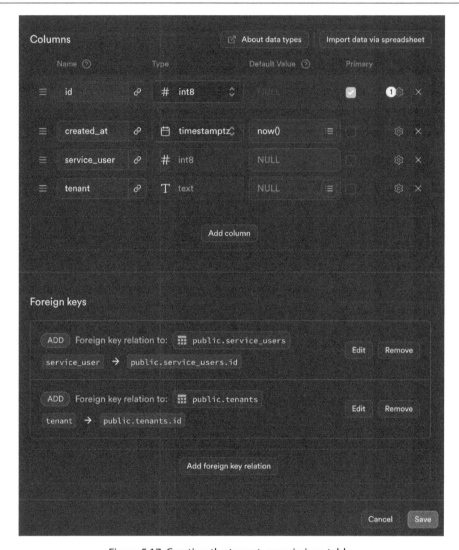

Figure 5.17: Creating the tenant_permissions table

Once saved, you'll have a table where you can identify which user has access to which tenant.

> **Reminder**
>
> We still don't have any permissions enforced in our application; we only created the database structure that will allow us to do so.

Your first service user row will have its `service_users.id` value set to 1. Let's use that user and define it so that it has access to two of our three tenants. Doing that will allow us to test our *n:n* multi-tenancy setup later. So, define your *service user* so that it has access to `packt` and `oddmonkey` by creating two rows in our new table, as follows:

service_user	tenant
1	packt
1	oddmonkey

Figure 5.18: Permission tuples

The result should look as follows:

Figure 5.19: Tenant permission rows

Semantically, your permission rows imply that the service user will be able to access the ticket system from **Packt Publishing** at the `packt.local` domain as well as the one from **OddMonkey Inc** at `oddmonkey.inc`. However, the user should not be able to access the ticket system from **activenode**.

At this point, you're all set – you have all the required tables and a minimal set of rows to be able to match a user permission for a tenant!

However, our application still doesn't check for or use anything of what we defined in the database. We'll connect it with actual permission checks in the next chapter. Back to this chapter, though, we will prepare our Next.js application so that it becomes multi-tenancy-capable. However, before doing so, we'll do a little refreshing side quest to help you keep your database changes intact.

Committing your database state (if you don't seed it, you lose it)

So far, you've put quite a lot of work into your Supabase instance, creating users and tables, and rows within those tables. Now, imagine that you restart your computer and run `supabase start` to spin up your local Supabase, but everything you've done on that local instance is unrecoverably lost and all the Terminal says is **Database unhealthy**. Usually, this shouldn't happen, but it did happen more than twice to me.

Your data in your local instance is stored in a **Docker volume** without you having to take care of it. But just as on a server, there are a thousand reasons why the volume can get damaged (for example, because you didn't stop the containers gracefully), and if that happens, no matter the reason, you have a problem if you don't have a backup. Instead of focusing on that problem, though, you should focus on avoiding it. Losing work means losing precious time, and that's never fun.

There's a simple fix for this: Supabase allows you to make a recoverable snapshot of the database structure and data. This snapshot is then committed as part of your git repository and is automatically loaded.

The underlying process is pretty easy but, as always, I want to give you proper background knowledge. To do so, let's look at the following questions:

- What happens when you run `supabase start`?
- What is the seed file and how can you create it?
- What is a migration file and how can you create it?

> **Note**
> We will now be using the `db diff` and `db dump` features from the Supabase CLI. These are going to be explained more extensively when we discuss production hardening in *Chapter 13*.

When you first ran `supabase start`, it had to create all services and your database from scratch. That's what you did in *Chapter 2*. In that chapter, I also mentioned the `seed.sql` file for the first time. It appeared in your Terminal stating the following:

```
Seeding data supabase/seed.sql....
```

After that, it never appeared again. When your database was freshly set up, the Supabase CLI looked for the `supabase/seed.sql` file and executed it inside the Postgres database after starting it. It doesn't check this file – it executes whatever is in there. However, nothing happened because that file is empty.

The idea behind that file is to be able to provide data instead of starting with an empty database, but only for a fresh setup. The next time you use `npx supabase stop`, it will only save your state in the Docker volume, and calling `npx supabase start` will just restore it. That's good because you don't want a fresh setup each time, but you can enforce it to start fresh and load the `seed.sql` file again by running `npx supabase db reset`.

These are just facts. Knowing that the `seed.sql` file is loaded when a fresh database setup is done makes it a candidate for being our backup file – so a full database snapshot, right? Technically, yes. But the seed file is supposed to contain only data (rows), not the whole database including structure. So, let's back up the data first and take care of getting the database structure after.

Getting your existing data from your running Supabase database into the `seed.sql` file is simple. Run the following command at the root of your project directory:

```
npx supabase db dump --local --data-only --file=supabase/seed.sql
```

This command will export (dump) the existing data into a file. By adding `--local`, you tell the CLI to export it from your local instance. With `--file`, you define a file in which to store it (replacing the existing file). After executing this command, you'll find a few SQL statements in that file, starting with `INSERT INTO`. That's the data export, so all went well.

Data is one thing, but what is data if you don't have the database structure preserved? As we mentioned earlier, structure should not be part of that `seed.sql` file.

For structure, Supabase uses so-called **migration files**. These files are also just SQL files and are technically the same thing as seed files. However, semantically, they're not the same. They are called migration files because they allow you to migrate from an empty database to one with structure and even from existing structures to modified structures.

Same as the seed file is not supposed to contain structure, a migration file is not supposed to contain data. This way, the separation is very clear and clean. There's a different command for creating migration files, though. Execute the following command to dump the structure into a migration file:

```
npx supabase db diff --local -f my_initial_structure
```

This command will magically create a file in `supabase/migrations/`. The filename that's created by the Supabase CLI is a timestamp that follows your given name after `-f`, so something like `YEARMONTHDATETIME_my_initial_structure.sql`. This will contain the commands to create the database structure – without the actual data.

After executing these two commands, you'll have two new files: a `supabase/seed.sql` file filled with data, and a `supabase/migrations/..._my_initial_structure.sql` file containing the structure. Those two files are your complete database snapshot! Now, a fresh setup will execute the migration files (structure) and then the seed file (data). So, simply commit those to your repository and you're super-safe.

Throughout this book, you'll be changing a lot of things in your database, both in terms of data and structure. The simplest solution for now is that whenever you want to back up your database, simply get rid of the existing `migration/*.sql` and `seed.sql` files, then run these steps again. Done!

There's an awesome side effect of what we just did. As you've just committed the database structure and some initial data as part of your repository, everyone working with you in that repository now gets the base setup when they run `npx supabase start` – out of the box. That's what makes it fun to work in a team.

Now that we're safe, let's head back to the tenant-matching logic.

Making our Next.js application tenant-aware

Looking at our database, we can tell which users have permission to a specific tenant. And yet our application is still just a basic login with mock data in the Ticket Management UI, ignoring our new shiny database changes. In this section, we'll make our Next.js app capable of handling multiple tenants.

> **Note**
>
> In this chapter, I'll be using two distinct terms: *tenant-aware* and *tenant-based*. They're not exactly the same but the second one needs the first one. The first one indicates that we have a structure that allows us to pass information about the tenant that we wish to show as part of our URL – for example, `https://some-url?tenant=tenantA`. But just because you opened that URL doesn't mean **tenantA** exists. Tenant-based refers to the fact that we have checked it against actual tenant information (that is, in the database). Let me give a quick example: If the login is tenant-aware, it means that we can differentiate if we should log in to **tenantA** or **tenantB** – we can't do this at the moment as everything is just single-tenancy. Then, if we take that information and check if that tenant exists, it's tenant-based.

There are two big questions to answer:

1. How can we get our application to differentiate between tenants (tenant-awareness)?
2. How can we use custom domains with that differentiation?

Answering the first question is simple: Make the tenant part of the URL! If we make the tenant's ID part of the pathname of the URL, then the differentiation problem is solved. So, instead of `/tickets/*`, the paths become `/TENANT_ID_HERE/tickets/*`, the login at `/` becomes `/TENANT_ID_HERE`, and so on.

The second question isn't trivial and will be solved in *Chapter 6*, but in essence, by knowing the domain of the request, we know the tenant and we can do internal rewrites.

For now, though, let's focus on the first step, which is making our Next.js application read tenants from the URL path. All you have to do to achieve tenant awareness via paths in the URL is to create a new folder called `app/[tenant]` in which we'll move all existing files that are in `app/` right now:

Figure 5.20: Path-based multi-tenancy

> **Note**
>
> The multi-tenancy approach we're taking for Next.js is also the one that's used by Vercel's Platforms Starter Kit (`https://github.com/vercel/platforms`).

Simply moving the files to the new dynamically parameterized folder makes the application tenant-aware because every route is a dynamic route containing the tenant ID. As part of this chapter, we will change a few things to make our code take the tenant parameter into account:

1.   We need to update our middleware to safeguard dynamic routes instead of static ones.

2.   We need to fix all statically coded paths we have in our files to be tenant-based paths.

3.   Lastly, we want to ensure the login is only shown for valid tenants.

Let's go through each of these options.

Enhancing the middleware to safeguard dynamic routes

Currently, our middleware checks if a user is authenticated and if so, redirects them to `/tickets`; otherwise, it heads back to the login page. But every single path now starts with a `/TENANT_ID_HERE` value, so we need to fix the redirection paths.

To be able to do that, we need to read the `tenant` value that is given in the path such that we can create proper redirects containing the tenant ID. In pages or components, it's very easy to access the parameters of dynamic routes, but because the middleware is the first point of touch of a request, dynamic route parameters are not yet resolved, so we have to read them on our own. To do that, we'll now read the full pathname of the requested URL and extract it from there.

In your current `middleware.js` file, you already have these lines:

```
const requestedPath = req.nextUrl.pathname;
if (requestedPath.startsWith("/tickets")) {
...
} else if (requestedPath === "/") {
...
```

We know that every route must now start with the tenant – for example, `/packt/tickets/new`. In that example, `packt` is the tenant, and everything behind it is considered the actual application path (`/tickets/new`). With simple JavaScript tricks, we can tear those two into two variables. The following snippet solves that:

```
const requestedPath = req.nextUrl.pathname;
// ...
const [tenant, ...restOfPath] = requestedPath.substr(1).split("/");
const applicationPath = "/" + restOfPath.join("/");
```

The preceding code splits the path by the slashes into an array (`['packt', 'tickets', 'new']`).
Then, it takes the first value as the tenant value and all of the rest as an array of strings
(`['tickets', 'new']`). The latter is converted back into a string by connecting each bit with
a slash and adding a leading slash, effectively creating `/tickets/new` as `applicationPath`.

Before using that tenant value for our redirects, let's also write a function that confirms that it doesn't
contain any characters that cannot be a tenant ID by enforcing a fixed set of characters. A tenant's ID
should not contain anything else but lowercase letters (a-z), any digits between 0-9, or underscores
and dashes. Let's write a check right after defining the tenant:

```
const [tenant, ...restOfPath] = requestedPath.substr(1).split("/");
if ( !/[a-z0-9-_]+/.test(tenant) ) {
  return NextResponse.rewrite(new URL("/not-found", req.url));
}
```

Now, if the tenant value isn't alphanumeric, it will show a Next.js error page. Plus, the best thing about
this is the fact that we don't have to check it in any other file anymore as the middleware will prevent
us from hitting any other file if the tenant ID isn't okay.

> **Note**
>
> Instead of using `NextResponse.rewrite` to rewrite a mismatch to the `/not-found` path
> internally, I'd rather have used return `NextResponse.error()` as it'll be a clearer solution.
> However, this is not (yet) supported. To learn more about this issue, go to `https://github.`
> `com/vercel/next.js/discussions/52233#discussioncomment-9274388`.

Awesome! Let's continue by adapting the existing `if` clause of the middleware, where we check if the
user tries to access a protected area or not.

I want to highlight that `applicationPath` is just the path of the URL without the tenant so that
you can easily check which part of the app is requested. With this, you can make your `if` conditions
and redirects tenant-based by replacing `requestedPath` with `applicationPath` and then
redirecting to routes containing the tenant identifier in front:

```
if (applicationPath.startsWith("/tickets")) {
  if (!sessionUser) {
    return NextResponse.redirect(new URL(`/${tenant}/`, req.url));
  }
} else if (applicationPath === "/") {
  if (sessionUser) {
    return NextResponse.redirect(new URL(`/${tenant}/tickets`,
      req.url));
  }
}
```

With this change, you restored the original logic while respecting the new folder structure. But don't rush into testing it just yet – we need to make sure we fix all paths, such as for the login, not just the ones in the middleware. That's what we'll do next.

Fixing all static routes in the application

The middleware was a special case where we had to decouple the tenant ID from the path of the URL manually to then be able to create proper URL paths with it. In all the other places where we're dealing with URL paths, we also need to make sure the tenant is contained within the URL as that's the basis of our multi-tenancy. We can use the utilities from Next.js to read the tenant ID so that we can create tenant-based paths.

Since you have to change URL paths to include the tenant ID – in multiple places now – you must build two utility functions. They will come in very handy in the next chapter as well. Create the `src/utils/url-helpers.js` file and fill it with the following code:

```
export function urlPath(applicationPath, tenant) {
    return `/${tenant}${applicationPath}`;
}

export function buildUrl(applicationPath, tenant, request) {
    return new URL(urlPath(applicationPath, tenant), request.url);
}
```

The `urlPath` function can be used everywhere you only need to generate the pathname, while the `buildUrl` function can be used in the backend where you want to create a full URL object.

Before we change anything, let me first give you an overview of the places where we used static routes so that we know where we have to put in work:

- Configuration files:
  - `next.config.js`
- Page files:
  - `error/page.js`
  - `magic-thanks/page.js`
- Component files:
  - The `Nav.js` component
  - The `Login.js` component
  - The `TicketList.js` component

- Route Handler files:

  - `auth/logout/route.js`

  - `auth/magic-link/route.js`

  - `auth/pw-login/route.js`

  - `auth/verify/route.js`

Let's go through them one by one, starting with `next.config.js`. Don't forget to import the `urlPath` function instead of hardcoding paths.

next.config.js

This file contains a static redirect from `/logout` to `/auth/logout`. We need to prefix it with the tenant, as shown in the following highlighted code, making it a tenant-based redirect:

```
redirects: async () => [
  {
    source: "/:tenant/logout",
    destination: "/:tenant/auth/logout",
    permanent: true,
  },
]
```

Error page and magic thanks page

In your current error page, the only static path is found at the **Go Back** link:

```
<Link role="button" href="/">Go back.</Link>
```

But to send someone back to the tenant's login page, we need the `/TENANT_ID/` path, something we can do with `urlPath('/', TENANT_ID_HERE)`. Fortunately, Next.js passes dynamic parameters as `params` to any page, allowing us to access `params.tenant`. That's why you can simply adapt your `ErrorPage` function like this:

```
export default function ErrorPage({ searchParams, params }) {
...
  <Link role="button" href={urlPath('/', params.tenant)}>
   Go back.
  </Link>
...
```

I want to remind you one last time that we don't have to check that `params.tenant` parameter – the middleware already did!

Now, the **Go Back** link works properly. The same problems occurs in `magic-thanks/page.js`, so go ahead and fix it there as well.

The Nav.js component

The `Navigation` component is an interesting one as it is a client component and not a page. Since we want the paths to not just be correct on the frontend but also on the backend, the only reasonable way to fix this is to pass the tenant parameter from the backend.

We cannot pass any parameters from a page down to the `Navigation` component because it doesn't sit in a page file but in the `[tenant]/tickets/layout.js` layout file. Luckily, the layout file gets the parameters of the rendered page, so you can also just access `params`.

Adapt the `TicketsLayout` function in that file as follows:

```
export default function TicketsLayout(pageProps) {
  return (
    <>
      <section style={{ borderBottom: "1px solid gray" }}>
        <TenantName tenantName="Packt" />
        <Nav tenant={pageProps.params.tenant} />
      </section>
      <section>{pageProps.children}</section>
    </>
  );
}
```

After that, adapt your `Nav.js` component so that it uses the `tenant` property:

```
export default function Nav( {tenant} ) { …
```

With that, you can now prefix all `href` attributes in the file, just as you did in the error page, like so:

```
<Link
  role="button"
  href={urlPath("/tickets", tenant)}
  {...(pathname === urlPath("/tickets", tenant) ? activeProps :
    inactiveProps)}
>
  Ticket List
</Link>
```

Also, don't forget to fix the path that you used within `onAuthStateChange` of this navigation file so that it becomes `router.push(`/${tenant}`)`.

Eventually, we'll also pass the tenant ID as `tenantName` to the `TenantName` component so that we don't have just always **Packt** there: `<TenantName tenantName={pageProps.params. tenant} />`.

The Login.js component

In the `Login.js` component, we have a similar case to the `Nav.js` component, but the difference is that the component is imported by a page, not a layout. So, we pass the `tenant` value from a page. Again, using the same principle as in the error page, make sure you get `params` from the page properties in `app/[tenant]/page.js` and then pass it to the `Login` component:

```
export default function LoginPage({ searchParams, params }) {
  ...
  return <Login tenant={params.tenant} ... />;
}
```

Inside the `Login.js` component, make sure you change not only the `href` attributes of the links (with the `urlPath` utility function) but also the form actions (for example, `'/auth/pw-login'`) and the `router.push()` path.

The TicketList.js component

In the `TicketList.js` component, we need to fix the links to the ticket details. Same as in the `Login.js` component, we get the `params.tenant` from the page where the component is used and pass it as a property to `TicketList`. So, `tickets/page.js` now looks like this:

```
export default function TicketListPage({params}) {
  return (
    <>
      <h2>Ticket List</h2>
      <TicketList tickets={dummyTickets} tenant={params.tenant} />
    </>
  );
}
```

Initially, the component code looked like this:

```
export function TicketList({ tickets }) {
  ...
    <Link href={`/tickets/details/${ticket.id}`}>
```

Now, the component code looks like this:

```
export function TicketList({ tickets, tenant }) {
  ...
    <Link href={urlPath(`/tickets/details/${ticket.id}`, tenant)}>
```

The Route Handlers

Your current Route Handler for logging out (`app/[tenant]/auth/logout/route.js`) is just three lines of code where, in the end, it redirects to `/`:

```
export async function GET(request) {
    const supabase = getSupabaseCookiesUtilClient();
    await supabase.auth.signOut();
    return NextResponse.redirect(new URL("/", request.url));
}
```

Redirecting it to `/THE_TENANT_ID` instead is super simple because we have the same access to `params` in the Route Handlers as we do in a page or layout. So, adapt the code like this and you're all set:

```
export async function GET(request, { params }) {
    ...
    return NextResponse.redirect(
      buildUrl("/", params.tenant, request)
    );
}
```

In the same way, you can now fix all the URLs in `magic-link/route.js`, `pw-login/route.js`, and `verify/route.js`.

When you've done this, all your files will be tenant-based. Now, we can make the login dependent on actual tenants.

Making the login tenant-based

Our paths are correctly set for multi-tenancy. However, you can still visit `localhost:3000/invalid-tenant` as the sign-in mechanisms don't care about the data in the database yet – ultimately, nothing is tenant-based yet. We want to change that now.

The login page should throw a 404 page instead of showing the login mask in case of an invalid tenant. We can achieve that directly with the login page itself. Our login page, `app/[tenant]/page.js`, runs on the server, which means it's safe to instantiate an admin Supabase client with which we check if the given tenant exists in the database or not.

At the top of your `LoginPage` function component, you can then search the tenants row using the Supabase client and the value of `params.tenant`. Given that the selection is asynchronous and needs to be awaited, our `LoginPage` must be defined as `async` as well. The result looks like this:

```
export default async function LoginPage({ searchParams, params }) {
    const supabaseAdmin = getSupabaseAdminClient();
    const { data, error } = await supabaseAdmin
```

```
        .from("tenants").select("*").eq("id", params.tenant).single();
    ...
}
```

Without `single()`, Supabase would return an empty list if no tenant was found. However, chaining `single()` forces Supabase to give us a single result (an object with the values) or, if none is found, end up with an error. Thus, we only need the `error` value to determine the tenant's existence. However, if the tenant exists, we want to show the tenant's name in the login box, so we also extract `data`.

First, let's take care of the error case. Having an error indicates that the tenant that's provided in the path doesn't exist. Fortunately, for showing an error page, there's a convenient function called `notFound()` in Next.js that can be imported from the `next/navigation` package and, when called, will show a 404 page. So, on your login page, after database selection, you want to check if an error occurred and trigger the `notFound` function:

```
const { data, error } = await supabaseAdmin.from("...
if (error) {
  notFound();
}
```

However, error-wise, we're not completely done. Calling the `notFound()` function requires us to define a `not-found.js` file. Create the `app/[tenant]/not-found.js` file and fill it with the following code:

```
"use client";
import ErrorPage from "next/error";

export default function NotFound() {
  return <ErrorPage statusCode={404} />;
}
```

This will show the default Next.js 404 error component. That's all there is to do to handle non-existing tenants. If you're signed out and try to access a tenant that doesn't exist at `http://localhost:3000/yummy`, it will show the error component:

Figure 5.21: The 404 error

For existing tenants, we can destructure the data and grab the tenant name:

```
if (error) {
  return notFound();
}
const { name: tenantName } = data;
```

Make sure you pass the `tenantName` value from this page to the `Login` component using the following command:

```
<Login tenantName={tenantName} ... />
```

Then, open the `Login.js` component file and add `tenantName` to the properties of the component:

```
export const Login =
  ({ formType = "pw-login", tenant, tenantName }) => { ...
```

Further below, there is the `<header>` section. Add `tenantName` inside, below the other elements, such that your code will look like this:

```
<header>
  ...
  <div style={{ display: "block", fontSize: "0.7em" }}>
    {tenantName}
  </div>
</header>
```

If you visit a valid login, such as `/packt`, then you will see the tenant name as well:

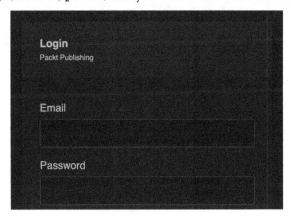

Figure 5.22: Login page with tenant name

But this is just the login page, which is shielded from being reached with the wrong tenant ID. This didn't solve the following problems:

- Someone can sign in with a valid tenant login URL and, after authentication, just switch the URL path to an invalid tenant URL (or one the user shouldn't have access to).

- The Route Handlers at `/auth/` are still technically reachable with a wrong tenant ID.

Both of those points aren't security flaws, but it feels wrong that they are doable. We will get rid of those in the next chapter – the first one will be fixed by actually checking the permissions of the user, while the second will be handled with a global check in our application.

Summary

In this chapter, you forged the backbone of multi-tenancy, sculpting a robust database structure and meticulously defining permissions for your users. Navigating the Table Editor, you established relationships between tables and fortified your project with your safety net of database snapshots. Shifting gears, you revamped the Next.js application, imbuing it with multi-tenant awareness, and adapted every file to align with the new structure. By reading the tenant ID, you strategically kept the login hidden from unauthorized territories, showcasing the power of your newfound structure.

As we move forward, the next chapter promises to build on this foundation, securing your application's authenticated realms by integrating tenant permissions. Get ready to fortify your digital stronghold!

6

Enforcing Tenant Permissions with RLS and Handling Tenant Domains

In the previous chapter, you made your Next.js application *tenant-aware* and implemented the necessary database structure for multi-tenancy. However, we cannot consider it *tenant-based* yet as the application isn't loading any specific tenant data or checking user-to-tenant permissions – it's just that the structure allows multi-tenancy.

Gear up to bring the application to another level as you'll make the application tenant-based in this chapter. Here, you'll explore the implementation of **row-level security** (**RLS**) policies for simple and secure access to user-bound data, as well as navigate the nuances of RLS dependencies, refine your policies, and introduce custom claims for a streamlined RLS experience while unlocking the secret of data access.

After gaining a good understanding of RLS, you'll continue to improve the existing authentication mechanisms; currently, they focus on having a valid session without considering the specific tenant context – which we want to change.

At the end of this chapter, you'll learn how to adapt the whole system so that it's domain-based so that no /tenant_id/ prefix is needed in the user-facing public URL. A great change!

At a glance, you'll learn about the following topics in this chapter:

- Learning to work with RLS
- Minimizing RLS complexity with custom claims
- Making the authentication process tenant-based
- Matching a tenant per domain instead of per path

Technical requirements

Compared to the previous chapter, where we mostly touched on the database with the help of the Supabase Studio dashboard, we will dig deeper and deal with SQL expressions, diving deeper into the database design requirements for completing our multi-tenancy. Although I'll explain everything, you need basic SQL knowledge to understand what's discussed in this chapter.

The related code for this chapter can be found within the Ch06_RLS_Permissions_Domains branch (https://github.com/PacktPublishing/Building-Production-Grade-Web-Applications-with-Supabase/tree/Ch06_RLS_Permissions_Domains).

Use the `reset` script (`npm run reset`) to get a fresh, hassle-free Supabase instance matching the progress of the branch.

Learning to work with RLS

Does it sound like lovely music when I tell you that you'll now fetch data based on the users' permissions from the database that you created in the previous chapter? Well, I hope so because as we step into the realm of RLS, you'll transform your multi-tenant application into a fortress of data integrity and peace of mind.

In this section, I'll show you what happens if we fetch data with the normal Supabase client, without any RLS policy set up. Then, we'll move on to understanding how a simple RLS can be set up and what effect it has on our data selection for signed-in users. From there, we'll create RLS policies that enforce the permissions we set up in the previous chapter and talk about RLS implications and potential pitfalls. We'll conclude by using so-called custom claims with RLS, allowing us to simplify tenant permission checks.

You will learn RLS by doing something that, UI-wise, sounds rather simple – we want to display the correct tenant name in the `TenantName` component by fetching it from the database. If you peek through the subsections quickly, you might get a feeling that there is a lot of content for just fetching a simple name. So, let me tell you one thing upfront – what we do here lays a foundational basis for working with the database based on defined permissions and will not just fetch a tenant name but also all future actions we do with the data from the database; that's why the topic seems bigger than the topic might suggest.

In the previous chapter, we changed `tickets/layout.js` to display the tenant's `id` as the tenant name, which we pass as part of the following URL: `<TenantName tenantName={pageProps.params.tenant} />`. That's showing the `id` value in the UI, not the tenant's real name. But can't we just instantiate the admin client that has full access to the data and fetch the tenant's name?

Well, to answer that question, it is important to focus on the fact that we're talking about the authenticated area right now and that our goal is to protect data from people who don't have the proper permissions to access it. Since our `layout.js` file is rendered on the server, we definitely could instantiate an admin client in it and fetch the tenant's name. But all you'd be doing would be ignoring all permissions to get data from the database. So, even if fetching a simple name with the admin client might not do any harm, even if acquired by faking a session, it's still not a clean thing to do and you should refrain from setting bad examples in your code.

Instead, wouldn't it be awesome if we could fetch the tenant name, and either the signed-in user has permission to do so or the tenant name isn't returned? And if all of this was controlled by the database? Well, that's exactly what RLS can do – without instantiating an admin client.

Fetching tenant data with the restrictive Supabase client

Let's get started by trying to select the data with the normal, safe Supabase client (the client using the Anonymous Key).

Currently, when logged in, in the header section, the `TenantName` component displays the tenant ID. That's because we simply pass through the ID (`<TenantName tenantName={pageProps.params.tenant} />`) from `tickets/layout.js`. But what it should display is the tenant's name, not the ID.

Given we have the ID, we can get the tenant's name from the database directly in the `TenantName` component. So, to avoid confusion, let's rename the property to `tenant` instead of `tenantName`: `<TenantName tenant={pageProps.params.tenant} />`.

As we can fetch data for that tenant from the database right in the `TenantName` component with the tenant ID, let's simply rename the `tenantName` property to `tenant` first. To do so, replace `<TenantName tenantName={pageProps.params.tenant} />` with `<TenantName tenant={pageProps.params.tenant} />`.

Now, we want to change the `TenantName` component so that it loads and displays the name. Let's start by initializing a `tenantName` variable that we'll render. It is initially set to **Unknown** as a fallback value. This fallback value will be overridden later when we load the tenant's name from the database successfully:

```
export default function TenantName({ tenant }) {
  let tenantName = "Unknown";
  ...
  return <strong ...>{tenantName}</strong>
}
```

Next, we want to grab the name from the database, based on the tenant `id` value, and fill this `tenantName` variable with the actual name. To do that, you'll need a Supabase client:

```
export default function TenantName({tenant}) {
  let tenantName = "Unknown";
  const supabase = getSupabaseCookiesUtilClient();
  ...
}
```

Similar to raw SQL statements, the Supabase client allows you to easily translate your normal sentences into database queries. Your sentence, in normal language, would be "I want to fetch data *from* the `tenants` table and *select* the `name` column value from it, in the instances where the row matches the current *tenant ID*." This sentence can be translated into a query, like so:

```
const supabase = ...;
const selection = await supabase
    .from("tenants")
    .select("name")
    .eq("id", tenant);
```

Getting the syntax for database selection requires defining the table from which we want to load data (`from(tableName)`). Here, you must provide a comma-separated string of columns you want to grab using `select(columnsString)`. In our case, it's just the `name` column. Finally, you can add conditions – here, I've used the `eq(column, value)` condition, which translates to "where [column] is equal to."

As this is an asynchronous request to the database, you need to wait for the result before using it. For that reason, we put `await` in front of it:

```
const selection = await supabase.from(...)
```

Awaiting only works inside of an asynchronous function, so you also need to add the `async` keyword to the TenantName component:

```
export default async function TenantName(props) {
```

For our selection, the Supabase client returns a single object containing the `data` and `error` keys. Let's deconstruct the object into two variables after awaiting:

```
const { data, error } = selection;
```

If the `error` value isn't `null`, it means that the selection failed; in that case, `data` will be null. Vice versa, if `data` is not null, the selection succeeded, and `error` will be `null`. We're not interested in handling a potential error here because the worst thing that will happen when we're not able to load the tenant data is that `tenantName` will still be `Unknown`.

Now, before doing anything else, add a `log` statement right after the deconstruction. This will help us analyze and debug the result:

```
console.log({
    tenant,
    data,
    error,
});
```

Next, open the URL to the **Packt Publishing** tenant, `http://localhost:3000/packt`, making sure to sign in if you aren't already. By doing so, you'll be forwarded to `http://localhost:3000/packt/tickets`. When running Next.js in your Terminal, you should see a log saying the following:

```
{ tenant: 'packt', data: [], error: null }
```

Figure 6.1: console.log output

The tenant is correct, no error occurred, and the data isn't null – it's just an array with no items. This means that the database request was successful, but it returned zero rows – hence, the array is empty.

But why? If you go to the database and check the `tenants` table, there's an entry with `id=packt`. So, shouldn't we get that as a result? Yes, we should, and that's where things get exciting.

RLS is activated on each of our tables. If you don't use the Superadmin client, then RLS will guard any table saying, "You can get all data from this table that you're specifically allowed to see, given by the policies you defined." Now, how many RLS policies have we defined so far? None! That means no policy would grant you access to any row, so you'll get an empty list.

This is very important to acknowledge as this doesn't mean that the request fails. It simply says, "Here are all the rows I can give you." And the number of rows the database can give you right now is zero. None. That's why it's simply an empty array, not an error.

To change this, we'll get our hands dirty by defining RLS policies to grant access to data – in other words, to return rows.

Defining RLS policies to access tenants based on permissions

We want to define policies that allow a user who has access to a tenant to also select the tenant's information from the database. To do this, we'll define RLS policies. The most convenient way of defining RLS policies is within the Supabase Studio UI. If you go to the **tenants** table, you'll find a button saying **Add RLS policy**:

Figure 6.2: RLS policies indicator

This button implies that RLS is enabled but there's no policy defined, so only admins can access data right now. If you click on that button, you'll be taken to a new view:

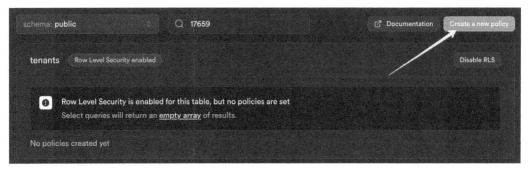

Figure 6.3: RLS policies status view for the tenants table

In this view, click the **Create a new policy** button to start creating a new policy. You will be taken to the **Create a new Row Level Security policy** modal, where you can write a policy or select one from the provided templates (*Figure 6.4*). We will write our own but the templates can be helpful for exploring samples:

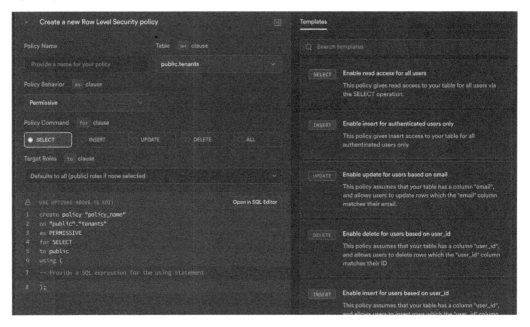

Figure 6.4: The button to create a new custom policy

Let's understand every part of the mask and fill it accordingly:

- **Policy Name**: Just type `my first policy` here. This name is for you and has no technical impact.

- **Table**: This is preselected to be the `tenants` (`public.tenants`) table, which is what we want.

- **Policy Behavior**: This determines if the policy we create will be a **Permissive** or **Restrictive** policy. The difference only comes into effect when there are at least two policies on the same table. This is best explained with some SQL pseudocode. Let's say you have two restrictive policies and two permissive policies. Internally, PostgreSQL checks `SELECT ... FROM tenants WHERE (restrictivePolicy1 AND restrictivePolicy2 AND (permissivePolicy1 OR permissivePolicy2);`. This can come in handy but most of the time (and in this book as well) you'll find only permissively defined policies. That's because there is no way to access data without RLS by default, so most of the time, we define permissive policies to give access based on certain conditions. Leave it set to **Permissive** and move on.

- **Policy Command**: This option asks you to select the operations for which the policy applies. You must select **SELECT, INSERT, UPDATE, DELETE,** or **ALL** (which includes all four). Since we want to grant selection access, choose **SELECT**.

- **Target Roles**: This defines the database roles for which the policy will apply. We haven't talked much about roles, but with Supabase, many developers don't care about the actual roles with policies. The only three roles you know by now are for unauthenticated users (`anon`), authenticated users (`authenticated`), and the admin user (`service_role`). If you leave this field empty, the policy applies to everyone (which is usually a good thing). Note that roles with admin rights (for example, `service_role`) can skip this. If you choose a role here, it just means that for every other role, it will be as if no policy was created (no policy = no data). But don't think about that too much for now; this field is the last thing you should be worrying about. Just choose **authenticated** since we want to create a policy for signed-in users.

- **SQL expression**: Here, you'll find the command for policy creation, where you can only edit the inner expression between `using ( ... )`. Nothing else can be edited because it's based on the other selections. Here, you must provide the actual policy condition in the form of a SQL expression. That may sound complicated but it's just like the condition of an `if` statement; it must resolve to a Boolean. If this expression is `true`, a user gets access to data; otherwise, no data is provided. What's the easiest thing you can put in an `if` statement for it to execute? `true`. So, let's start simple and give every signed-in user access to all the data in the `tenants` table by putting `true` in that field:

```
 🔒  USE OPTIONS ABOVE TO EDIT

 1   create policy "my first policy"
 2   on "public"."tenants"
 3   as PERMISSIVE
 4   for SELECT
 5   to authenticated
 6   using (
 7   |  true
 8   );
```

Figure 6.5: Creating your first policy

If you're thinking forward, yes, this is just the first step to understanding policies. Later, we will improve the policy based on the permissions we've defined in the tenant_permissions table. But for now, click **Review**, then **Save policy**. You should see a summary of the existing policies in the tenants table, as shown in *Figure 6.6*:

Figure 6.6: RLS policies summary of the tenants table

Again, go to http://localhost:3000/packt/tickets to see what your Terminal says now. There shouldn't be an empty array inside data anymore, but one with the selected row data. Since we only selected the name column, it will also only return an object with the name property, as shown in *Figure 6.7*:

```
{
  tenant: 'packt',
  data: [ { name: 'Packt Publishing' } ],
  error: null
}
```

Figure 6.7: Non-empty row data

Awesome – you just received your first data from the database by creating an RLS policy! There's only one last improvement to make. At the moment, we get an array with one row. This is correct, but we don't need an array if we already know that it's going to be one row. The Supabase client has an additional chaining method to say, "I expect exactly one row," and that method is `.single()`. Hence, your selection becomes as follows:

```
const selection = await supabase
  .from("tenants")
  .select("name")
  .eq("id", tenant)
  .single();
```

The interesting part about `single()` is that it will fail when there are zero or multiple results – in such a case, the returned object would be `{ data: null, error: 'some error message here' }`. But since we use eq with the primary id value, we search for one specific (or single) entry, so `single()` is perfect for us. For other use cases, there is also a `.maybeSingle()` method, which will allow zero or one row (but not multiple).

With our added `.single()` method, the log output in the Terminal is without the array, as shown in *Figure 6.8*:

```
{ tenant: 'packt', data: { name: 'Packt Publishing' }, error: null }
```

Figure 6.8: Data result after using .single()

Finally, to render the name, assign it with `tenantName = data.name;`. Your header UI will show the correct name. Although we will make sure that only valid tenants can be accessed later, I'm a big fan of designing for failure to make the application robust. So, let's make sure the application doesn't break if the data is `null` (so the selection resulted in an error for whatever reason). We'll use the following approach, which falls back to the previous value (**Unknown**): `tenantName = data?.name ?? tenantName;`.

Go ahead and try it out with `http://localhost:3000/packt/tickets` (as shown in *Figure 6.9*), as well as `http://localhost:3000/oddmonkey/tickets`. You should see the actual name in the header section now:

Figure 6.9: The header displaying the tenant's name

You can also see the name at `http://localhost:3000/activenode/tickets`, albeit you'll have no permissions because our current policy doesn't respect those permissions yet.

Additionally, you can test a non-existing tenant such as `http://localhost:3000/foobar/tickets` (as shown in *Figure 6.10*), for which it couldn't fetch data from the database, so it sticks with showing **Unknown**:

Figure 6.10: The header displaying the fallback string

With that, you've fetched your first data based on an RLS policy. However, it's not permission-based yet. Let's change that!

Creating a permission-based RLS policy

The next step is to constrain our RLS policy so that we can only get data for the two tenants for which we have permissions in the database defined (for our user, we added `packt` and `oddmonkey` permissions in the previous chapter, but not `activenode`).

> **Attention**
>
> This is a very important section as it explains how RLS works.

On the `tenants` table, the button previously shown in *Figure 6.2* will now say **1 Auth policy**. Click this to get back to the RLS overview again. From there, click **Edit** to edit the existing policy:

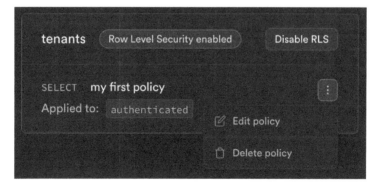

Figure 6.11: Editing your RLS policy

The same mask that was shown in *Figure 6.5* will open, with the only difference being that the Policy Command option cannot be edited (it's hidden away as you cannot alter that, which is fine). Instead, we want to change **USING expression** so that it resolves to `true` if the user has access to that tenant – but which tenant? For people who have never used RLS (which is the majority of people), this is often very confusing as it's not clear which parameters will be available in this expression.

Imagine RLS as something that goes through every single row of the table (`tenants`), like a `for` loop, and executes the expression to check if it resolves to `true`. Only then will it give you access to that row, and then continue to check the other rows. So, with our existing expression being just `true`, it will go through each row and give you access to each row.

This raises a new question: If we're writing a policy that checks rows inside the `tenants` table, how would we know anything about the permissions that are in another table, such as the `tenant_permissions` table? The short answer is that in SQL expressions, you can execute sub-expressions to retrieve additional information from other tables. That's what we'll do in just a moment.

But first, let's understand the policy we need to define in human language by using an example row. Say the tenant row that RLS is checking access for is the `packt` row:

id	name	domain
packt	Packt Publishing	packt.local

Figure 6.12: The packt row

Let me ask you a question: When should a service user (for example, a service user with `id=1`) have access to that row? The answer is if there is a `tenant_permissions` row where `tenant` equals `packt` and the `service_user` column matches the `id` value of that service user (`1`).

If you translate this answer into a SQL expression with hardcoded values, it would look like this:

```
EXISTS (
   SELECT FROM
      tenant_permissions tp
   WHERE
      tp.tenant = 'packt'
      AND
      tp.service_user = 1
)
```

Let me explain. Here, `EXISTS` makes the whole expression a Boolean, resolving to `true` if there is any data inside or `false` if the inside expression didn't return any data. Now, let's look at what's inside the parentheses. We're selecting rows from the `tenant_permissions` table, which we alias as `tp`, and selecting those rows where the tenant `id` value matches `packt` and the service user `id` value matches `1`.

Although this statement is syntactically correct SQL, it's not the statement we need. For the sake of understanding it better, let's assume this was our actual RLS expression and elaborate on what this would mean. Say the service user with `service_user.id = 70` is signed in and tries to access the `tenants` table's data. Semantically, RLS would go through each row of the `tenants` table as follows:

1. Can this user access the `packt` row? Yes, because a permission row with `packt` and service user `1` exists.

2. Can this user access the `oddmonkey` row? Yes, because a permission row with `packt` and service user `1` exists.

3. Can this user access the `activenode` row? Yes, because a permission row with `packt` and service user `1` exists.

So, using hardcoded values doesn't make sense, but this example makes it clear that a given RLS expression will be checked against each row. What you want to do instead is refer to the dynamic tenant value per row and include the signed-in user instead of a hardcoded one, like this:

```
EXISTS (
  SELECT FROM
    tenant_permissions tp
    WHERE
    tp.tenant = tenants.id
    AND
    tp.service_user = THE_SERVICE_USER_ID_OF_CURRENT_USER
)
```

If this was our RLS policy expression for the `tenants` table, it would translate into normal language – for example, "The user has access to a tenant row if a row in the permissions table exists with the same tenant and the current service user. Here, `tenants.id` refers to the `id` value of the tenant of the row that is checked." This process is visualized in *Figure 6.13*:

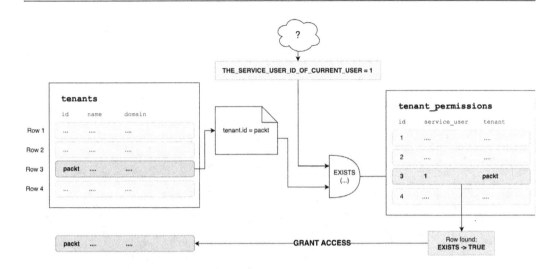

Figure 6.13: RLS policy check per row

The first condition, `tp.tenant = tenants.id`, works, but the second one, `tp.service_user=...`, doesn't because `THE_SERVICE_USER_ID_OF_CURRENT_USER` isn't a variable that exists. So, we need to resolve the service user `id` value differently.

That leaves us with a question: How would the database even know who's signed in? As explained in *Chapter 1*, there's a special connection between the authentication service and the database within the Supabase instance. When we execute any action with the Supabase client, such as selecting data with `supabase.from()...`, the user data is securely exchanged with the database service. This means that for each action, no matter who calls it, the database can tell if it's an authenticated user, and also knows the user ID of that user.

But how can we get access to that user data inside the database? Supabase not only stores users in the `auth` schema of the database but also has special functions in that schema that can be called at any point in time. One of them is `auth.uid()`, which returns the UUID that is stored with a user alongside the `auth.users` table. That's the ID that you used to build a relationship in our `service_users` table (*Figure 6.14*):

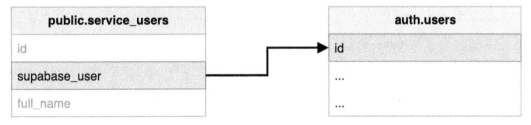

Figure 6.14: The service_users/auth.users relation

This `auth.uid()` function is a normal PostgreSQL function and can be called as part of any SQL expression. So, let's try it. Open a new browser window, go to Supabase Studio (`http://localhost:54323`), and click on the **Terminal** icon.

This opens the SQL editor of the database, where you can execute any SQL command. Let's execute the aforementioned function by entering `SELECT auth.uid();` and then pressing **Run**:

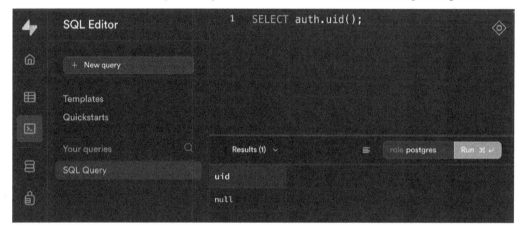

Figure 6.15: The SQL Editor view

As shown in *Figure 6.15*, the result is `null`, and there's a simple reason for this. When you're inside Supabase Studio, you're acting like you're directly connected to the database, with an admin role (the `postgres` role), without anything in between, Hence, you can do what you want but there is no authentication information and you're not a user of the `auth.users` table – you're simply a database admin. That's why running `auth.uid()` will simply return `null`.

But if `auth.uid()` is used as part of an RLS policy, it will only return `null` if the user is not authenticated. Otherwise, it will give the UUID (`auth.users.id`) of the user. With that knowledge at hand, a naïve approach would be to use the following code:

```
EXISTS (
  SELECT FROM
    tenant_permissions tp
    WHERE
     tp.tenant = tenants.id
     AND
     tp.service_user = auth.uid()
)
```

Have a close look. Can you spot why this wouldn't work? Open your tables in Supabase Studio to compare the columns and their types before continuing – I'm sure you can catch the problem!

Our condition, `tp.service_user = auth.uid()`, compares the left-hand value, which is an integer-based service user ID, with the right-hand value, which is the UUID of a signed-in user. In other words, this code tries to match not just two different – hence unmatchable – types but also tries to match the ID of a service user against the ID of a Supabase user. We need a different condition to solve that.

So, we have the ID of the auth user, but we need to get the ID of the service user. We can get this via the ID of the auth user by using a nested `EXISTS` statement:

```
EXISTS (
  SELECT FROM
    tenant_permissions tp
    WHERE
    tp.tenant = tenants.id
    AND
    EXISTS (
      SELECT FROM service_users su
        WHERE
        su.id = tp.service_user
        AND
        su.supabase_user = auth.uid()
    )
)
```

With this expression, the policy is asking: "Does a row with the given tenant ID exist? If yes, is there also a row (`AND EXISTS`) within `service_users` that matches the service user ID from the permission and has a reference (`supabase_user`) to the user that is signed in?"

That sounds like a reasonable policy. Let's see if it works. Go back to the **Editing policy** area, rename the policy `can read tenant if has permission`, enter the expression we created in the mask (*Figure 6.16*), and save it:

```
 🔒  USE OPTIONS ABOVE TO EDIT                                    Open in SQL Editor
 1    alter policy "can read tenant if has permissions"
 2    on "public"."tenants"
 5    to authenticated
 6    using (
 7      EXISTS (
 8        SELECT FROM
 9          tenant_permissions tp
10        WHERE
11        tp.tenant = tenants.id
12        AND
13        EXISTS (
14          SELECT FROM service_users su
15            WHERE
16            su.id = tp.service_user
17            AND
18            su.supabase_user = auth.uid()
19        )
20      )
21
22    );
23    alter policy my first policy
24    on "public"."tenants"
25    rename to "can read tenant if has permissions";


                                                          Cancel    Save policy
```

Figure 6.16: Editing our policy

If you visit your application now, you'll notice that the header will show the fallback name of **Unknown**:

Figure 6.17: The header showing the fallback name

Our expression is correct, both semantically and syntactically. Yet, we're not getting the data. You could feel sad, but this is the point where you should get excited. In the next section, we're going to solve this puzzle!

Understanding and solving RLS implications

This part is extremely exciting as it doesn't just resolve the issue of not getting data, but it will also open your eyes to the magic world of RLS and its awesome security implications.

Our existing RLS policy controls read access (SELECT) in the tenants table. Inside this policy, we're accessing two other tables to make our RLS check complete: the tenant_permissions table and the service_users table.

We have a proper policy defined but still, just like when we didn't have any policies, we don't get any data. The reason isn't immediately obvious, so let me explain. The RLS check inside the tenants table runs, but as soon as it tries to find data in the other two tables, inside of the EXISTS statement, it returns zero rows for tenant_permissions and zero rows for service_users. That's because both have no RLS policy defined. This is why the policy we defined in the tenants table cannot match anything.

That is awesome because, by design, RLS is extremely secure! I like to use an analogy here: Imagine that there's a castle with a front door and an inner door. Even if you have the key to the inner door, you cannot reach the inside if you don't also have the key to the front door.

> **Key learning**
> RLS policies run with the permissions of the signed-in user. When one policy in table1 accesses data from table2, the RLS policies in table2 are checked. If there is no policy defined in table2, it means that table2 will return an empty set of rows, and the RLS policy in table1 can't match anything with table2, so it will resolve to false.

To get data, hence making our existing policy allow data, we must create reasonable policies in the other two tables, service_users and tenant_permissions. We will start with service_users. To which rows within service_users should a signed-in user have read access? The most obvious choice is that a user should have access to its own service user row.

I want to mention that sometimes, it can make sense to give broader permissions – for example, giving a user read access to all service user rows that are part of the same tenant can make sense to be able to read their names (full_name). But we won't go down that path as broader permissions can easily become a security risk. We'll see this in the upcoming sections when we cover RLS implications. Here, we'll stick with the most restrictive policy – which is also much easier to implement than the broader one.

You already know how to create an RLS policy, so go to your service_users table and create a new policy – name it access own user data, choose the **authenticated** role select the **SELECT** operation, and add the following expression:

```
supabase_user = auth.uid()
```

But why are we not using `EXISTS` in the statement? Since the only check we need to do is a value comparison within the row itself, `service_users.supabase_user = auth.user()` sufficiently returns a Boolean.

You can see the **Adding new policy** mask in *Figure 6.18*:

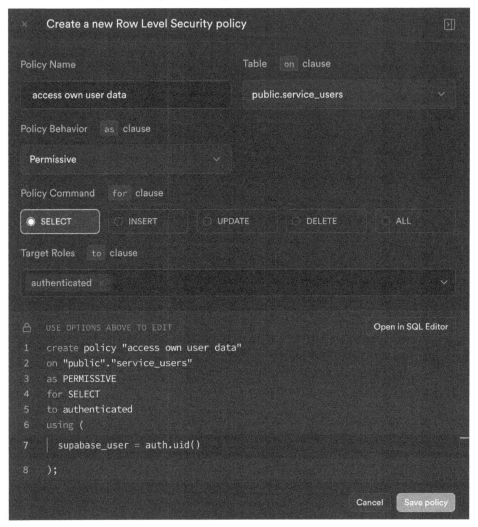

Figure 6.18: Defining the service user's RLS policy

> **Note**
>
> I mentioned earlier that you shouldn't think too much about the role of an RLS. Although it's an essential feature, in both of our RLS policies, it wouldn't make a difference if you chose no role at all (instead of **authenticated**). Once you make your policy dependent on matching a user with `auth.uid()`, it will anyways fail if the user is not authenticated as `auth.uid()` would return `null`. However, it is best practice to always add a role such that the RLS doesn't consume performance for roles for which it shouldn't run anyway.

Once the policy has been saved, the signed-in user will have access to its user data within `service_users`.

Now, we need to write a policy for the data in the `tenant_permissions` table. The most reasonable policy is to constrain access to `tenant_permissions` rows where the respective signed-in service user matches with `service_user` – so the user will only be able to read its own permissions.

Go to the `tenant_permissions` table and add a new RLS policy – name it **allow reading own permissions**, then choose the **SELECT** operation and the **authenticated** role.

Now, let's make a sentence from which we derive the expression "You should be able to access a row within `tenant_permissions` if there's a row in `service_users` where the currently signed-in Supabase user (`auth.uid()`) is given and where, on that same service user row, the service user ID matches with the service user ID value of the permission row." The resulting expression should feel familiar:

```
EXISTS (
  SELECT
    FROM service_users su
    WHERE su.id = tenant_permissions.service_user
    AND su.supabase_user = auth.uid()
)
```

It's the same expression we used in our nested `EXISTS` statement in the `tenants` table. Save it and test your application again. Now, it will retrieve the tenant's name for all tenants you have access to. This means that for **activenode**, at `http://localhost:3000/activenode/`, the access will now be correctly denied as we don't have a permission row matching our user.

With that, your RLS policies have been set up to work seamlessly together. In the next section, we will re-evaluate those policies based on our recent decisions.

Shrinking RLS policies based on the implications

You already know that RLS policies are dependent on other policies if other foreign tables are included. Here, we want to use that knowledge to optimize the complexity and readability of the queries. Let's have another look at our existing RLS policies, from smallest to biggest.

In `service_users`, we have the following:

```
supabase_user = auth.uid()
```

In `tenant_permissions`, we have the following:

```
EXISTS (
  SELECT FROM service_users su
    WHERE su.id = tenant_permissions.service_user
    AND su.supabase_user = auth.uid()
)
```

Finally, in `tenants`, we have the following:

```
EXISTS (
 SELECT FROM tenant_permissions tp
  WHERE tp.tenant = tenants.id
   AND
   EXISTS (
     SELECT FROM service_users su
      WHERE
       su.id = tp.service_user
       AND
       su.supabase_user = auth.uid()
   )
 )
```

Do you notice anything?

- The third – and biggest – policy is the most complex one because it requires matching data from two other tables
- The second one only checks data from one other table and is the second-biggest
- The first one only references data from within the row, so it's the smallest

But that's not all there is to notice. The biggest policy essentially repeats the condition from the second biggest one, while the policy one size down repeats the condition from the smallest one.

In the smallest one, there's nothing to optimize – it's as small as it can be. But the second one checks if there is a matching entry in the `service_users` table where the following check is also made: `supabase_user=auth.uid()`. However, given the policy inside the `service_users` table, we cannot access any other row but rows of our own. This means that with our current policy design, the second condition is redundant.

So, we can reduce the policy in `tenant_permissions` to the following:

```
EXISTS (
  SELECT FROM service_users su
  WHERE su.id = tenant_permissions.service_user
)
```

Going one step further, knowing that we can only access permissions of our own, we can make the `tenants` RLS policy pretty compact now:

```
EXISTS (
  SELECT FROM tenant_permissions tp
  WHERE tp.tenant = tenants.id
)
```

Take a break and let this sink in. In the `tenants` table, we now only check for the existence of a row from `tenant_permissions` that matches the tenant. But we're not checking against the signed-in user because we know that `tenant_permissions` will only give us rows that belong to the signed-in user. What we did has the upside that we can have RLS policies depend on one another so that we can build a reliable chain of RLS policies, similar to stacking LEGOs. This comes with a few problems, all of which I will highlight in the next subsection.

Learning about RLS implications

We just removed the complexity of RLS expressions in terms of size. But now, when someone opens the `tenants` policy, that person might think "What? Checking for the tenant ID isn't enough; the user must be checked as well." So, the person analyzes more of the policies and needs to put them together in their head, piece by piece, to understand that everything has been set up. This means reducing expression complexity can increase mental complexity. This would be different if an RLS policy would immediately show its dependent policies but, unfortunately, such a feature doesn't exist.

But there's another potential pitfall. With the most minimal expression, every policy is only as secure as the one it depends on. I mean, that's why programmers try to avoid dependencies when there's no necessity to add them – to have less mental load. No one likes to be dependent, but that's why you should test your applications.

Let's play this through. Let's assume we change the `service_users` policy to `true`. This would imply that a user could not only access the data of all service users but would have access to the data of all `tenant_permissions` rows because our simplified expression says that we can read all permissions where we have access to the matching user – which are all users, so all permissions. It gets worse because the `tenants` policy says that we have access to all tenants to which we find rows in `tenant_permissions` – which are all rows. Long story short: flick one wrong switch and we're screwed.

So, what's the deal? Should we bring back the complex expressions instead so that they provide more safety? Yes and no. The first principle you should keep in mind is to always have as few dependencies as possible per your database design. So, if you're maybe thinking about adding another table that could also simply be a column in an existing table, maybe think about which serves you better. But I don't want to go into the endless world of database possibilities.

I want to propose a compromise: If we only revert the `tenant_permissions` policy, we gain the independence of the `service_users` table with a rather cheap column comparison:

```
EXISTS (
  SELECT FROM service_users su
   WHERE su.id = tenant_permissions.service_user
   AND su.supabase_user = auth.uid()
)
```

Now, the `tenants` policy still has reduced complexity but if the `service_users` policy was just `true`, it would still provide the same safety due to the additional check in our `tenant_permissions` policy. I'm not saying you should make compromises – the only way to leakproof your application is testing – but I wanted to give you more clarity about RLS.

Also, it's worth noting that having a lot of RLS dependencies will naturally slow down your queries as there is more to process.

At this point, you've gained a really solid understanding of RLS policies. However, at the same time, you might feel a little bit vague about connecting them, fearing that growing projects will become overly complex and slow. Well, let me tell you this: In a production project where I wasn't yet aware of RLS optimizations in general, I implemented many RLS policies with a lot of cross-referencing activating additional policies from other tables and the project was still fast enough for hundreds of users. Nevertheless, I'm here to show you how you can do even better. So, in the next section, we will make our RLS policies simpler both in size and complexity.

> **Note**
> Since we will change the policies in the next section again, I have created a file containing the current policies inside `supabase/book/Ch06_base_rls.sql`.

Minimizing RLS complexity with custom claims

In this section, we'll reduce the RLS complexity for our existing policies and future policies by using something called a **custom claim**. This sounds very special but it's just values that we add to the user object that are easily accessible.

> **Note**
>
> What you'll learn in this section is part of RLS optimization, which is more extensively discussed in *Chapter 10* and *Chapter 13*. The difference is that, in this section, we are only optimizing for the sake of having less overall complexity and hence more safety while later, you'll learn about further optimization tricks.

In *Chapter 5*, I told you that there is a column in the auth.users table of Supabase that's tailored for custom data that cannot be edited by the user on its own, but only by us, or more specifically by an admin Supabase client. It's the raw_app_meta_data column (though when dealing with it technically, such as through the Supabase client or helper functions, it's always just named app_metadata). We will bring the tenant IDs to which a user has access into that app_metadata column of the user.

Here's a visualization of what we have right now (top) and what we plan to achieve (bottom):

Figure 6.19: Visualization of our planned addition

As you can see, we don't plan to get rid of existing tables or existing references and instead plan to include tenant_permissions right in the user. Now, you might be thinking: "But David, didn't you say in *Chapter 5* that it's best practice to create tables instead of filling up the custom metadata in the auth.users table?"

Yes, and that is still true! The tables are our source of truth. The additional information we can add to `app_metadata` can only hold unstructured data – it's just a JSON object – and cannot be used to build relationships with other tables. But it will allow us to make fast permission checks.

Your next question might be, "But if we add that, don't we have to take care of keeping the metadata in sync with the actual table as we will then have the permission information in two places?" Yes and no. If add data statically or hardcode it in `raw_app_meta_data`, which we can, then we have to take care of syncing it. Fortunately, there's a solution from Supabase that allows us to have the tenant information inside `app_metadata` without us having to statically add it.

Let me summarize what the plan is once more: Wherever we deal with the user object, we want to be able to read the tenant permission from `app_metadata` of a user. Let's see how we can achieve that.

Extending app_metadata with tenant permissions

Now, I want to show you how you can implement tenant permissions as part of the user data from `auth.users`. Go to Supabase Studio, navigate to the Table Editor, select the `auth` schema, and then select the `users` table. Find the user for which you created the permissions for `oddmonkey` and `packt` and scroll to the `raw_app_meta_data` column. There, you should find a JSON object with some data that Supabase added for internal usage:

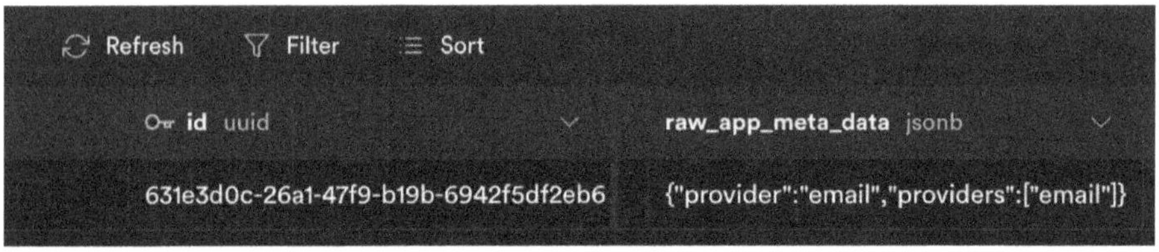

Figure 6.20: raw_app_meta_data in auth.users

Here, `raw_app_meta_data` is a JSON filled with data that Supabase uses internally.

We could change the `raw_app_meta_data` object so that it also holds the tenant IDs directly as part of the user app metadata. Initially, we have the following:

```
{
  "provider": "email",
  "providers": [
    "email"
  ]
}
```

We want to change this so that it looks like this:

```
{
  "provider": "email",
  "providers": [
    "email"
  ],
  "tenants": ["packt", "oddmonkey"]
}
```

To avoid accidental manipulation, the Table Editor doesn't let you edit the values in the actual interface for the auth.users table – it's in read-only mode – but you could use a simple SQL expression that you execute in Supabase Studio's SQL editor to update the value. Let's do that.

Manually extending app_metadata

To update a specific user within auth.users, you need to know which row to update, so copy the ID of the user for which you want to set the tenants first. In my case, it's 631e3d0c-26a1-47f9-b19b-6942f5df2eb6:

Figure 6.21: Copying the id value from an auth.users row

With that id value, you can move to the SQL editor, then enter and run the following:

```
UPDATE auth.users
  SET raw_app_meta_data = '{"provider": "email","providers":
["email"],"tenants": ["packt", "oddmonkey"]}'
  WHERE id='THE_AUTH_USER_ID_YOU_COPIED';
```

When you run this and check the raw_app_meta_data field afterward, you'll see that it successfully updated the value of the raw_app_meta_data column of that user row as defined.

Even more so, since raw_app_meta_data is in JSON form, we can now easily check whether the user has the tenant defined at the user level by getting the tenants property from that JSON and using the ? operator to check whether the value contains a specific tenant value. For instance, to check if the packt value is inside of the tenants array of that user, you can execute the following code:

```
SELECT * FROM auth.users
WHERE (raw_app_meta_data->'tenants') ? 'packt'
AND id = 'THE_AUTH_USER_ID_YOU_COPIED';
```

This should return true in the output of your SQL editor, as shown here:

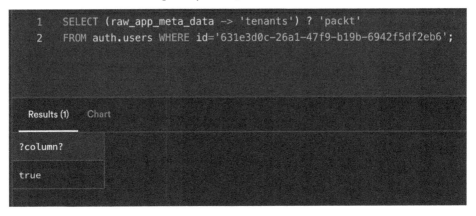

Figure 6.22: Checking if packt is inside the tenants array

That's a good first step. Here, you manually added tenant access information to a user. Let's move on and understand auth.jwt() as well as how to optimize our RLS policies before we learn how to replace the manual approach of setting the tenants array value with a more dynamic solution.

Understanding the auth.jwt() function

Understanding the ease of working with JSON objects within Postgres is the first step. Before changing anything in our RLS policies, I need you to understand that we don't just use the auth.uid() function, which we use in our policy in the service_users table, but also the auth.jwt() function.

Essentially, auth.jwt() contains the row data of the currently signed-in user (or null if not authenticated) parsed as a JSON object. This JSON object given by auth.jwt() contains an app_metadata key that we can use to check the (manually added) tenants value inside of it. In other words, with auth.jwt(), we can easily check defined app_metadata in RLS expressions without the need to query other tables.

But before we use it in an RLS, let's test what we get in both scenarios (unauthenticated/signed-in) when calling auth.jwt() directly in the Supabase Studio SQL editor. If you execute SELECT auth.jwt();, it returns null:

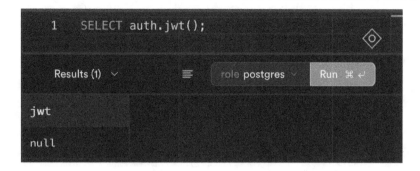

Figure 6.23: Executing auth.jwt() in the editor

That's because you execute it as the `postgres` role and the `postgres` role is not a *Supabase user* – it's a database user.

However, you can also simulate being signed in. In Supabase Studio, you can impersonate an account by clicking on the **role** dropdown, selecting **authenticated role**, and then choosing the user you wish to simulate/impersonate:

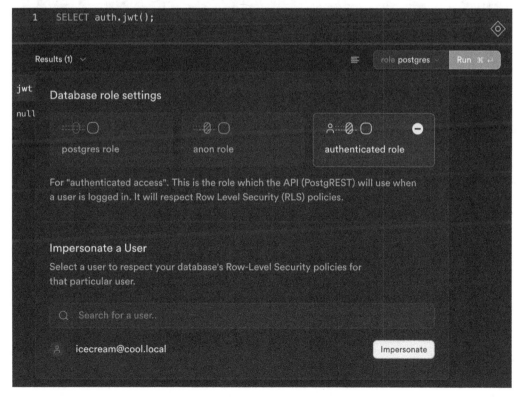

Figure 6.24: Simulating an authenticated user

When choosing my user – that is, **icecream@cool.local** – and running the same command, I got the following output:

Figure 6.25: Executing the function with a simulated user

In the output, you'll find a lot of values inside the returned object that we aren't interested in.

However, if you copy the output, you'll find the `app_metadata` entry. As the whole result of `auth.jwt()` is a JSON object, we can specifically and only select that metadata by using the `->` operator with the `SELECT auth.jwt() -> 'app_metadata';` expression. This is shown in *Figure 6.26*:

Figure 6.26: Grabbing app_metadata from the user object

As this is an object itself with a `tenants` property, you can add `-> 'tenants'` to get the value. From there, we can check the existence of a tenant string within the array by using a question mark, as we did earlier. Hence, to check if the simulated user has access to `packt`, we can write the following:

```
SELECT (auth.jwt() -> 'app_metadata' -> 'tenants') ? 'packt';
```

Executing the preceding code will resolve to `true`, as shown in *Figure 6.27*:

```
1    SELECT (auth.jwt() -> 'app_metadata' -> 'tenants') ? 'packt';



Results (1) ∨                          ☰    role authenticated  ⚲ icecream@co...  ∨   Run ⌘ ↵

?column?

true
```

Figure 6.27: The app_metadata tenants check resolves to true

Finally, we have what we need for RLS: an expression with which we can check permissions, and which returns a Boolean value. Let's get this into our RLS expression.

Using auth.jwt() checks in RLS expressions

In this section, we'll look at each of our existing RLS policies to decide how the new `tenants` value of `app_metadata` and what we learned in the previous subsections will help us optimize RLS policies:

- The current `service_users` RLS policy is `supabase_user = auth.uid()`. There's nothing to improve here; it's as simple as it can be.

- In `tenant_permissions`, we currently have the following policy:

```
EXISTS (
  SELECT FROM service_users su
    WHERE su.id = tenant_permissions.service_user
    AND su.supabase_user = auth.uid()
)
```

  Take a close look and note that no tenant is being checked, only the user ID. This makes sense because being able to access permissions is solely dependent on the user to whom they are assigned. As it's requesting data from `service_users`, it will still trigger the policy from `service_users`. Long story short, there's nothing we can optimize here either.

- In the `tenants` table, our RLS policy is as follows:

```
(EXISTS ( SELECT
   FROM tenant_permissions tp
  WHERE (tp.tenant = tenants.id)))
```

It's a small policy. However, it has the aforementioned problems of high mental load and multiple dependencies. For this table, we only care about whether a user can access a tenant. Since we now have the permission data at hand with the user data, we can replace this policy with the following RLS policy:

```
COALESCE(
    (auth.jwt() -> 'app_metadata' -> 'tenants') ? tenants.id,
    false
)
```

Here, COALESCE is like an if-else statement – if the first argument is null, it'll return the second argument (false). Our policy is independent of other tables – if a user has the tenants.id value of the currently checked row defined in its app_metadata, the user must be granted access. If there is no match, then the first argument will be null and everything will resolve to false. You can go ahead and save this one as your new policy for the tenants table.

You might not be able to foresee why this isn't just a small improvement but a major win, so let me elaborate.

Everything in our application revolves around the users and their tenant permissions. Selections are the most frequent operations in the database and, similar to the tenants table, data access from future tables such as the one that will be holding the tickets is bound to tenant permissions. If we were to use unoptimized RLS, as we did earlier, we would have a lot of unnecessary load on the database. But this isn't the only win here. By enriching app_metadata with the tenant IDs, the user data that's fetched from the Supabase client will now always be enriched with the tenant IDs as part of the app_metadata value that's returned with it. Also, since app_metadata is also stored in the session, it can come in handy in use cases such as the one shown later in the *Rejecting to visit invalid and forbidden tenant URLs when signed in* section.

In short, the benefits of this custom claim will become clear throughout this book.

There's just one more theoretical improvement to achieve: We wouldn't necessarily want to set tenants in the app_metadata manually. Well, we won't anyway because we'll do that with application code within Next.js. But what I mean is that it would be great if the database could set this value for us for the user's app_metadata object. Let me show you the custom access token auth hook and let me explain why we won't be using it.

Setting the tenants value on app_metadata with an auth hook

In this subsection, you'll learn about the **custom access token auth hook**. This is a function that you can define to change what auth.jwt() returns. But how will that work? Will such a function be called each time we call auth.jwt()? No. Such a custom access token function will be called when a user session is created (when the user signs in or the session token is refreshed due to expiry). That's great because that also means that whatever we run in such a function will not unnecessarily be called multiple times.

OK, but that's theory. How can we use it and implement it? Here's a simplified visualization of what we have (top) and what we can do with a custom access token function (bottom):

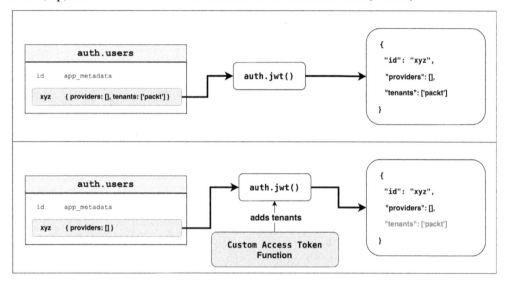

Figure 6.28: A simplified visualization of the custom auth hook

To achieve what's shown in the lower half of the diagram, we must write a SQL function that takes user object data (event data) as a JSON-based input and returns the modified version of it. This is explained at https://supabase.com/docs/guides/auth/auth-hooks?queryGroups=language&language=add-admin-role#hook-custom-access-token.

Let me give you a sample definition of a custom access token function and then tell you why we won't use this approach. The following SQL code looks rather complex since we haven't talked about SQL functions yet (we will deal with database functions for the first time in *Chapter 8*), but I'll explain afterward:

```
CREATE OR REPLACE FUNCTION
   public.custom_access_token_hook(event jsonb)
RETURNS jsonb
LANGUAGE plpgsql
AS $$
   DECLARE
      claims jsonb;
      user_id uuid;
   BEGIN
      user_id := (event->>'user_id')::uuid;
      claims := event->'claims';
```

```
    -- Check if 'app_metadata' exists in claims
    if jsonb_typeof(claims->'app_metadata') is null then
        -- If 'app_metadata' does not exist, create an empty object
        claims := jsonb_set(claims, '{app_metadata}', '{}');
    end if;


    -- Set a claim 'user_id_test' with the value of the user_id
    claims := jsonb_set(claims, '{app_metadata, user_id_test}',
      to_jsonb(user_id::text));

    -- Update the 'claims' object in the original event
    event := jsonb_set(event, '{claims}', claims);

    -- Return the modified or original event
    RETURN event;
  END;
$$;

GRANT EXECUTE
  ON FUNCTION public.custom_access_token_hook
  TO supabase_auth_admin;

REVOKE EXECUTE
  ON function public.custom_access_token_hook
  FROM authenticated, anon, public;
```

If the given SQL code is executed, it creates a function inside the Postgres database that takes a parameter named `event` in the form of a JSON object that contains two properties – `claims` and `user_id`. Here, `user_id` is saved in a local variable of that function with `user_id :=` `(event->>'user_id')::uuid;`. Then, `->>'user_id'` reads the value as text from the given `event` object and `::uuid` converts it into a `uuid` database type (because every Supabase user ID is a `uuid` type).

With `claims := event->'claims';`, we store the value of the `claims` property in our local `claims` variable. The next `if` statement only checks that, within the `claims` object, there is already an `app_metadata` object. If not, it adds an empty one.

Then, with `claims := jsonb_set(claims, '{app_metadata, user_id_test}',` `to_jsonb(user_id::text));`, we set the `user_id_test` property on `claims.app_metadata` with the value of `user_id`.

Finally, we set the changed `claims` value in the event object with `event := jsonb_set(event,` `'{claims}', claims);` and return that.

The GRANT command makes sure that the database role that executes this function has access to this function (`https://supabase.com/docs/guides/database/postgres/roles#supabaseauthadmin`). The REVOKE command revokes permission to call this function on the default roles.

If we were to use this function as our custom access token function, it would mean that there would be always a `user_id_test` field that we could access via `auth.jwt() -> 'app_metadata' -> 'user_id_test'`. In the same way I've added `user_id_test` to the `app_metadata` object, you could enrich it with `tenants` permission data by selecting it from `tenant_permissions`. However, I just wanted to show a simpler code sample to explain how the hook works.

Locally, you could enable this hook via the following entry in `config.toml` within the `[auth]` section:

```
[auth.hook.custom_access_token]
enabled = true
uri = "pg-functions://postgres/public/custom_access_token_hook"
```

On `supabase.com`, you can manage auth hooks by going to `https://supabase.com/dashboard/project/_/auth/hooks`.

It was important to me to explain this idea as this book is about enabling you to make your own decisions in any future Supabase projects. But now, I'd like to clarify why we won't use Auth Hooks for this project in the book.

Why we won't use the custom access token auth hook

In theory, I'd like to use the auth hook as a best practice model as we can generate user object information as part of a dynamic function and hence don't have to sync data in multiple places.

However, at the time of writing, the custom access token auth hook has a few major problems:

- The `app_metadata` information that's generated by the auth hook function is not available in the `app_metadata` object of the `getSession()` or `getUser()` function. Instead, you would have to use the approach described at `https://supabase.com/docs/guides/database/postgres/custom-claims-and-role-based-access-control-rbac#accessing-custom-claims-in-your-application`.

- The auth hook's custom claims are not reflected when using Supabase Studio's impersonation (issue link: `https://github.com/supabase/supabase/issues/27841`). However, this is very confusing when you're trying to test permissions within the central UI, especially within a bigger team.

- Since the custom auth hook is only called on token generation, it could lead to stale data. In other words, if we change something that should be reflected within `app_metadata`, we would have to force a session regeneration.

With these problems at hand, I find it will bring more complexity and issues than it will help us advance, which is why we will disregard using auth hooks in this book for as long as issues 1 and 2 remain.

In the next subsection, we'll briefly discuss how to keep our additionally stored `app_metadata.tenants` data in sync with the table data.

Keeping custom claims in sync with the table data

So far, you've created your user in Supabase Studio and, with the help of an UPDATE SQL expression, added the tenant IDs to a user's `app_metadata`, mirroring the tenant permissions. The problem is that you did that *manually*. That's not just extra effort, but it's also error-prone. Also, if we change a permission row within `tenant_permissions`, it doesn't match the tenant IDs in `app_metadata` anymore. That sounds like a potential security leak. Let's briefly discuss if keeping it in sync is even a real problem or not.

If you want to keep both in sync, that's doable and not a hard problem to solve. However, if something cannot get out of sync, there's certainly no need to handle such a state. In one of my jobs consulting frontends, someone asked, "But what if the server delivers the wrong [JavaScript] file?" I responded: "Then you have a much more severe problem than caring about the frontend code because then we don't control it."

So, asking "What if...?" questions is one thing; facing reality is another. You control the app. Once you have the registration set up, at the end of the next chapter, the registration will do what you did manually first: making sure the permissions are set in the table as well as in `app_metadata`. Hence, at the point of registration, it's definitely in sync.

Sure, that only solves the problem at the time of user creation. But with a user deletion, you wouldn't even have to think about syncing as the user would be gone and everything related to it as well (thanks to Cascade). So, what happens if the permissions of a user change because a row within `tenant_permissions` is deleted or added, for example?

That's easy to answer for this project: Once we've implemented the registration, we're not planning to do that, especially not manually. Yes, our application supports that a single user can have multiple tenants assigned. That's technically feasible because it's an awesome project. But nowhere in our application will `tenant_permissions` be modified after registration – except when we're manually *manipulating* the database. That means that with the existing project, if you don't manually manipulate the database rights, there's no need to sync the permissions.

However, in the future, who knows what you're about to do with this application. So, your next question might be, "But what if I want to sync?" Well, in that case, you have three scenarios to which I'd like to give an answer:

1. Do you want to casually do manual database manipulations on `tenant_permissions`? In that case, just casually do the manual update, as we did previously, setting the `app_metadata` fields with a SQL statement. For casual, individual updates, this shouldn't be much of a burden.

2. Do you want to build an admin panel that allows admins to assign users to more than one tenant? In that case, you can write application logic (similar to what we do in our registration) with which you can update `app_metadata` with your JavaScript code on the backend with a Superadmin client.

3. Do you want the `app_metadata` field of a user to automatically synchronize with the `tenant_permissions` table? In that case, you have two good options:

    A. **Database triggers** can execute SQL functions whenever a table's data changes – in this case, `tenant_permissions`. You'll see how triggers work in *Chapter 8*.

    B. Instead of saving the custom claims directly in the `auth.users` database row, you can enrich the `app_metadata` object with an auth hook, as explained previously.

With this at hand, we won't dive deeper into syncing `app_metadata.tenants` with `tenant_permissions` as we won't need it. However, if you need automatic syncing, consider the options I've mentioned.

Optimizing RLS with custom claims is now something that you're aware of and your Supabase journey backpack is now packed with a pretty solid case of RLS knowledge. That'll help a lot in implementing ticket management. In the next section, we'll make the actual authentication process tenant-based, similar to what we did for the login page, already where we display the login details based on their existence in the database.

Making the authentication process tenant-based

At this point, we have a setup that respects the tenant permissions when selecting data from the database. Our current setup differentiates tenants by a path in the URL and the login page is shown for valid tenants only.

But something's bugging me: I can visit the login of the **activenode** tenant at `http://localhost:3000/activenode/` to sign in with my credentials, even though the account I'm signing in with only has permissions set on `packt` and `oddmonkey`, not `activenode`. When I sign in and I'm forwarded to `/activenode/tickets`, I will see **Unknown** as the tenant name in the header as I have no permission for it.

And here's another thing: even if I sign in on the path of a tenant that I have permission for, I can freely jump around to other tenants by changing the URL because the middleware only checks for a valid session.

Both are not a security concern as our RLS policies will shield the data, but it's just nonsense to be able to do that. So, let's fix both of those issues.

Preventing password login on a foreign tenant

If someone signs in via the frontend, our `Login.js` will listen to a successful `SIGNED_IN` event with `onAuthStateChange(..)` and immediately forward it to the `/tenant_id/tickets` page. As we use the default Supabase method of `signInWithPassword()`, we cannot prevent the process of getting a valid session. But we can prevent forwarding and sign out immediately if the user doesn't have access to the tenant. We can do that without any delay thanks to our previously implemented `app_metadata.tenants` addition.

As shown in the code samples of *Chapter 4*, `onAuthStateChange` not only passes the event as the first parameter but also the valid user session as the second parameter. Here, `app_metadata` is part of the session object. It can be found at `session.user.app_metadata`, which contains our `tenants` array. Also, the current tenant ID is passed to the `Login` component as a tenant variable anyway (we did that in the previous chapter). Therefore, we have everything at hand to check if the user matches with the tenant. What we can do here is adapt our `onAuthStateChange` listener, as follows:

```
onAuthStateChange((event, session) => {
  if (event === "SIGNED_IN") {
    if (session.user.app_metadata.tenants?.includes(tenant)) {
      router.push(`/${tenant}/tickets`);
    } else {
      supabase.auth.signOut();
      alert("Could not sign in, tenant does not match.");
    }
  }
})
```

This prevents us from signing in on foreign tenants on the frontend. However, signing in with a password on the backend will hit the `pw-login/route.js` file. For the sake of simplicity, we can pretty much do the same process there as we're also just using `signInWithPassword`. In the `login/route.js` file, we already read the `.user` object into the `userData` variable from which we can read the `app_metadata` values. We only have to adapt the `if` clause for the error case. Here, we have the following `if` statement:

```
if (error || !userData) {
  return NextResponse.redirect(...);
}
```

We need to change this so that it looks like this:

```
if (
  error ||
  !userData ||
  !userData.app_metadata?.tenants.includes(params.tenant)
) {
```

```
await supabase.auth.signOut();
return NextResponse.redirect(...);
}
```

Now, even before a session in the browser is created, it's deleted when the tenant-user combination doesn't match. A user isn't forwarded to a tenant to which there is no matching permission anymore, at least for the password-based login. Plus, by checking against the allowed tenants, we automatically exclude non-existing clients (although we don't even show the login page for non-existing clients).

In the next section, things will get a bit trickier as we want to prevent the same issue when using the magic link login.

Preventing the magic link login for foreign tenants

As opposed to the password-login variant, where we could already check the session value and compare the tenant ID against `app_metadata.tenants`, with the magic link, there will be no session as the session is only initiated after the user has clicked the link. But that's not helpful. We don't even want to send a link to the user if either the user doesn't have access to that tenant or if there is no such tenant.

Thankfully, we get the related user object as soon as the token is generated. To get a token within `auth/magic-link/route.js`, we must make use of the following code:

```
const { data: linkData, error } =
    await supabaseAdmin.auth.admin.generateLink(...);
```

Here, `linkData` contains token information but also the user object for which it is created. As we already learned, the user object contains `app_metadata`, which contains information about the assigned tenants.

Thus, given the user object from the link generation, we simply don't send the magic link and redirect the user to the error page if we don't have a tenant match within `app_metadata` of the user object. Let's do that. For error handling in the magic link generation, there's currently the following code within the route handler:

```
if (error) {
    return NextResponse.redirect(...);
}
```

Let's extend it the same way we did in the password-login route handler, but we'll access `linkData.user` this time:

```
const user = linkData.user;
if (
    error || !user.app_metadata?.tenants.includes(params.tenant)) {
    return NextResponse.redirect(...);
}
```

That wasn't hard, was it? Go ahead and try it. Move to `http://localhost:3000/activenode` and request a magic link with a user who doesn't have access to that tenant. You should be directed to the error page.

We solved the issue of being able to sign in on a non-matching tenant. Next, we'll ensure that a signed-in user with a valid session will be shown an error page when they switch to a tenant URL to which no permission is given.

Rejecting to visit invalid and forbidden tenant URLs when signed in

I've said it before but it's worth emphasizing that there is no security issue if a user visits a page from another tenant as we secure our application data via RLS. However, it doesn't make sense to show any page, so we want to prevent this. This makes for a good application design.

Unlike the login page, it's not helpful to execute a tenant-user match check at the page level as we aren't talking about a single file that we wish to adapt but multiple files, all being part of the Ticket Management UI at `/tenant/tickets/*`. So, we need a more overarching solution. That's what the middleware is for. Thankfully, the middleware already contains everything we need. The middleware has a `tenant` variable defined containing the tenant ID from the URL, it reads an existing session, and it has an `if` clause to check if the Ticket Management UI path has been accessed (`.startsWith('/tickets')`).

So, if the Ticket Management UI path is accessed with a given session but the session doesn't contain the matching tenant inside `app_metadata.tenants`, we will simply show an error page.

Let's do that. Right now, `middleware.js` contains the following code:

```
if (applicationPath.startsWith("/tickets")) {
  if (!sessionUser) {
    return NextResponse.redirect(new URL(`/${tenant}/`, req.url));
  }
} else ...
```

So far, we haven't been interested in the `else` case of that `if` statement as this just meant a user session being given and we simply allowed it to continue with the request. But now, we want to check `app_metadata.tenants`. If we don't find a match, we'll simply return an error:

```
if (applicationPath.startsWith("/tickets")) {
  if (!sessionUser) {
    return NextResponse.redirect(new URL(`/${tenant}/`, req.url));
  } else if (!sessionUser.app_metadata?.tenants.includes(tenant)) {
    return NextResponse.rewrite(new URL("/not-found", req.url));
  }
} else ...
```

That's all there is to do. Now, if you sign in and navigate from a tenant you have access to (for example, **packt**) toward a tenant you shouldn't have access to (for example, **activenode**), you should face an error page!

Figure 6.29: Error page upon accessing a "forbidden" tenant

> **Reminder**
>
> There's a point about reading values from the session that I have mentioned multiple times now but will repeat because it should be stored in your memory: The session that we read with `getSession()` inside `middleware.js` isn't trustworthy as it's not validated against the server. It's simply the session that's part of your browser storage (cookie); if someone changes that, they can also change the `app_metadata` data that is part of it. What I want to point out is the fact that if someone does that, our application will still be secure – it's just that they will see some broken layouts with missing data.

Our authentication mechanisms are now tenant-bound and no longer generic. A user can only visit tenants with proper permissions given and can only sign in on such tenants. In the next section, we'll make another big leap by assigning tenants to be matched via domain instead of the URL path.

Matching a tenant per domain instead of per path

Now comes another very exciting part that I've been waiting for. In this section, we will match our tenants via custom domains (hostnames). I'd like to say upfront that if you're happy with the solely path-based multi-tenancy choice, you can skip this part; mapping domains isn't a necessity for the rest of this project but it's quite interesting to learn how to achieve it.

> **Note**
>
> We are changing the way the paths are resolved in this section as we want to, for example, be able to visit `packt.local:3000` instead of the `localhost:3000/packt` URL. That means the URL localhost:3000/packt will not be available after this change. However, you will still be able to use localhost for convenience as shown in the section *Bringing back localhost with mapped domains*.

Here, the goal is that each tenant's login page can be reached via `tenant-domain.ending/` and its Ticket Management UI can be reached at `tenant-domain.ending/tickets`. So, when the domain is used, no additional pathname with the tenant ID is going to be needed.

Just as a reminder, here are the domains we have set up in our `tenants` table:

id	domain (hostname)
`packt`	**packt.local**
`activenode`	**activenode.learn**
`oddmonkey`	**oddmonkey.inc**

Figure 6.30: The tenants table data

Let's make sure we have those domains working on our computer.

Adding custom domains via the hosts file

Before we go into implementing a solution, we need to make sure to have the domains mapped locally to our computer. Regardless of whether you're using Windows, Mac, or Linux, it's done in the related `hosts` file.

The format of the `hosts` file is the same for all systems. It's a plaintext file where each line represents a mapping from host to IP in the following format:

```
127.0.0.1 localhost
1.2.3.4 some.host
234.243.342.222 foobar
```

The first line is found in all common `hosts` files. It states that the hostname's `localhost` points to the computer itself (`127.0.0.1`).

On Windows, you can find the file to edit at `\Windows\System32\drivers\etc\hosts`. On Linux or macOS, you'll find it at `/etc/hosts`. In either case, you need to edit the file as an administrator to be able to save your changes.

To be able to run domain-mapped tenants locally, we need to add the three tenant domains to point to our computer, like so:

```
127.0.0.1 packt.local
127.0.0.1 activenode.learn
127.0.0.1 oddmonkey.inc
127.0.0.1 doesnot.exist
```

Once you've added the three lines to your `hosts` file, you can test whether your computer maps it properly by opening your browser and entering `http://activenode.learn:3000/` (make sure your application is started with `npm run dev`). Seeing the 404 page from Next.js means that the host-mapping works:

Figure 6.31: 404 error from Next.js on a custom-mapped host

Right now, all we've done is point `activenode.learn`, `oddmonkey.inc`, `packt.local`, and `doesnot.exist` to `127.0.0.1` (so to your own machine). For now, they're only aliases for `localhost`, which means you can access the `packt` tenant by opening `http://activenode.learn:3000/packt` or the `activenode` tenant with `http://odmonkey.inc:3000/activenode`. But that's going to change in the next section, where the domain will uniquely target the correct tenant without the need to specify the path anymore.

Mapping domains in our application

There are two variants of mapping domains in our application. Both can be replaced with each other with not much effort and both lead to the same result. However, both have their advantages and disadvantages:

- The first variant is the cached, file-based variant. It is extremely fast as no connection to the database will be required at the point of resolving the domain-to-tenant mapping. However, there is a slight delay when adding a new tenant domain.

- The second variant is the fully dynamic variant, which can handle a new tenant, including its custom domain, immediately. However, it comes with the downside that an additional database request is needed for each request.

As a heads-up, as both variants work toward the same thing, you can implement whichever you like more. In the project repository, I'll use the first – that is, the file-based variant. The second one builds upon the first. So, you need to follow the steps of the first one either way. Let's get into the nitty gritty bits of both.

Ultra-high-performing domain mapping

Knowing that we have three tenants and we know the matching hostnames, we can rewrite the internal path of Next.js so that it becomes `/activenode/*` when the URL matches `activenode.learn/*`.

In theory, Next.js allows you to create such rewrite rules as part of the `next.config.js` file, as explained at `https://nextjs.org/docs/pages/api-reference/next-config-js/rewrites#header-cookie-and-query-matching`. But practically, rewrites via `next.config.js` don't make sense in our application. There's a very simple reason for this: the middleware runs before a rewrite rule in `next.config.js` is checked – as opposed to defined `redirects`, which are activated without the middleware becoming active.

Long story short, the rewrites would collide with the middleware logic. But the middleware itself can execute such rewrites, and that's exactly what we will do now!

First, you want to create a file next to the middleware file – for example, `src/tenant-map.js` – in which we map tenant IDs with their domain (I know, that's static, but I need your patience here; by the end of this variant, I'll hint at how to make it semi-dynamic). Then, fill in the file, as follows:

```
export const TENANT_MAP = {
  "activenode.learn": "activenode",
  "packt.local": "packt",
  "oddmonkey.inc": "oddmonkey",
};
```

At the top of your middleware file, import TENANT_MAP with `import { TENANT_MAP } from './tenant-map.js';`.

It's important to focus here as there are quite a few things to consider. With the help of TENANT_MAP, we want to check if the hostname with which our application was requested matches one of the keys of TENANT_MAP. If it does, then we want to rewrite the path internally by adding the tenant prefix. If it doesn't, then we want to show an error page.

Let's start by checking if the tenant exists. If so, we'll output a simple JSON stating that we found a match; if not, we'll return an error.

In your current middleware, you'll find the following section of code:

```
...
const requestedPath = req.nextUrl.pathname;
const sessionUser = session.data?.session?.user;
const [tenant, ...restOfPath] = requestedPath.substr(1).split("/");
const applicationPath = "/" + restOfPath.join("/");
...
```

Above the `requestedPath` definition, you can do the following check to show an error page for an unmatched tenant:

```
const hostnameWithPort = req.headers.get("host");
const [hostname, port] = hostnameWithPort.split(":");
if (hostname in TENANT_MAP === false) {
```

```
    return NextResponse.rewrite(new URL("/not-found", req.url));
  } else {
    return NextResponse.json({ tenantMatch: true });
  }

  const requestedPath = ...
```

From the request headers, we read the host header into `hostNameWithPort`. Afterward, we derive the hostname by splitting away the port part. In the `if` clause, we check if the host exists as a key in our `TENANT_MAP` or not. If it exists, we just show a JSON message to prove our code is working.

Now, if you visit `activenode.learn:3000`, you should see a JSON message:

Figure 6.32: Opening the app with a valid hostname

However, opening `doesnot.exist:3000` should show the 404 page:

Figure 6.33: Opening the app with an invalid hostname

This means our host check is working. However, instead of showing a JSON message, we want to show the page for the valid tenant. So, let's remove the `else` part with `NextResponse.json({ tenantMatch: true })` so that we end up with the following:

```
  ...
  if (hostname in TENANT_MAP === false) {
      return NextResponse.rewrite(new URL("/not-found", req.url));
  }
  const requestedPath = ...
```

This is where things get interesting. Earlier, when we created the path-based solution, we had to remove the tenant prefix to find out the actual application logic path – that is, `/tickets` instead of `/tenant_id/tickets`. But from the domain perspective, the pathname will be the application logic path again. Don't worry – we're not changing our path-based application structure. However, we need to rewrite it internally properly by prefixing it with the tenant ID that we derive from TENANT_MAP with the hostname as the key. This means you can remove the following code:

```
const [tenant, ...restOfPath] = requestedPath.substr(1).split("/");
const applicationPath = "/" + restOfPath.join("/");
```

Replace it with the following:

```
const tenant = TENANT_MAP[hostname];
const applicationPath = requestedPath;
```

Now, when we open `packt.local:3000/tickets`, the tenant ID will resolve to `packt` and the application path will be `/tickets`.

When a user isn't signed in but tries to access `/tickets`, the user is currently forwarded to `/tenant_id/`. But with the domain-based approach, it should be just `/`, so we must construct the URL accordingly.

Usually, we could use something like `new URL('/', req.url)`, but unfortunately, this trick doesn't necessarily give us the correct hostname with our locally rewritten domains, as described in the following Next.js issue: `https://github.com/vercel/next.js/issues/37536#issuecomment-1154191063`.

Instead, we must construct the URL on our own by creating a `tenantUrl` variable that incorporates `protocol` (either HTTP or HTTPS), `hostname`, and `port`. The `hostname` and `port` variables already exist. The `protocol` can be grabbed from the request's `nextUrl` object:

```
const { protocol } = req.nextUrl;
const tenantUrl = `${protocol}//${hostname}:${port}/`;

if (applicationPath.startsWith("/tickets")) {
...
```

Although this is correct, in a production environment, you'll deal with ports that are not sent with the URL as they're default ports (for example, 443 for HTTPS). In that case, the port value is empty. So, let's incorporate that in our composited URL variable:

```
const portSuffix = port && port != '443' ? `:${port}` : '';
const tenantUrl = `${protocol}//${hostname}${portSuffix}/`;
```

Now, if a port is passed (for example, 3000 locally), it gets attached properly; otherwise, it will be just the protocol and the hostname, as expected.

With this constructed URL string, we can now comfortably fix our redirects so that they include the tenant domain. Your new `if` block should look like this (the changes are highlighted):

```
if (applicationPath.startsWith("/tickets")) {
  if (!sessionUser) {
    return NextResponse.redirect(new URL("/", tenantUrl));
  } else if (!sessionUser.app_metadata?.tenants.includes(tenant)) {
    return NextResponse.rewrite(new URL("/not-found", req.url));
  }
} else if (applicationPath === "/") {
  if (sessionUser) {
    return NextResponse.redirect(new URL(`/tickets`, tenantUrl));
  }
}
```

Now, you have proper redirection handling in place for special cases. But the crucial part is still missing: rewriting the internal URL for everything that is not an error and is not redirected.

That's easy because if none of our `if` clauses take action, it means that there is no special case to handle and we're letting the user pass to the page that was requested. Since the `/tickets` path or even `/` does not exist since we changed to the multi-tenancy structure, we must prefix it with a tenant.

And that's what we can do with `NextResponse.rewrite()`. Instead of using `return response.value` at the end of the middleware function, we can set the internal path to resolve with the prefixed tenant:

```
const rewrittenResponse = NextResponse.rewrite(
  new URL(`/${tenant}${applicationPath}`, req.url),
  {
    request: req,
  }
);
const cookiesToSet = response.value.cookies.getAll();
cookiesToSet.forEach(({ name, value, options }) => {
  rewrittenResponse.cookies.set(name, value, options);
});

return rewrittenResponse;
```

What we do here looks complex but is rather simple. First we create a rewritten response (`https://nextjs.org/docs/pages/api-reference/functions/next-response#rewrite`) where we tell Next.js to internally route to the prefixed route with the tenant id that we derived from the domain (for example, requesting `/tickets` with the host `packt.local` is rewritten to `/packt/tickets`). As the `.rewrite` function creates a new `NextResponse`, we then make sure to pass the `request` object onwards. Then, since this new response doesn't have cookies attached, we take the cookies from our existing `response.value` and apply it to the new response object. Finally, we return this `rewrittenResponse` object. However, there's still something missing. In this case, it's not sufficient to only change the path of the URL – by doing so, you also throw away attached query parameters (`?foo=bar`). But since we're making use of query parameters, we would break some functionality. Let me give you an explicit example.

The login's `applicationPath` is `/`. If you have a URL like `http://packt.local/?magicLink=yes`, it's supposed to show the magic link form. However, with the rewrite code given in the previous code block, it shows the normal login. This happens because this code ends up rewriting `packt.local/?magicLink=yes` internally to `/packt/`. That's because the tenant is `packt` and `applicationPath` is `/`, so we're missing query parameters as they aren't part of the pathname (`applicationPath`) of the URL. You can easily fix that by also passing the URL's query string as it was parsed and given to you in `req.nextUrl.search`:

```
const rewrittenResponse = NextResponse.rewrite(
  new URL(`/${tenant}${applicationPath}${req.nextUrl.search}`
  { request: req }
);
```

That's it. If you're asking yourself if you need to also fix the `buildUrl` function to include query parameters everywhere where we construct URLs, you don't. This is simply because, in all our existing use cases where we used `buildUrl`, we don't need or want to pass query parameters.

Now, if you open `http://doesnot.exist:3000/`, you'll still see that it correctly fails to load. However, if you open `http://packt.local:3000/` or `http://activenode.learn:3000/`, it will route you to the correct tenant-based page.

That's lovely! However, there's a new, old problem now: In *Chapter 5*, we changed all static links to tenant-based links by prefixing the URL path with the tenant ID. Now, with the domain approach, we don't want that prefixing anymore. This is easy to change, so let's do it!

Fixing all the links so that they're domain-based

Earlier, when we went from single-tenancy to path-based multi-tenancy, we had to fix all the links to have the tenant ID as a prefix in it. Now, internally and structurally, we still have path-based multi-tenancy but the domain-based approach hides the tenant ID in the path away by the rewrite. So, for link generation, we have to revert all the tenant prefixing again. Thanks to our existing work, this won't take much effort.

First, we need to change the `next.config.js` logout redirect back to the unprefixed path by getting rid of `/:tenant`:

```
const nextConfig = {
  redirects: async () => [
    {
      source: "/logout",
```

```
      destination: "/auth/logout",
      permanent: true,
    },
  ],
};
```

Next, with our `url-helper.js` utility functions, we can change all existing paths by changing the `urlPath` function to this:

```
export function urlPath(applicationPath, tenant) {
  return applicationPath;
}
```

This function might seem redundant as we're passing two variables of which we simply return the first, but don't make the mistake of getting rid of the function solely because of that. By having it abstracted instead of hardcoded, you can easily switch back to path-based routing at any point without refactoring all of your code.

But changing the `urlPath` function isn't enough, unfortunately. We also need to make sure that the created URLs with `buildUrl`, such as in the magic link route handler, contain the actual domain. Let me remind you that in the middleware, we used the header information to get the domain name safely. We need to do that with the `buildUrl` function as well.

Thankfully, we have the required code for deriving the hostname and port in the middleware. We're now moving that code to our `url-helper.js` file inside the `buildUrl` function. With that, we'll also create a `getHostnameAndPort` utility function that gives us the hostname and port based on a given request object:

```
export function getHostnameAndPort(request) {
  const hostnameWithPort = request.headers.get("host");
  const [hostname, port] = hostnameWithPort.split(":");

  return [hostname, port];
}

export function buildUrl(applicationPath, tenant, request) {
  const [hostname, port] = getHostnameAndPort(request);

  const portSuffix = port && port != "443" ? `:${port}` : "";
  const { protocol } = request.nextUrl;
  const tenantUrl = `${protocol}//${hostname}${portSuffix}/`;

  return new URL(urlPath(applicationPath, tenant), tenantUrl);
}
```

This ensures that your generated URLs, wherever you use `buildUrl`, include the correct domain. None of this code logic of deriving the host is new; it's the same code where we use middleware already. And because of that, you can now even clean up your middleware by using those functions instead of having duplicated code. Open the `middleware.js` file and find the following block:

```
const hostnameWithPort = req.headers.get("host");
const [hostname, port] = hostnameWithPort.split(":");
```

Exchange it with this block (we only need the hostname in the middleware itself now):

```
const [hostname] = getHostnameAndPort(req);
```

Then, further down, you can completely get rid of the following code:

```
const portSuffix = port && port != "443" ? `:${port}` : "";
const { protocol } = req.nextUrl;
const tenantUrl = `${protocol}//${hostname}${portSuffix}/`;
```

The previous code's only task was to help us construct a correct URL for which we can now use `buildUrl`. Previously, we had this:

```
return NextResponse.redirect(new URL("/", tenantUrl));
```

After deleting the previous code, it becomes this:

```
return NextResponse.redirect(buildUrl('/', tenant, req));
```

We also had this:

```
return NextResponse.redirect(new URL(`/tickets`, tenantUrl));
```

This now becomes this:

```
return NextResponse.redirect(buildUrl("/tickets", tenant, req));
```

The last bit of our middleware, the `NextResponse.rewrite` part, must not be changed – it's rerouting internally, not externally!

Awesome, we're done with domain mapping. In the next section, I'll explain how this seemingly static approach can be made dynamic.

Making the static TENANT_MAP dynamic

With many B2B applications, there is a contract between you and the tenant. That process leaves sufficient time to deploy another tenant configuration. Changing the `tenant-map.js` file to add another tenant, committing it, and with that triggering an automatic deployment (if used on platforms such as Vercel or Netlify, which automatically deploy when you push) is a thing that takes less than a minute.

However, maybe in the future you'll add a **Register a new tenant** option and you're at the beach drinking a Mojito. You don't want whoever registered a new tenant to wait until you're back to find the time to edit that file, that's for sure!

What you can do is make use of deployment pipelines in combination with events, such as those from webhooks (we will discuss such events in the *Diving Deeper into Security and Advanced Features* part). There are so many ways to achieve this. We won't dive deep into it, but let me sketch it for you quickly. When a new tenant is added to the `tenants` table, you can let Supabase trigger an HTTP request to whatever you like. That could be a GitHub webhook link that causes your project to be built. In that case, you could create a step that reads the existing tenants from the database and creates the `tenant-map.js` file as part of the build process. Voila, it's dynamic!

So, by integrating it into a clever pipeline, you have both a high-performing solution and the comfort of not having to edit the file manually. So, once a tenant is created, the process will take typically less than five minutes for it to be domain-mapped (this is just a random guess; that depends on your pipeline, though most of my pipelines run in less than one minute).

You're now equipped with the knowledge of how to implement domain-based multi-tenancy, which performs extremely well, and you even know how to automate the process so that you don't have to manually generate that file. The only downside is that you either have to change the file on your own for new tenants (which is fine if you grow one client per month, for example) or create some kind of automated file creation, as described here.

In the next section, you'll learn how to map a domain instantaneously and solely through the database data, without having to create a `tenant-map.js` file.

Fully dynamic domain mapping

Say you have a fast-growing multi-tenancy application. Every other minute, a new tenant is created. When that tenant is created, it should be available immediately, not one minute later. The following solution solves that. The additional effort of achieving that is near zero, but it comes with slower load times for each request.

We will do the tenant check dynamically, so you don't need the `tenant-map.js` file. What you need instead is a connection to Supabase to read the tenant. The code logic we need to achieve that already exists in your **Login** page (`/[tenant]/page.js`), which checks for the existence of a tenant:

```
const supabaseAdmin = getSupabaseAdminClient();
const { data, error } = await supabaseAdmin
  .from("tenants")
  .select("*")
  .eq("id", tenant)
  .single();
```

To dynamically resolve the domain, we need to use that database selection in the middleware. In the current middleware, we have the following code block where, at the end, the hostname is checked with the help of `TENANT_MAP`:

```
const supabase = getSupabaseReqResClient({ req, res });
const session = await supabase.auth.getSession();
const [hostname] = getHostnameAndPort(req);

if (hostname in TENANT_MAP === false) {
  return NextResponse.rewrite(new URL("/not-found", req.url));
}
```

We need to initialize an admin client to get access to all tenants. Then, with that admin client, we must execute the same command shown on the login page. If it returns an error, it means the tenant does not exist. So, the extracted code section becomes as follows:

```
const supabase = getSupabaseReqResClient({ req, res });
const session = await supabase.auth.getSession();
const [hostname] = getHostnameAndPort(req);

const supabaseAdmin = getSupabaseAdminClient();
const { data: tenantData, error: tenantError } = await supabaseAdmin
  .from("tenants")
  .select("*")
  .eq("domain", hostname)
  .single();

if (tenantError) {
  return NextResponse.rewrite(new URL("/not-found", req.url));
}
```

Now, as we derive the tenant ID from the database, not `TENANT_MAP` anymore, we need to switch `const tenant = TENANT_MAP[hostname];` with `const tenant = tenantData.id;`. And that's all there is to do – now, every single request will trigger and wait for the tenant domain to be confirmed.

As I mentioned previously, I don't use this variant if there's no definite need as I think the downside of having to do a database selection before the actual request happens (not in parallel) isn't worth the extra fun. Still, in today's fast-paced world, more often than not, you'll need instantaneous availability.

Next, I will show you how to use the localhost version again for a fluid development experience, despite all of the domain-mapping changes that we did.

Bringing back localhost with mapped domains

You might be thinking, "Yeah, mapping domains is awesome for deploying them, but I don't want to map the same domains for local development – I just want to develop with the localhost version." There's an easy way to do that and a more complicated one. The easy way is to override the tenant for local development – which is what I'll show you now.

Since you derive the hostname via the domain, you can also simply override it for local development. For that, create another environment variable in your .env.local file and add a configuration that defines the tenant that you want to run. Here's an example:

```
OVERRIDE_TENANT_DOMAIN=packt.local
```

Then, go to url-helpers.js and edit getHostnameAndPort so that if an override value exists, it takes it as the hostname; otherwise, you derive the hostname as usual:

```
export function getHostnameAndPort(request) {
  const hostnameWithPort = request.headers.get("host");
  const [realHostname, port] = hostnameWithPort.split(":");

  let hostname;
  if (process.env.OVERRIDE_TENANT_DOMAIN) {
    hostname = process.env.OVERRIDE_TENANT_DOMAIN;
  } else {
    hostname = realHostname;
  }

  return [hostname, port];
}
```

That's it. Now, you can use localhost:3000 and it will show the login of the tenant you've overridden. Pretty helpful, right?

The more complicated way would be to bring back the path-based variant when we're on localhost. That's not wizardry – it's definitely feasible! However, we would need to check the hostname being localhost and switch up the middleware logic so that it doesn't rewrite paths and not redirect based on domains. Also, we would need to adapt url-helpers.js so that it would use tenant prefixes on localhost. As there's no massive benefit over the easy solution I've shown you, we won't mess up our beautiful code logic with that.

Summary

In this chapter, you transformed your application into a robust multi-tenant system – that's a reason to celebrate!

With a solid understanding of RLS principles, you've enabled user-specific data access and learned how RLS dependencies work, as well as how to remove their complexity and connect them with untamperable custom claims. By gluing authentication mechanisms to user permissions, you've ensured that the application only allows access to authorized individuals. Additionally, the ability to adapt to a domain-based system offers a seamless user experience without the need for a `/tenant_id/` prefix in public URLs, which gives our project an improved enterprise-like behaviour.

Armed with these improvements, your application is now geared to provide a customized and secure experience for each tenant.

In the next chapter, we'll explore how to integrate registration, which is intricately tied to the tenant, and even cover signing in with Google (OAuth). I'm looking forward to this next phase, and I hope you are too!

7

Adding Tenant-Based Signups, including Google Login

At this point, you have created a secure multi-tenant setup. However, signing up by oneself is currently not implemented – we have only created users manually in Supabase Studio so far.

In this chapter, we will now introduce a registration process where users with an email matching the tenant domain can sign up, gaining automatic access to the corresponding tenant. That means that if I own the `bearclaw@activenode.learn` or `rabbit@activenode.learn` email, I'll be able to register and, after email confirmation, sign into the **activenode** tenant. Likewise, if I own `nicebooks@packt.local`, I'll be able to register an account that is legitimized on the **packt** tenant. With that, you'll also gain an understanding of what it means to disable the default signup functionality – despite implementing a signup.

After having learned how email registration can be integrated within our application, I will introduce you to the OAuth mechanism and how it works by adding an OAuth login mechanism to allow signing in with Google. This will allow you to also use other providers to log in, such as GitHub.

Finally, you'll learn why there might be cases of invalid user registrations and how to prevent such.

So, in this chapter, we will cover the following topics:

- Understanding the impact of disabling signups
- Implementing the registration page
- Processing the registration with a Route Handler
- Enabling OAuth/Sign-in with Google
- Dealing with invalid user registration

Technical requirements

To add registration features, there's no additional knowledge required compared to the previous chapters. You should remember, though, that for any changes in the `config.toml` file, you must restart your local instance for them to take effect.

The related code for this chapter can be found within the `Ch07_Registration` branch (`https://github.com/PacktPublishing/Building-Production-Grade-Web-Applications-with-Supabase/tree/Ch07_Registration`).

Understanding the impact of disabling signups

In this section, I'll explain to you the potential benefits of disabling signups in your instance, even though, for implementing a signup, this might sound paradoxical.

The Supabase client has a publicly callable `supabase.auth.signUp(..)` function (`https://supabase.com/docs/reference/javascript/auth-signup`) that creates a Supabase user and sends a user a login link via the built-in mailing system. The function also returns the newly created user. "Publicly" here also means that anyone can call this method with the Anonymous Key – which is exposed via the frontend.

Now, you may be thinking: is creating a Supabase user enough to use our application? It's not. Our application requires custom claims (see *Chapter 6*) to be set on the user's `app_metadata` – which only admin powers can do. However, it's not just that; our application also requires a service user and at least one tenant permission to be in the database. Besides that, we don't want `signUp()` to send any emails. We want to use our own custom emails that can be tailored to each tenant.

So, knowing that one could call `signUp()` from the outside and create an unusable user, having `signUp()` can lead to maliciously created unusable users (so-called orphan users) as they don't have the required related data. From an application security standpoint, you can ignore having orphan users, as they can't reach any database data of our application.

> **Note**
> The experienced database users among you are already asking yourselves at this point: "Can't we solve this with database triggers that create the related table entries when a new user is created?" Yes, it is technically possible, but it moves the registration complexity to SQL. An example of how this can be done is shown at `https://supabase.com/docs/guides/auth/managing-user-data`. However, even if you opt for the trigger-based variant later, I strictly urge you to follow everything in this chapter to have an easy time following along. Also, such a trigger obviously requires trigger knowledge, which is first discussed in *Chapter 8*.

So, with the `signUp(..)` method, you have a tool at hand that creates a Supabase user and, with the built-in mail templates, sends a login link. That's good to know and useful for either a simple application or one that wants to use the built-in functionality for sending emails.

But we want to create our own, fully controlled signup! For that, we don't need the public email signup possibility and, hence, can disable it. In Supabase, we can disable signups generally or disable specific authentication methods. Although I would love to disable email signups specifically, in Supabase's current state, you can only disable a whole authentication feature, including signing in. This is not what we want, so we disable signups generally – this will still allow you to log in as usual.

Having signups disabled while implementing signup functionality will improve your understanding of Supabase authentication – even though, spoiler alert, we will re-enable it when implementing the Google login. However, let's disable signups for now.

Disabling signups generally

For disabling signups in your local instance, open the `supabase/config.toml` file, scroll to the `[auth]` section, and change `enable_signup` to `false`.

For hosted instances, on `www.supabase.com`, go to **Project Settings | Configuration** and click on **Authentication** to open the authentication options. There, you'll see an activated **Allow new users to sign up** toggle, which you want to deactivate:

Figure 7.1: Disabling signups generally

Now that you have signups disabled, no one can use the public signup methods anymore.

Let me additionally tell you how you'd disable specific authentication methods.

Disabling specific signup methods

As I mentioned, you can also disable specific signup methods in Supabase. However, it disables the whole authentication method including signing in with that specific method. So, if you were to disable the email signup method, we couldn't use the `signInWithPassword()` function for it anymore.

But let me still tell you how you can do it in case you need it (e.g., if, for a specific project, you only wanted LinkedIn or Slack authentication).

For the local instance, all authentication options can be configured within the `[auth.method_name]` sections of the `config.toml` file. In each of those, you find an `enable_signup=` flag set to `true` or `false`. So, for example, the SMS-based authentication method can be disabled by doing this:

```
[auth.sms]
enable_signup=false
```

By default, in your local instance, `email` and `sms` signups are enabled. As we don't use SMS authentication, you can leave it configured as shown in the previous block.

On your hosted instance, you can disable specific authentication methods by going to **Authentication | Providers**, where you'll find a list of all possible providers. There, usually, only **Email** is activated by default. You can disable the authentication by switching off the **Enable Email provider** toggle:

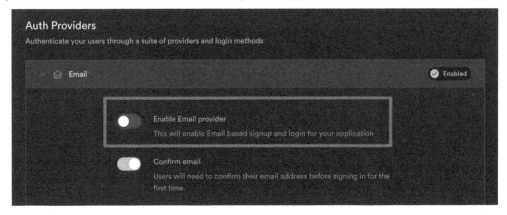

Figure 7.2: Disabling email authentication

You've disabled signups and learned how you can disable specific authentication providers. Let's move on to implementing the registration flow next.

Implementing the registration page

As we don't have a registration UI yet, we'll fast-forward to creating one. We want the registration page (called **Create account**) to look pretty much the same as the **Login** page: a box where you can enter your email and a password of your choice to create a new account. It shall be reachable at `tenant.domain/register`.

First, you want to make sure that the registration is reachable with a click in the UI, so simply add a `Link` component to your `Login.js` component below the existing buttons:

```
<Link
  href={urlPath("/register", tenant)}
  style={{
    textAlign: "center",
    display: "block",
    marginTop: "1em",
  }}
>
  Create account
</Link>
```

My login screen will now look like this:

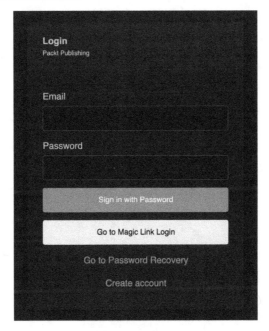

Figure 7.3: Login screen

Design note

If your login page looks slightly different, that's normal. I just tweaked the styles as the project grew to differentiate the buttons better. If you want yours to look exactly the same as mine, you can just copy the same code from the `Login.js` file of the `Ch07_ Registration` git branch.

Now we need the registration page. Create the `src/app/[tenant]/register/page.js` file and fill it with dummy code, just to check that everything works as expected:

```
export default async function Registration({ params }) {
  const { tenant } = params;
  return <strong>Registration Page of Tenant-ID={tenant}</strong>;
}
```

If you open it using `activenode.learn:3000/`, for example, and click on **Create account**, then you should see your registration page mentioning the **activenode** tenant ID.

From here, we'll jump immediately to replacing it with the final registration page code. I won't explain the UI code itself as it's just a slightly changed, more minimal version of what we've used for the login page. It's the same box showing the tenant's name at the top and containing a form with the same input fields, but with an additional field for the user's name. I highlighted the important parts in the following code:

```
import { getSupabaseAdminClient } from "@/supabase-utils/adminClient";
import { urlPath } from "@/utils/url-helpers";
import Link from "next/link";
import { notFound } from "next/navigation";

export default async function Registration({ params }) {
 const { tenant } = params;

 const supabaseAdmin = getSupabaseAdminClient();
 const { data, error } = await supabaseAdmin
   .from("tenants")
   .select("*")
   .eq("id", tenant)
   .single();

 if (error) notFound();

 const { name: tenantName } = data;

 return (
  <form method="POST" action={urlPath("/auth/register", tenant)}>
    <article style={{ maxWidth: "480px", margin: "auto" }}>
     <header>
      <strong>Create account</strong>
      <div style={{ display: "block", fontSize: "0.7em" }}>
       {tenantName}
      </div>
```

```
    </header>

    <fieldset>
     <label htmlFor="name">
      Your name <input type="text" id="name" name="name" required />
     </label>
     <label htmlFor="email" style={{ marginTop: "20px" }}>
      Email <input type="email" id="email" name="email" required />
     </label>
     <label htmlFor="password" style={{ marginTop: "20px" }}>
      Choose a password{" "}
      <input
         type="password" id="password" name="password" required />
     </label>
    </fieldset>

    <button type="submit">Register now</button>

    <Link
      href={urlPath("/", tenant)}
      style={{
       textAlign: "center",
       display: "block",
       marginTop: "1em",
      }}
     >
      Go back to login
    </Link>
   </article>
  </form>
 );
}
```

So, there's a form on your registration page that is being sent to /auth/register for which we'll need a Route Handler. When the form is submitted, it will send the email and password as part of a POST request. At the bottom of the form, there's a link to allow users to go back to the login page. That's all. There's no frontend part because, just like with the magic link, we need admin rights and hence can just simplify our lives by sending the form directly to the Route Handler, where it's processed on the backend. Here is the result:

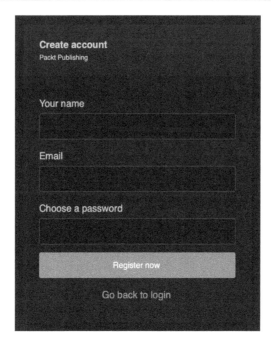

Figure 7.4: Registration page

Now we must create the matching Route Handler that processes the form data, leading to a successful or rejected signup.

Processing the registration with a Route Handler

Now we will process the sent form data (email and password) and match the given email with the tenant's domain, or else reject it. With a successful registration, we will send an activation link. To do this, let's start by reading and validating the data.

Reading and validating the form data

To process the registration, we need a Route Handler at app/[tenant]/auth/register/route.js. As our form sends POST data, the function needs to be named POST. Let's add some Route Handler code to read the form data and the tenant, and output it as a JSON for us to check whether everything works as expected:

```
export async function POST(request, { params }) {
  const formData = await request.formData();
  const name = formData.get("name");
  const email = formData.get("email");
```

```
  const password = formData.get("password");
  const tenant = params.tenant;

  return Response.json({ email, password, tenant });
}
```

Having done that, go to /register and enter some@mail.supa as the email and 1234 as a password, and you should see the same output as shown in *Figure 7.5*:

Figure 7.5: The Route Handler showing the form data and the correct tenant

Having confirmed that the Route Handler and the form work well together by checking the JSON, we now want to introduce a few validations to the sent data, as data that is being sent from a browser to our backend must never be trusted. We want to check whether the provided email address is even a technically correct email address and whether the password and name fields are not empty. If one of these is not the case, then it wouldn't make sense to process the request further.

To check whether the email is a string, we use a very simple regex to make sure it follows a basic email syntax. Other than that, we introduce a minimal function, isNonEmptyString, to confirm we didn't receive empty values. So, in the Route Handler, below const tenant = params.tenant, add the checks as follows:

```
const isNonEmptyString = (value) =>
  typeof value === "string" && value.trim().length > 0;

const emailRegex = /^\S+@\S+$/; // simple front@back regex
if (
  !isNonEmptyString(name) ||
  !isNonEmptyString(email) ||
  !emailRegex.test(email) ||
  !isNonEmptyString(password)
) {
  return NextResponse.redirect(
    buildUrl("/error", tenant, request),
    302
  );
}
```

> **Recommendation**
>
> For big projects or complex validations, I recommend using the Zod validation library, since it is simple yet mighty at the same time (`https://zod.dev/`).

After that, we want to split the email address, finding the back part that contains the hostname. Let's add the derived host to the JSON output as well:

```
const [ , emailHost] = email.split('@');
return Response.json({ email, password, tenant, emailHost });
```

When I submit the registration form again with `some@mail.supa`, I see this:

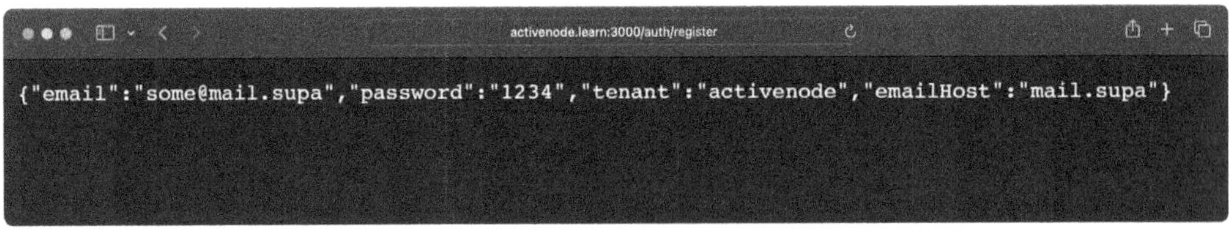

```
{"email":"some@mail.supa","password":"1234","tenant":"activenode","emailHost":"mail.supa"}
```

Figure 7.6: The Route Handler parsing the emailHost

As a next step, we need to grab the tenant domain from the database to match it against `emailHost`. We instantiate the Supabase admin client, and request the tenant from the database where the tenant id matches the current tenant and the domain matches the one from the email:

```
const supabaseAdmin = getSupabaseAdminClient();
const { data, error } = await supabaseAdmin
  .from("tenants")
  .select("*")
  .eq("id", tenant)
  .eq("domain", emailHost)
  .single();
```

With that done, let's first handle the error case and then move on to handling the successful case.

Rejecting registration

When the database returns no data, it means no matching tenant was found with the same host as the email. In that case, we'll forward to an error page. We do so by passing a specific type, `register_mail_mismatch`, as well as the email itself to the error page. Then, to safely append the email and its potential special characters to the URL, we encode it with the `enocdeURIComponent` standard JS function.

So, right below your previous `select()` statement, add the following code to handle the failed selection:

```
const safeEmailString = encodeURIComponent(email);
if (error) {
  return NextResponse.redirect(
    buildUrl(
      `/error?type=register_mail_mismatch&email=${safeEmailString}`,
      tenant,
      request,
    ),
    302,
  );
}
```

Now, let's make sure that this specific type, `register_mail_mismatch`, is handled with a specific error message by going to `error/page.js` and adding the highlighted parts of the following code block:

```
const knownErrors = [
  ...
  "register_mail_mismatch"
];
...
return (
  ...
  {type === "login-failed" && ...}
  ...
  {type === "register_mail_mismatch" && (
    <strong>
      You are not legitimated to register an account with
      <u>{searchParams.email}.</u>
    </strong>
  )}
  ...
)
```

If you test the registration again on `activenode.learn:3000/register` and enter an email that matches the tenant domain – for example, `foo@activenode.learn` – you'll see the JSON response (like back in *Figure 7.6*). However, if you enter an email such as `does@not.exist`, you'll be forwarded to the error page shown in *Figure 7.7*:

Figure 7.7: Email tenant mismatch error

That's a good start, but instead of outputting a JSON, we now want to create an actual account.

Handling account creation

Despite disabled signup, you can create a Supabase account with the admin client using the following `createUser` function (`https://supabase.com/docs/reference/javascript/auth-admin-createuser`):

```
supabaseAdmin.auth.admin.createUser({ email, password });
```

This would create the `auth.users` entry with the email and password we got from the form.

However, don't forget that we want to add the tenant access to the user's `app_metadata` object within the `app_metadata.tenants` array (see the custom claim explanation in *Chapter 6*). We can do so by directly passing the `app_metadata` property to the `createUser` function. It will then return the user data object when everything went well, or else an error. Add it now to the `/auth/register/route.js` Route Handler code as shown:

```
if (error) {
  ...
}
const { data: userData, error: userError } =
  await supabaseAdmin.auth.admin.createUser({
    email,
    password,
    app_metadata: {
      tenants: [tenant],
    },
  });
```

You already know from other functions that if Supabase returns a non-empty error, the data will be empty, and vice versa. However, what we don't know is how to handle specific errors, such as when the user couldn't be created because it already exists.

For user creation, we can derive this information from the error string as it will contain the following phrase: **A user with this email address has already been registered**. We can add the following code to catch that specific error and any other unknown errors too:

```
const { data: userData, error: userError } = ...

if (userError) {
  const userExists =
    userError.message.includes("already been registered");
  if (userExists) {
    return NextResponse.redirect(
      buildUrl(
        `/error?type=register_mail_exists&email=${safeEmailString}`,
        tenant,
        request,
      ),
      302,
    );
  } else {
    return NextResponse.redirect(
      buildUrl("/error?type=register_unknown", tenant, request),
      302,
    );
  }
}
```

As you can see, we first check whether an error occurred at all. If so, we check whether it was because the user already exists in the system or because of any other unknown error such as a database connection failure. Depending on that, we send the person to the error page with either `register_mail_exists` or `register_unknown`.

Both types need to be added to the `knownErrors` variable in the error page and be rendered accordingly. Here's what I am using to render the fitting messages within `error/page.js`:

```
{type === "register_mail_exists" && (
  <strong>
    There is already an account registered with  
    <u>{searchParams.email}</u>.
  </strong>
)}
```

```
{type === "register_unknown" && (
  <strong>
    Sorry but an unknown error occurred when trying to create
    an account.
  </strong>
)}
```

In `register/route.js`, you now have code that creates a Supabase user, including `app_metadata`, as well as handling potential failures.

We must now create the `service_users` entry and the `tenant_permissions` entry immediately after user creation to ensure the user receives the required permissions for the tenant.

Adding the service user and permission rows

Although we have only selected data with the Supabase client so far, it's easy to insert data too, especially with the admin client, which doesn't have any limitations. Our `userData` object from the `createUser` result contains the full `auth.users` object from which we get the respective Supabase user's ID (`userData.user.id`). This ID we need for creating an entry in `service_users` and with the service user ID, we create the `tenant_permissions` row. The process is pretty straightforward.

Let's first handle the service user creation:

```
if (userError) {
  ...
}

const { data: serviceUser } = await supabaseAdmin
    .from("service_users")
    .insert({
      full_name: name,
      supabase_user: userData.user.id,
    })
    .select()
    .single();
```

As you can see, we use `insert` in our `service_users` table, passing `full_name` given by our name variable and `supabase_user` given from our previous `createUser(..)` call. The other values (`id` and `created_at`) of the row are automatically filled by the database.

But why do we chain all of it again with `.select().single()`? If we don't, the row will be inserted but its data will not be returned. By chaining a simple `select()` method call, it will select the previously created row and return its data.

You might also ask, why don't we handle the error value of this statement here? Well, you can; it's not like there's no possibility that it couldn't fail. However, knowing that the previous createUser call succeeded (hence, the user didn't exist, and the database was available) and knowing our existing database architecture, it's extremely unlikely that it would fail at this point, so I didn't want to bother you with more error handling code here.

Also, if the service user creation fails, our tenant_permissions insert will fail as well as it expects a valid service user id. We will check the error case for inserting the tenant_permissions row and, by that, we also implicitly catch potential previous errors.

Now that you have the service user, you want to make sure to create the proper tenant_permissions row for that user. For that, you simply execute an insert on the tenant_permissions table like so:

```
const {error: tpError} =
  await supabaseAdmin.from("tenant_permissions").insert({
  tenant,
  service_user: serviceUser?.id,
});
```

If you are wondering why we use the serviceUser?.id conditional chaining and not serviceUser.id, well, we didn't check the service user insert for failure so, though unlikely, the serviceUser object could be null. By using conditional chaining, we would forward an empty value to the tenant_permissions insert, which would correctly cause this insert statement to fail. If tpError is empty, it means the data was added successfully.

But if not, we want to revert the registration by deleting the user:

```
if (tpError) {
  await supabaseAdmin.auth.admin.deleteUser(userData.user.id);
  return NextResponse.redirect(
    buildUrl("/error", tenant, request), 302
  );
}
```

As you can see, when tpError is provided, we call the auth.admin.deleteUser function, which forces the deletion of the Supabase user (not the service user).

You might wonder why we don't need to delete anything from the service_users table in that case. That's because the entry in the service_users table has a relation to the Supabase users table (auth.users) in which we selected the **CASCADE** option for deletion (we did this in *Chapter 5*). This means that when the Supabase user is deleted, the service user is deleted with it automatically.

For a successful registration, let's now send an activation link via email and redirect the user to a success page.

Sending the activation email

Question: shouldn't the user be able to sign in with the password after registration? Well, yes, but the user isn't activated yet so the sign-in will fail. An inactive user becomes active when they log in using an OTP (or a magic link).

> **Note**
>
> With a local instance, the additional, required activation of a user is often disabled by default. You can enable it in the `config.toml` file by setting `enable_confirmations = true` in the `[auth.email]` section.

The activation ensures that the given email really exists. Otherwise, I could just enter any non-existing email matching the tenant domain and get access to a tenant, which would be a huge security failure.

When talking about the password recovery feature in *Chapter 4*, I told you that it's really just the same as sending a magic login link – it's a link that contains a one-time valid token to sign you in. The activation link does the exact same thing.

The only difference between a magic link, a recovery link, and a signup activation link is that each one carries additional internal information about where it originated from – from a magic link request, from a recovery request, or from a signup. However, effectively, you could send just a magic link – when the user clicks it, it would also activate the account as the email was proven to exist.

We've already sent OTP links; however, this time, we will use the `type: 'signup'` option instead of `'magiclink'` or `'recovery'` when generating the link.

As we would now have to copy/duplicate the magic link generation code from the `auth/magic-link/route.js` file inside our `auth/register/route.js` file and only change a few things, such as setting `type` to `'signup'`, we'll instead abstract it into a new `src/utils/sendOTPLink.js` file, exporting a reusable function for sending an OTP link. This makes it cleaner and easier to maintain than code duplication.

This is what the new `src/utils/sendOTPLink.js` file looks like:

```
import { getSupabaseAdminClient } from "@/supabase-utils/adminClient";
import nodemailer from "nodemailer";


import { getSupabaseAdminClient } from "@/supabase-utils/adminClient";
import nodemailer from "nodemailer";
import { buildUrl } from "./url-helpers";

export async function sendOTPLink(email, type, tenant, request) {
  const supabaseAdmin = getSupabaseAdminClient();
  const { data: linkData, error } = await supabaseAdmin.auth.admin.
generateLink(
```

```
  {
    email,
    type,
  },
);

const user = linkData.user;

if (error || !user.app_metadata?.tenants.includes(tenant)) {
  return false;
}

const { hashed_token } = linkData.properties;

const constructedLink = buildUrl(
  `/auth/verify?hashed_token=${hashed_token}&type=${type}`,
  tenant,
  request,
);

const transporter = nodemailer.createTransport({
  host: "localhost",
  port: 54325,
});

let mailSubject = "";
let initialSentence = "";
let sentenceEnding = "";

if (type === "signup") {
  mailSubject = "Activate your account";
  initialSentence = "Hi there, you successfully signed up!";
  sentenceEnding = "activate your account";
} else if (type === "recovery") {
  mailSubject = "New password requested";
  initialSentence = "Hi there, you requested a password change!";
  sentenceEnding = "change it";
} else {
  mailSubject = "Magic Link requested";
  initialSentence = "Hey, you requested a magic login link!";
  sentenceEnding = "log in";
}
```

```
await transporter.sendMail({
  from: "Your Company <your@mail.whatever>",
  to: email,
  subject: mailSubject,
  html: `
  <h1>${initialSentence}</h1>
  <p>Click <a href="${constructedLink.toString()}">here</a> to
    ${sentenceEnding}.</p>
  `,
});

return true;
}
```

Please note that none of this code logic is new! I simply copied it from the existing `auth/magic-link/route.js` file and the only things I've changed are as follows:

- In the error case, I return `false` instead of redirecting, and in the success case, I return `true` instead of redirecting – this simply allows us to use this function in our Route Handlers to handle the redirections

- I added an `if` clause to switch between the content and subjects of the email, depending on the provided type

With that, we can now easily make both the magic link route code and the signup (register) route code. Let's start with adapting the magic link route code within `auth/magic-link/route.js`:

```
import { NextResponse } from "next/server";
import { buildUrl } from "@/utils/url-helpers";
import { sendOTPLink } from "@/utils/sendOTPLink";

export async function POST(request, { params }) {
  const formData = await request.formData();
  const email = formData.get("email");
  const type =
    formData.get("type") === "recovery" ? "recovery" : "magiclink";

  const errorUrl =
    buildUrl(`/error?type=${type}`, params.tenant, request);
  const thanksUrl = buildUrl(
    `/magic-thanks?type=${type}`,
    params.tenant,
    request,
  );
  const otpSuccess = await sendOTPLink(
```

```
    email,
    type,
    params.tenant,
    request
  );

  if (!otpSuccess) {
    return NextResponse.redirect(errorUrl, 302);
  } else {
    return NextResponse.redirect(thanksUrl, 302);
  }
}
```

That looks much cleaner!

In our `auth/register/route.js` file, we will now follow a similar approach calling the `sendOTPLink` function that is sending our activation link. All we have to add is one line of code before our final `return`:

```
if (tpError) {
  ...
}

await sendOTPLink(email, "signup", tenant, request);
return Response.json(...);
```

Again, I'm not interested in the return result as I'm confident it must succeed as we've properly set up the user.

This will generate a link with the `signup` type, which means that our verification route for that link must make sure that the type is forwarded when using `verifyOtp()`, or else it will fail, even if the token is valid. So, in the verification route within the `auth/verify/route.js` file, you need to add this type (new code is highlighted):

```
...
const isRecovery = searchParams.get("type") === "recovery";
const isSignUp = searchParams.get("type") === "signup";
let verifyType = "magiclink";
if (isRecovery) verifyType = "recovery";
else if (isSignUp) verifyType = "signup";
...
const { error } = await supabase.auth.verifyOtp({
  type: verifyType,
  token_hash: hashed_token,
});
```

Awesome. A proper magic link will be sent and it can be activated with our verification Route Handler. Let's finalize the registration flow by creating a success page and redirecting the user to it.

Redirecting the user to a success page

We want to let the user know that the registration succeeded and that they should check their inbox to activate the account. Hence, immediately after `sendOTPLink(..)`, we want to add a redirection to the registration success page:

```
return NextResponse.redirect(
  buildUrl(
    `/registration-success?email=${safeEmailString}`,
    tenant,
    request
  ),
  302
);
```

Then, we create the matching page by creating `app/[tenant]/registration-success/page.js` and filling it with some user-friendly content, as follows:

```
import { urlPath } from "@/utils/url-helpers";
import Link from "next/link";

export default function RegistrationSuccessPage({
  searchParams,
  params,
}) {
  const { email } = searchParams;
  const { tenant } = params;

  return (
    <div style={{ textAlign: "center" }}>
      <h1>Registration succeeded!</h1>
      <p>Check your email ({email}) for a link to activate your
      account.</p>
      <br />
      <Link role="button" href={urlPath("/", tenant)}>
        Login
      </Link>
    </div>
  );
}
```

At this point, when I visit `activenode.learn:3000/` and register with the `supercool@activenode.learn` email, I'll be forwarded to the registration success page:

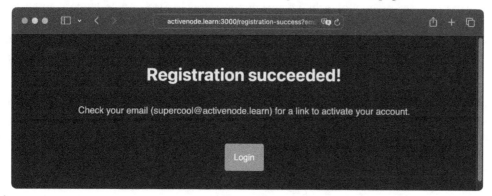

Figure 7.8: Registration success page

Then, when I go to my local Inbucket postbox (`http://localhost:54324/monitor`) and click the activation link (the word **here**), it will immediately sign me in:

Figure 7.9: Activation mail in Inbucket

The registration is complete. You can see that everything works by going to Supabase Studio and checking the created entries in the tables, alongside the data stored. How awesome is that?

In the next section, we'll discuss the integration of foreign login providers such as Google.

Enabling OAuth/Sign-in with Google

Supabase has a massive amount of login provider options: Google, Slack, Apple, Discord, Bitbucket, and many more, with the list quickly growing. It's impossible to explain them all, but luckily, they all follow the OAuth standard, which means that if we show you one implementation, you're prepared to implement the others. If you would like to learn more about the OAuth standard, you can dig deeper at `https://en.wikipedia.org/wiki/OAuth`.

To enable you to use whatever login provider you want, we'll implement the Google login and play it through with the `oddmonkey.inc` host. Be aware that due to security constraints, making this run on your local development system requires dealing with HTTPS, but that's not a problem as we'll solve it together.

Here's what we need to do:

1. Obtain Google OAuth credentials.

2. Configure our Supabase with the OAuth credentials.

3. Add a "Sign in with Google" option, triggering the OAuth process.

4. Solve the HTTPS security problem.

5. Build a verification route to finalize the registration (we will also ensure here that we match the email domain with the tenant domain).

> **Important task**
> Before you start with the first step, you must re-enable signups in your instance as the OAuth login will otherwise not work (we will discuss this further in *Preventing invalid user registration*).

Now, let's start with obtaining Google OAuth credentials.

Obtaining Google OAuth credentials

This is not a book about Google, so I'll keep the explanation about obtaining the credentials concise yet sufficient. To be able to identify users in your application with Google, as the admin of your application, you need a Google Cloud project. You can create one in the Google Cloud console (`https://console.cloud.google.com/`). There, you need to click the dropdown box, which, when unselected, says **Select a project**:

Figure 7.10: Project dropdown on the Google Cloud console

This will open the option to create a new project. Click on **New project** and enter a name of your choice, then save and create it:

Figure 7.11: Projects modal

Next, go to **APIs & Services**. There, you'll find a **Credentials** button leading you to the page with a + **CREATE CREDENTIALS** button. Click that and then choose **OAuth client ID**:

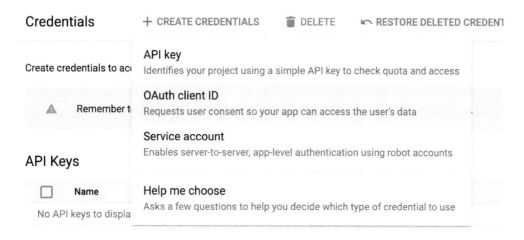

Figure 7.12: Requesting new OAuth credentials

If you click it, it will first ask you to set up some OAuth consent screen settings (the screen that the user sees when trying to sign in with Google). Follow the consent screen setup by choosing **External** as the application type:

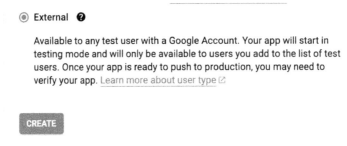

Figure 7.13: Consent screen type choice

Then, it will ask you for an app name and an email – feel free to use whatever you want. You can leave the other fields empty and skip to the end of the wizard, finalizing your choices by saving them.

Now, in the **APIs & Services** navigation, you can click on **Credentials** again to start generating your OAuth credentials. Click again on + **CREATE CREDENTIALS**, then **OAuth client ID**. In the **Application type** selection, choose **Web application** and give an arbitrary name to it.

In the **Authorized redirect URIs** section, you'll need to add the URL where the OAuth sign-in will be processed – the so-called OAuth callback URL. When triggering an OAuth login with Supabase, it must follow Supabase's flow and return to the Supabase Auth Service to complete sign-in. For that, you need the appropriate callback endpoint URL.

On your local Supabase instance, the Auth Service sits at `http://localhost:54321`. It has an endpoint that processes OAuth callbacks, which is at `http://localhost:54321/auth/v1/callback`. This is the URL we need to provide to Google in **Authorized redirect URIs**:

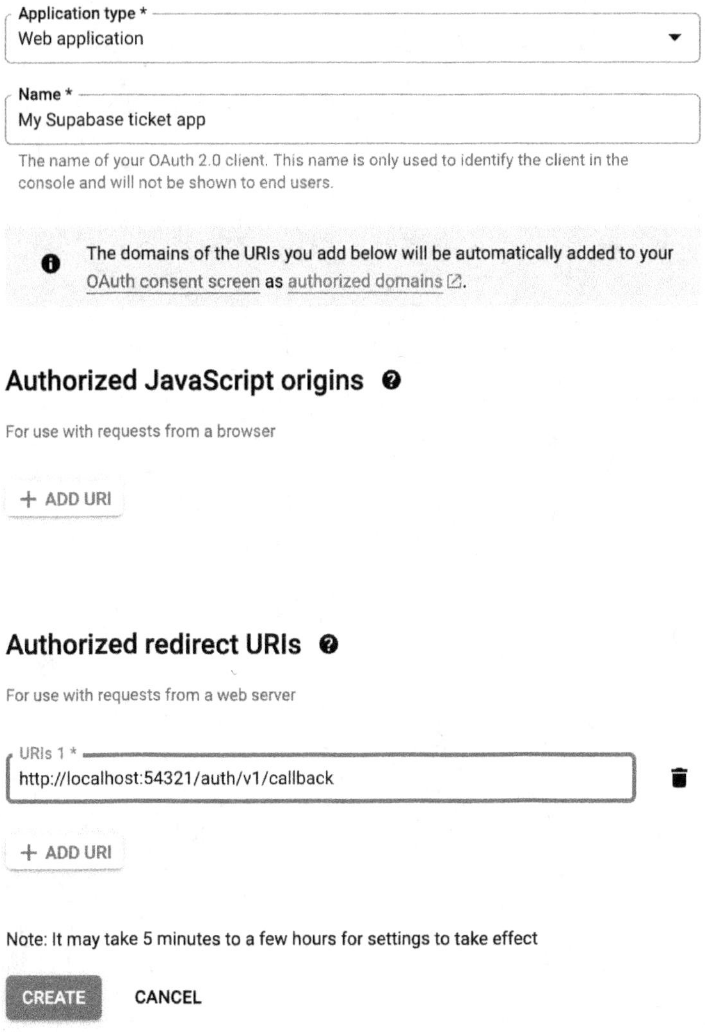

Application type *

Web application ▼

Name *

My Supabase ticket app

The name of your OAuth 2.0 client. This name is only used to identify the client in the console and will not be shown to end users.

ⓘ The domains of the URIs you add below will be automatically added to your OAuth consent screen as authorized domains ☑.

Authorized JavaScript origins ❷

For use with requests from a browser

＋ ADD URI

Authorized redirect URIs ❷

For use with requests from a web server

URIs 1 *

http://localhost:54321/auth/v1/callback   🗑

＋ ADD URI

Note: It may take 5 minutes to a few hours for settings to take effect

CREATE   CANCEL

Figure 7.14: OAuth redirect URI setup

If you're connecting it with your real www.supabase.com instance, you can find the callback URL by going to **Authentication | Providers**, where you can activate the Google provider. In that Google provider mask, it will show the **Callback URL** field (*Figure 7.15*). You can have both the local callback URL and the one from your www.supabase.com instance added to the **Authorized redirect URIs** in your Google setup.

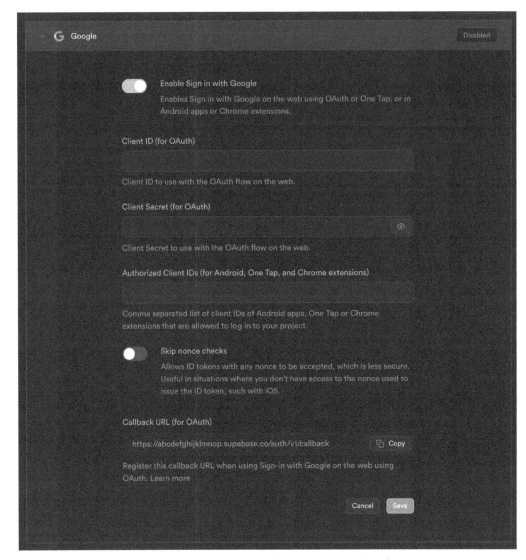

Figure 7.15: Setting up Google OAuth provider in Supabase

After setting the callback URIs, complete the last step in the Google credentials mask and you'll be given a **client ID** and a **client secret** – the client ID identifies this application to be used with the OAuth process, while the secret allows Supabase to process the registration with the Google API. To be able to do so, we have to configure Supabase with the credentials given to us.

Configuring our Supabase instance with the OAuth credentials

As you can see back in *Figure 7.15*, on www.supabase.com, you need to paste both the **Client ID** and **Client secret** values in the provider mask and save it.

To allow your local instance to handle Google logins, go to the supabase/config.toml file and add the following code below the [auth] section:

```
[auth.external.google]
enabled = true
client_id = "env(SUPABASE_AUTH_EXTERNAL_GOOGLE_ID)"
secret = "env(SUPABASE_AUTH_EXTERNAL_GOOGLE_SECRET)"
redirect_uri = "http://localhost:54321/auth/v1/callback"
```

With that, you're telling your local Supabase instance that it will find the client ID and the secret for Google in a .env file, as well as that Google will need to redirect to the Auth Service callback URL.

Unfortunately, the config.toml file doesn't read our existing environment variables from the .env.local file. Instead, it reads variables defined in a .env file. So, next to your .env.local file, create a .env file that contains two lines like these:

```
SUPABASE_AUTH_EXTERNAL_GOOGLE_ID=PUT_THE_CLIENT_ID_HERE
SUPABASE_AUTH_EXTERNAL_GOOGLE_SECRET=PUT_THE_SECRET_HERE
```

As we'll work on a "secure" (HTTPS) localhost connection, you have to add https://oddmonkey.inc:3000/** to the list of additional_redirect_urls in the [auth] section of the config.toml file (you can surely also add the other tenant URLs if you want to test it with those).

Finally, stop the local Supabase instance with supabase stop and start it again with supabase start for the new values to become active.

Now, let's make sure the user can initiate sign-in with Google.

Adding a "Sign in with Google" option triggering the OAuth process

In the Login.js file, we need a button for the user to initiate Google sign-in. You can add a button in the position of your choice, but I'll do it right below the **Submit** button:

```
<button
  type="button"
  onClick={() => {
    //...
  }}
>
  Sign in with Google
</button>
```

Inside the `onClick` function, we now add the code that triggers the OAuth process. To do that, we call `signInWithOAuth()`. In there, we tell Supabase the provider to use (`google`) and give it a URL to which our Auth Service shall redirect when signing in is done – that's a similar process to what our magic link uses:

```
supabase.auth.signInWithOAuth({
  provider: "google",
  options: {
    redirectTo: window.location.origin + "/auth/verify-oauth",
    queryParams: {
     access_type: "offline",
     prompt: "consent",
    },
  },
});
```

`window.location.origin` is just the protocol and the hostname together, so we can add our path to it. When signing in is done, go to `/auth/verify-oauth`, where we will soon create another Route Handler. `queryParams` are just parameters that are going to be added to the Google OAuth URL. The first one, `access_type: "offline"`, allows us to refresh the user token on behalf of the user (`https://stackoverflow.com/a/30638344/1236627`), and the second one, `prompt: "consent"`, just shows the user the consent when signing in.

Navigate to `http://oddmonkey.inc:3000/` and click the new **Sign in with Google** button. Unfortunately, the button won't do anything. Instead, you'll see an error in your browser console, stating that it cannot access `window.crypto.subtle`:

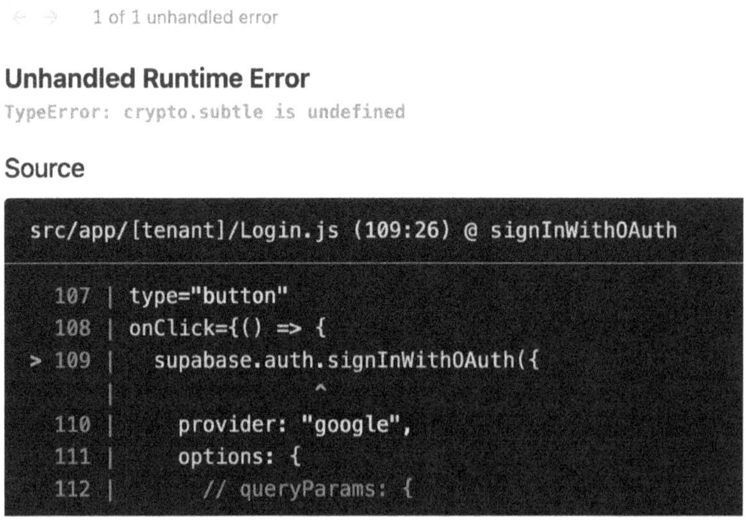

Figure 7.16: The crypto.subtle error

That's what we'll solve next.

Solving the crypto/HTTPS security problem

On sites that do not use a secure HTTPS connection, the `crypto.subtle` function is not available due to security restrictions in the browsers – the only exception is localhost. However, we can enable HTTPS locally with a simple HTTPS proxy tool instead.

Stop running your Next.js application and run `npx local-ssl-proxy --source 3000 --target 3001` in your terminal. This will proxy everything that runs on port 3001 through a secured connection on port 3000. However, right now, nothing runs on port 3001.

Then, run Next.js on port 3001 by using `npx next dev -p 3001` instead of `npm run dev`. Once both commands are running, you can visit the secured `https://oddmonkey.inc:3000/` site (you need to accept the certificate). You can also visit any other tenant on HTTPS now.

> **Tip**
> Having to call two long commands is cumbersome; however, you can install an additional package that can handle two scripts with one call. You simply execute `npm i concurrently` in your terminal and then add a script in your `package.json` file as follows: `"ssl": "concurrently --kill-others 'npx next dev -p 3001' 'npx local-ssl-proxy --source 3000 --target 3001'"`. This allows you to run both commands by just running `npm run ssl`.

If you click the **Sign in with Google** button now, it will work and you're forwarded to the Google authentication process, where it asks you to confirm signing in. If you accept, you'll end up at the `/auth/verify-oauth` path throwing a 404, as the Route Handler doesn't exist yet. However, as the authentication will have passed the Supabase Auth Service already, the Supabase user will already be created (you can see it in the user overview in Supabase Studio). This user, in its current state, is an orphan user as it doesn't have any connection to our application – no service user, no permissions.

Go ahead and test it with any Google account. Afterward, you can delete the user in the dashboard, though you don't have to.

In the next section, we will build a verification route that takes care of OAuth logins and can even handle incomplete ones – such as this one, which is missing `app_metadata` and the related table data. There, we will make sure each OAuth user is properly set up, the same as when one registers with an email and password at `/register`.

Building a verification route to finalize the registration

When the Supabase authentication has done its job, the user will be forwarded the URL provided in `redirectTo`, which contains a special code that will allow us to create a session for that user. So, in our example, it will go to `https://oddmonkey.inc:3000/auth/verify-oauth?code=SOME_CODE_TO_INITIATE_A_SESSION`.

We have to process that code to initiate a session. Let's create the matching Route Handler and see how we can initiate a session with the given code, as well as how we complement the user with the required data for our application (the `app_metadata.tenants` data, the `service_user` table row, and the `tenant_permissions` table row). We'll also make sure that only tenant-matching emails are accepted.

Create a Route Handler file at `src/app/[tenant]/verify-oauth/route.js`.

Similar to our `register/route.js` file, we need to match the email against the tenant domain, but to be able to do that, we need the email – at the moment, we only have a code in the URL. However, this code can initiate a session and, as part of that, return all of the user data that belongs to it, including the email. This is done by sending the `code` part of the URL to the `exchangeCodeForSession(code)` function of Supabase:

```
import { getSupabaseCookiesUtilClient } from "@/supabase-utils/
cookiesUtilClient";
import { buildUrl } from "@/utils/url-helpers";
import { NextResponse } from "next/server";

export async function GET(request, { params }) {
  const url = new URL(request.url);

  const supabase = getSupabaseCookiesUtilClient();
  const { data: sessionData, error: sessionError } =
    await supabase.auth.exchangeCodeForSession(
      url.searchParams.get("code"),
    );

  if (sessionError) {
    return NextResponse.redirect(
      buildUrl("/error?type=login-failed", params.tenant, request),
    );
  }

  return NextResponse.json({ session: sessionData.session });
}
```

You'll notice that the `getSupabaseCookiesUtilClient()` function is used, not the admin client. That's crucial. We need a client that is user-bound, as we're initiating a user session with it with a code that is valid once, for that user. Our admin client doesn't handle cookies and must not be used to handle data bound to the user, such as when initiating a session for a user.

For now, we're just outputting the session data as JSON when everything has gone well, but we want to read the email instead and match it with the tenant host.

You can grab the email by accessing `sessionData.user.email`. Again, we split the email into its front and back parts, as we did in `register/route.js`, and then initiate an admin client to check whether the current tenant matches with that domain:

```
const { tenant } = params;
const supabaseAdmin = getSupabaseAdminClient();
const { user } = sessionData;
const { email } = user;
const [, emailHost] = email.split("@");
const { error: tenantMatchError } = await supabaseAdmin
  .from("tenants")
  .select()
  .eq("id", tenant)
  .eq("domain", emailHost)
  .single();

if (tenantMatchError) {
  await supabase.auth.signOut();
  return NextResponse.redirect(
    buildUrl(
      `/error?type=register_mail_mismatch&email=${email}`,
      params.tenant,
      request,
    ),
  );
}
```

Now, the first thought of most of us is: why do we only sign out again and not simply delete that Supabase user if it doesn't match anyway? That's something very dangerous to do here. If we were to now delete that user, based on the fact that the user doesn't match the current tenant, it would mean that a potentially pre-existing user, trying to log in on the wrong tenant, would be eradicated, and with it, all of the data the user created (tickets, ticket comments, etc.). So, without further checks, don't delete the user here; we'll handle this scenario in the last section.

The code below the `tenantMatchError` if-clause only gets executed if the user session was initiated successfully. In that case, we must complete the registration by enhancing the user with the missing table data and setting the `app_metadata.tenants` array. However, the problem is that we are not able to determine whether the user just signed up and we need to set up the missing data, or whether it's a usual login for which we already did the registration setup – our OAuth process does not differentiate between those two.

Logically, what we need is some kind of indicator that tells us whether we need to add additional data for that user-tenant combination or not. When the user does not have an `app_metadata.tenants` array with the current tenant ID inside, we need to complete the registration. Otherwise, we can just skip it as it's already there.

To do this, we create a helper Boolean variable called `needsInitialSetup` that checks whether a tenant either is not defined on the user's `app_metadata` value or does not contain the current tenant id – in those cases, it will resolve to `true`:

```
const needsInitialSetup =
    !user.app_metadata.tenants?.includes(tenant);
```

Now, if `needsInitialSetup` is true, we want to set `app_metadata` to contain the current tenant id, as well as create a service user and a tenant permission row in the database. The latter two inserts are the same thing as we've done in `register/route.js`. Changing the `app_metadata` value of an existing user, however, is done with `updateUserById(userId, options)`. The final code looks as follows:

```
if (tenantMatchError) {
    ...
}

const needsInitialSetup = ...;
if (needsInitialSetup) {
  await supabaseAdmin.auth.admin.updateUserById(user.id, {
    app_metadata: {
      tenants: [tenant],
    },
  });

  const { data: serviceUser } = await supabaseAdmin
    .from("service_users")
    .insert({
      full_name: user.user_metadata.full_name,
      supabase_user: user.id,
    })
    .select()
```

```
    .single();

  await supabaseAdmin.from("tenant_permissions").insert({
    tenant,
    service_user: serviceUser?.id,
  });
}
...
```

You see that we get the full name from the existing user object as part of the `user_metadata` value. That is because the name was given to us by the Google OAuth process and Supabase saved it as part of the `user_metadata` value.

We're nearly done, but there's a small design failure that you'd only notice in a very special case, yet it's still important to address – our application allows the assignment of multiple tenants; we designed it like that.

Yes, the registration itself wouldn't allow for that as it's bound to the domain of the email. However, I can manually assign a user to another tenant (which we did in *Chapter 5*). That means that the user I'm signing in with could've existed before – on a different tenant. Since it would not have the new tenant assigned, it would still make `needsInitialSetup` be `true`. Then, with the given `updateUserById(...)` call, I'm overwriting the `app_metadata.tenants` entry to contain just that one new tenant. Instead, we want to take a potentially pre-existing value into account.

That's as easy as changing it to the following:

```
await supabaseAdmin.auth.admin.updateUserById(user.id, {
  app_metadata: {
    tenants: [tenant, ...(user.app_metadata.tenants ?? [])],
  },
});
```

With this code, we're expanding any pre-existing `tenants` array into the updated `tenants` array, falling back to expanding an empty array (using `?? []`) if none existed before.

Cool! We have integrated OAuth with Google into our application and now we're super-safe! The process of adding any other OAuth-based provider such as GitHub can just reuse the same verification route due to how the OAuth authentication process with Supabase works.

Next, I want to have a brief, last look at dealing with potentially incomplete registrations.

Dealing with invalid user registration

As we had to generally re-enable signups for the intuitive OAuth login, there's a possibility that someone creates user accounts solely in `auth.users`. It can happen for two reasons at this point:

1. Someone uses our OAuth sign-in process but is not legitimated to access the tenant. Still, a Supabase user will be created as this action takes place before we can check it.

2. Someone uses the Anonymous Key exposed in the frontend and triggers the public `signUp` method on their own, hoping to breach the application security and gain access beyond our registration methods.

In both cases, the user won't get far. They will be an orphaned Supabase user with no access to our application whatsoever. However, it's not nice to have orphan users. So, how can we deal with it?

What you'll ideally want is to delete such orphan users as soon as possible. As this is not a definitive necessity, we won't go into implementation details for this one. Instead, I'll give you a few tips such that you are able to implement your own cleanup or prevention methods.

For Google OAuth, one way to avoid accidental signups is by telling Google the allowed email domain upfront. We can do that by passing the `hd` parameter to the `queryParams` property of the `signInWithOAuth` function (`https://developers.google.com/identity/openid-connect/openid-connect#sendauthrequest`). As we do have the tenant domain already (we fetch it in `app/[tenant]/page.js`), we can pass it on to the `Login.js` file and set the `hd` parameter as follows:

```
supabase.auth.signInWithOAuth({
  provider: "google",
  options: {
    queryParams: {
      access_type: "offline",
      prompt: "consent",
      hd: tenantDomain,
    },
    redirectTo: window.location.origin + "/auth/verify-oauth",
  },
});
```

That's a tactic to avoid accidental orphan users for this specific case. However, as this happens in the frontend, you can still get orphan Supabase users if one removes that domain from the request.

Also, that's just for our Google login. One can still create orphans with `auth.signUp()`.

My personal favorite when dealing with cleanup tasks, such as orphan removals, is using the cronjob extension (which will be shown in *Chapter 13*) that will, for example, run every two minutes and execute an SQL expression that will delete users that don't have tenants defined in their `app_metadata. tenants` value. That's an easy win.

However, long story short: if you really need to prevent orphans upfront, there is no way other than putting registration logic into the database with triggers.

Summary

Fantastic job! You've nailed tenant-bound registration.

You've learned about the `signUp()` method to create Supabase users. With that, you were made aware of the issue of having orphan users and implemented a signup with the help of `createUser`, albeit a disabled signup.

In the process of implementing the signup flow, you gained knowledge about sending account activation links, which are just magic links with a twist.

You also added the cool feature of signing in with Google accounts, and with that, you've set up Google credentials for Supabase and learned that external providers require signups to be enabled, which highlighted the problem of orphan users once more.

On top of that, you were shown the tools at hand to test OAuth with local domains by using HTTPS locally.

In the final stretch, you gained insights into the options to consider to handle orphan users.

With all of these essential foundations squared away, let's shift gears and explore the exciting realm of fetching data within the ticket system. Ready up for the next exhilarating phase!

Part 3:
Managing Tickets
and Interactions

In this part, we will bring your application to life with dynamic ticket management and enhanced user interactions. You'll learn how to create, manage, and display tickets, implement real-time comments, and enable secure file uploads. This part is all about making your app interactive and engaging for users, so are you ready to see your app in action?

This part includes the following chapters:

- *Chapter 8, Implementing Dynamic Ticket Management*
- *Chapter 9, Creating a User List with RPCs and Setting Ticket Assignees*
- *Chapter 10, Enhancing Interactivity with Realtime Comments*
- *Chapter 11, Adding, Securing, and Serving File Uploads with Supabase Storage*

Implementing Dynamic Ticket Management

You have put a lot of work into creating a shielded, multi-tenant application. Now it's time to pivot to its core functionality, breathing life into the application: the ticket management itself.

In this chapter, we will implement the creation and deletion of tickets, as well as showing the ticket details and their related data.

You'll also enhance the ticket list to show tickets from the database and implement paging, sorting, and searching features, as well as learn how to create enums, how to cache values with triggers, and why and when caching within the database values can be useful.

In this chapter, you will dive into the following topics:

- Creating the tickets table in the database
- Creating tickets and using triggers
- Viewing the ticket details
- Listing and filtering tickets
- Deleting tickets

Technical requirements

There are no additional technical requirements, unlike in the previous chapter. It can be helpful, however, if you have experience with SQL functions.

You'll find the related code of this chapter here: `https://github.com/PacktPublishing/Building-Production-Grade-Web-Applications-with-Supabase/tree/Ch08_Tickets`.

Don't forget that you can get a fresh setup with clean database data for that same branch with npm run reset.

Creating the tickets table in the database

To be able to manage the tickets, we first need to create a table for its data. However, which columns will we need to create? Let's have a look at our current UI, the Ticket Details page, that is shown when you click on one of the mock tickets on our ticket list:

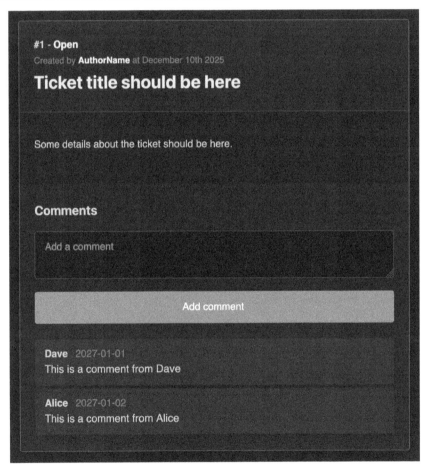

Figure 8.1: Our current ticket creation form

Given the existing UI, we can derive the minimal set of fields we need in the table (and in parentheses, I have included intuitive column names):

- A unique ticket id (`id`)
- The state of the ticket, for example, **Open** or **Done** (`status`)
- The author of the ticket (`created_by`)
- The creation date (`created_at`)
- A ticket title (`title`)
- A ticket description (`description`)
- The tenant ID to which the ticket belongs (`tenant`)

> **Note**
>
> Forget about the comments section shown in *Figure 8.1* for now. This involves additional database work that we'll handle separately in *Chapter 10*.

Let's convert this requirement into a database table. Go to the Supabase Studio, click on the Table Editor, and then the + **New Table** button.

Choose `tickets` as the table name and leave the two pre-created columns, `id` and `created_at`, unchanged. Then add the following new columns:

- Add `title` and `description` with the **text** type. Ensure that neither column can be null.
- Add `created_by`, also making sure that it's not nullable. Then click on **Add foreign key relation** and choose `public.service_users.id` as the reference to the `created_by` column. This is analogous to when you referenced the service user when creating the `tenant_permissions` table.
- Add `tenant` to contain a relation to the `tenants` table (`public.tickets.tenant -> public.tenants.id`). As with the `created_by` column, make sure to uncheck **Is Nullable** here.

You can see these columns in *Figure 8.2*:

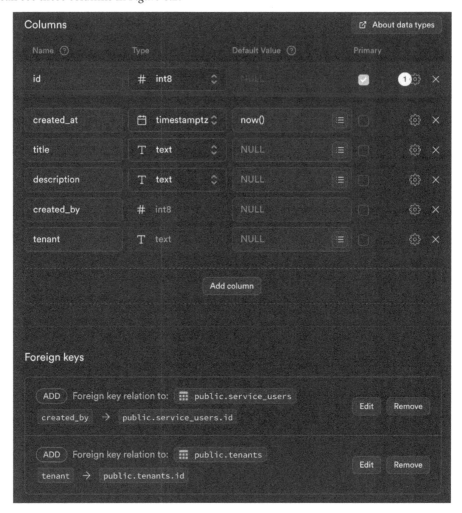

Figure 8.2: The table creation mask without a status column

Next, you'll want to add the ticket's state column. However, for that, we will not allow arbitrary text values but define an **enum**. This is a fixed, defined set of possible values. To be able to do that, save your table, then create a state type and add the state column afterward.

In the **Studio** sidebar, click on **Database** and navigate to **Database Enumerated Types**.

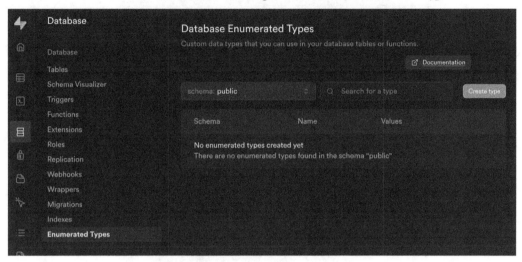

Figure 8.3: The Enumerated Types section in Studio

Here, you can create types with predefined values, which can then be used as column types. Click on the **Create type** button, then in the mask that appears, enter the **ticket_status** name for our new type. In the **Values** section, add five possible ticket status values: open, in_progress, canceled, information_missing, and done, as shown in *Figure 8.4*.

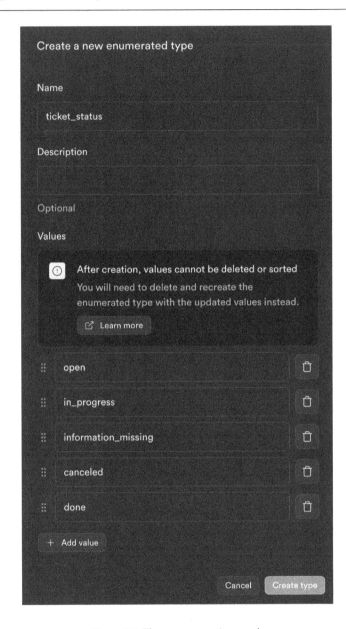

Figure 8.4: The enum creation mask

Finalize the creation of your `ticket_status` enum type by clicking **Create type** at the bottom of the mask. Now, your first enum shows up in the list of enumerated types:

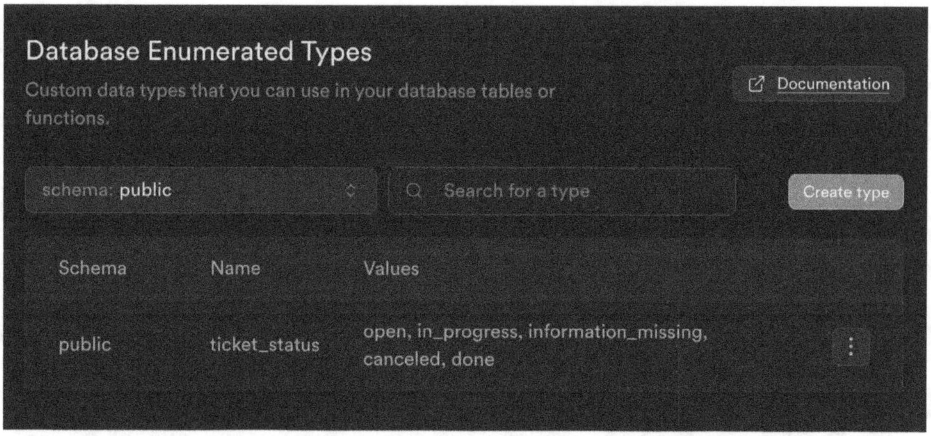

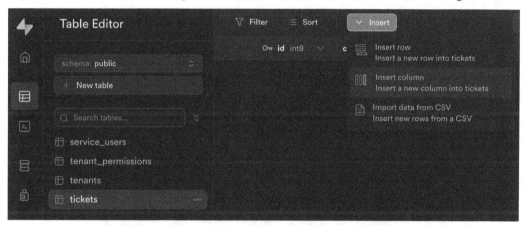

Figure 8.5: The enum list

This newly created type is now available alongside the other database types and can be used for columns. So, let's add the ticket status column to our existing tickets table. Click on the Table Editor and move to your `tickets` table. At the top of the table, click on **Insert Column**, as shown in *Figure 8.6*.

Figure 8.6: Adding a new column in the tickets table

A mask opens where you need to enter the name for the new column – we'll name it `status`. Then, in the **Type** dropdown, you can choose your previously created `ticket_status` type, and below that, you can now choose a default value. It makes sense to set a newly created ticket's status to `open`; hence, you should choose **open** as the default ticket value, like in *Figure 8.7*.

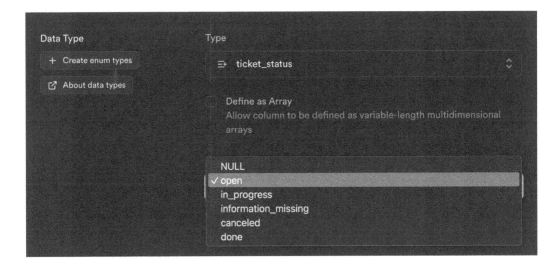

Figure 8.7: Setting the default value of the ticket

Choosing a default value doesn't explicitly prevent one from choosing null as a column value, so make sure to deactivate **Is Nullable**. Then click the **Save** button on the mask to add the status column to your table.

You now have your minimal tickets table ready to be used and you've learned how to create your own enum types. Next, you'll learn what you need to do to allow the convenient creation of tickets with your UI and what problems you need to solve to show all required ticket details.

Creating tickets and using triggers

In this section, you'll implement and extend the existing form for ticket creation and learn which RLS policies you need. If you sign in and go to https://packt.local:3000/tickets/new, it will show the following UI, which we created in *Chapter 3*:

Figure 8.8: The current ticket creation form

It looks like we are missing fields, as our table contains more than just two columns, but in fact, we're not missing anything. The other column values are implicitly given. The author (`created_by`) is the one who submits the ticket, the `created_at` date is automatically filled at ticket creation, and the `status` value is set to `open` by default. So, the only two custom fields, right now, are `title` and `description`.

This form will get more fields when we are extending its features but first, we want to allow the creation of a ticket with the existing two fields and show the respective ticket details on the Ticket Details page afterward.

> **Note**
>
> You've mastered building a full-fledged multi-tenant Next.js application with Supabase, conquering both the frontend and the backend! Now we'll primarily leverage frontend or backend code – whatever is more reasonable. However, we won't add a backend fallback for each frontend implementation. Why? Duplicating code logic wouldn't provide you with any learning benefit from here on and would make the app harder to maintain.
>
> Think of it like this: backend fallbacks are vital for apps where a failed request has a real impact, such as losing a sale in e-commerce. For most apps, such as our ticket management system, prioritizing frontend interaction code and using backend where reasonable streamlines development and the user experience. Next.js will still render anything it can on the server, it's just that, for example, the form for ticket creation wouldn't be doing anything if JavaScript in the browser was disabled – which is fine because we don't expect that to happen.

Implementing the ticket creation logic

We now want to implement the code that will insert the form data as ticket data to the database.

To do that, let's head to the `tickets/new/page.js` file where the ticket creation form is located. There you'll find the form submission listener as follows:

```
<form
  onSubmit={ (event) => {
    event.preventDefault();
    alert("TODO: Add a new ticket");
  }}
>
```

When the form is submitted, we want to make sure that both input fields contain proper values and then insert the ticket into the `tickets` table with the Supabase client and wait for the result. While waiting, we want to show a little loading spinner and disable the form inputs for a good user experience.

Inserting data into the database without checking the form first wouldn't be reasonable. That's why we will add a little bit of validation and provide user feedback if it fails.

Let's say that a ticket title should at least be over five characters, and the description be at least over 10 characters. We want to enforce that on the frontend first as immediate feedback, and then also in the database itself.

In the frontend, we can do a simple string-length check as follows:

```
onSubmit={ (event) => {
  event.preventDefault();
  const title = ticketTitleRef.current.value;
  const description = ticketDescriptionRef.current.value;

  if (title.trim().length > 4 && description.trim().length > 9) {
    // process form
  } else {
    alert(
      "A title must have at least 5 chars and a description must at
least contain 10"
    );
  }
}}
```

In place of the `// process form` comment, we now want to show a loading spinner and disable the form elements. We'll introduce another React state at the top of the component, directly below the `const ticketDescriptionRef = ...` definition, like so:

```
const [isLoading, setIsLoading] = useState(false);
```

Then replace `// process form` with `setIsLoading(true);`.

With the `isLoading` variable, we can now disable the input fields and the button:

```
<input disabled={isLoading} ...
<textarea disabled={isLoading} ...
<button disabled={isLoading} ...
```

Now you can enhance the UI by also showing a loading spinner inside of the button in that case by adding `aria-busy={isLoading}` to the button. `Pico.css` will automatically add a spinner animation to an element that has `aria-busy="true"`.

At this point, when you now enter the proper values and submit the form, it should look like *Figure 8.9*.

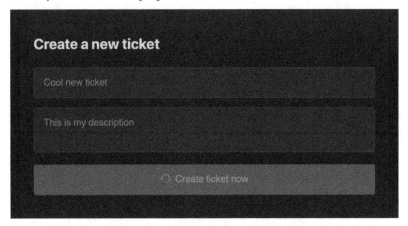

Figure 8.9: The form after submission

What's missing now is actually adding the data to the database. As part of the registration implementation, you've already used the `.insert()` command from Supabase, so you're aware of that, but you've only used it with the admin client up until now.

Here, in the ticket creation form, we need a secure, user-bound Supabase client respecting RLS. Let's instantiate one right below the useState line we added in `tickets/new/page.js`:

```
const [isLoading, ...
const supabase = getSupabaseBrowserClient();
```

Then, when inserting, the ticket needs to know the tenant it is added to. Fortunately, the tenant information is internally passed to every page as it's in the internal path structure. Hence, we read it by using the `params` value that is passed to the page:

```
export default function CreateTicket({ params }) {
  const { tenant } = params;
  ...
```

Now, in the onSubmit function, right below executing setIsLoading(true), we want to use the Supabase client to trigger an insert and fetch the result again by calling insert() and chaining it with select() and single():

```
supabase
  .from("tickets")
  .insert({
    title,
    description,
```

```
    tenant
})
.select()
.single()
.then(({ error, data }) => {
 if (error) {
  setIsLoading(false);
  alert("Could not create ticket");
  console.error(error);
 } else {
  alert("Successfully created ticket");
 }
});
```

As you can see, in the error case, we show a simple error alert and log the error in the browser console; otherwise, we display a message to the user saying **Successfully created ticket**.

When you submit the form, it will go into the error case and tell you about violated RLS policies:

Figure 8.10: The error when trying to insert a row

That's because we haven't added RLS policies for that table and hence, we can neither read from it nor write to it. Let's change that and add policies that allow us access to both reading and writing policies on tenants that we have access to.

Go to the Table Editor, navigate to the `tickets` table, click **Add RLS Policy**, and open the policy creation mask (we did this back in *Chapter 6*).

In the opened mask, we need to add two policies – one for creating rows (`INSERT`) and one for reading them (`SELECT`). Just because you can write a row doesn't mean that you can read it, which is why we need to add both.

We will start with the `SELECT` policy and name it **allow reading tickets of allowed tenant**. As we want to allow the signed-in user to read all tickets of tenants for which permissions are provided, it's the same rule as the one we added in the `tenants` table in *Chapter 6*, with just one difference. In that chapter, in the `tenants` table, we checked whether `tenants.id` was inside of the `app_metadata.tenants` array. This time, we need to check whether `tickets.tenant` is inside of `app_metadata.tenants`. So, it's the same thing with a different name.

We have the following policy in the `tickets` table:

```
COALESCE (
    (auth.jwt() -> 'app_metadata' -> 'tenants') ? tenants.id,
    false
)
```

Now, this becomes the following RLS policy in our `tickets` table:

```
COALESCE (
    (auth.jwt() -> 'app_metadata' -> 'tenants') ? tickets.tenant,
    false
)
```

Go ahead and add that as a policy in the **USING expression** of the policy mask now and make sure that the **SELECT** operation is selected (see *Figure 8.11* for reference).

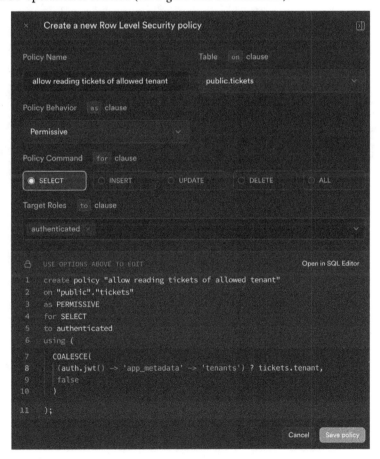

Figure 8.11: Adding an RLS policy in the tickets table

Save that policy and immediately add the exact same policy again with the **INSERT** operation selected. Once done, it will show your two existing policies for the `tickets` table:

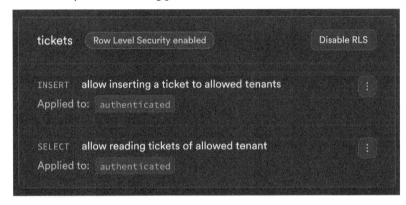

Figure 8.12: The tickets table's RLS policies

Move back to your form and try to submit a new ticket again. Even though you have proper RLS policies created, you'll now face a new error:

```
Object { code: "42501", details: null, hint: null, message: 'new row violates row-level
security policy for table "tickets"' }
    code: "42501"
    details: null
    hint: null
    message: 'new row violates row-level security policy for table "tickets"'
  <prototype>: Object { … }
```

Figure 8.13: The error message about the missing created_by value

The error states the obvious: we haven't provided a `created_by` value, although we must. We could execute a database selection to fetch our own `service_users.id` value and then pass it on to `insert()`. That would work. The thing is, I don't want to do that. `created_by` shouldn't be something the user sets; it should be implicit set by whoever is signed in. This is what database triggers are tailored for, and we'll solve that in the next subsection.

Using triggers to derive and set the user ID

Database triggers aren't unique to Supabase; they're found among all major databases. A trigger within a database is a function that is executed depending on defined events. For example, a trigger can be executed before a new row is inserted and can edit or adapt the inserted data, or even reject the insertion altogether.

Triggers in PostgreSQL consist of two parts: the function definition that the trigger will call and the trigger definition itself. They are easiest to understand if we implement one, so let's create a trigger function that sets the correct service user ID for the signed-in user for us.

First, you need to create a new database function that a trigger can execute. You can do this by going to **Database** in Supabase Studio and then navigating to **Functions**.

Figure 8.14: The database functions overview

There, you'll find a **Create new function** button. Click it and a mask like the one in *Figure 8.15* opens. Enter a technical name for the function (I will choose **set_created_by_value**), leave the schema as **public** and, for **Return type**, choose **trigger**.

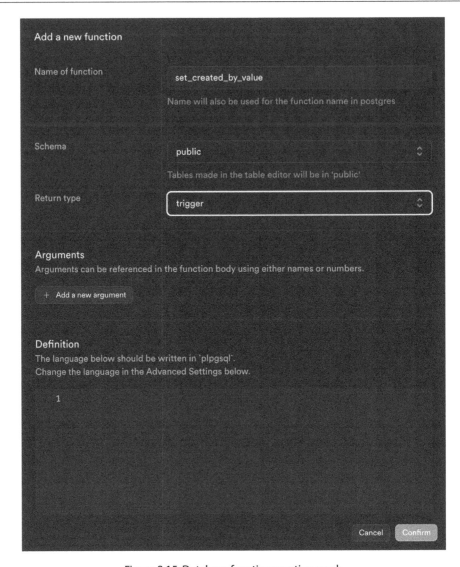

Figure 8.15: Database function creation mask

For triggers, you can ignore the **Arguments** section of the mask – you'll learn about this in *Chapter 9* – but moving on, you'll be faced with the **Definition** section. There, we will write the expression that shall be executed before a new row is inserted. A function definition in Postgres always needs the following surrounding it:

```
BEGIN
  -- the sql code to execute for the row
END;
```

> **Note**
>
> - - starts a comment in SQL.

Between BEGIN and END, we now write the SQL code to be executed when the row is added. The logic we write must put the service user ID value into the `created_by` column value of a newly inserted row. For that, you need to know that a trigger that runs before an INSERT operation will automatically have a variable named NEW defined, containing the to-be-inserted row data.

You can hence set or change the value of a column of that row by writing NEW.`column_name` = `value`;. What we'll need is something like this:

```
BEGIN
  NEW.created_by = THE CODE THAT FETCHES THE SERVICE USER ID OF THE
    SIGNED IN USER;
END;
```

You already know that you can use auth.uid() to get the Supabase user ID of the signed-in user and you also know that your `service_users` table has a `supabase_user` column where that ID is stored. That means that we can simply write a SELECT statement on the right side of the assignment as follows:

```
NEW.created_by = (SELECT id FROM service_users WHERE supabase_user =
auth.uid());
```

The only thing left to do to complete the trigger function is to return that row, so the final definition is as follows:

```
BEGIN
  NEW.created_by = (SELECT id FROM service_users WHERE supabase_user =
    auth.uid());
  RETURN NEW;
END;
```

That's it. Save the function by clicking on **Confirm** in the mask (*Figure 8.15*).

Before continuing, let's quickly answer two open questions you may be thinking about:

- What happens if it cannot find the ID, for example, because a user is not signed in? Well, then it correctly fails – a non-authenticated user shouldn't reach this statement or else it was a malicious request anyway, so everything works as expected.

- With which rights is the SELECT inside of the function executed? Does it respect existing RLS policies? Here, with the way in which we defined the function, and without further advanced settings, it will always respect RLS. So, if there wasn't a policy on the service_users table that allowed the signed-in user to fetch their own row from it, the function would fail as well.

At this point, the trigger function is defined but the trigger itself has not been defined yet. That means that, right now, your function does nothing, it's just ready to be used. To deal with this, navigate to the **Triggers** section:

Figure 8.16: Database triggers view

Click on **Create a new trigger**, and in the mask that opens, enter the technical name for the trigger. A good convention is to start with tr_, so let's use tr_tickets_autoset_created_by.

In the **Conditions to fire trigger** section of the mask (see *Figure 8.17*), for **Table**, choose **tickets**, and for **Events**, choose **Insert**. This makes sure that it will only execute when an insert on the tickets table happens.

For **Trigger type**, ensure that **Before the event** is selected, and for **Orientation**, ensure it fires per **Row**.

The final conditions look like this:

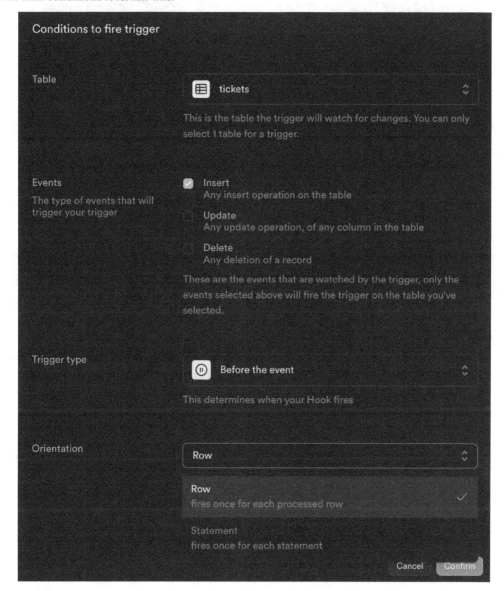

Figure 8.17: Trigger conditions

Finally, below the **Conditions to fire trigger** section, you'll find the **Function to trigger** section. Here, you'll need to add the function that you want to execute – this should be the **set_created_by_value** function that you created earlier.

Figure 8.18: The Function to trigger section

The trigger is ready, so save it by clicking on **Confirm**. Now, before a new row is inserted into the tickets table, it will execute the function and set the correct `created_by` value.

You can prove that by going back to the ticket creation UI, such as for `packt` at `https://packt.local:3000/tickets/new`. When you enter the fields correctly, it should show **Forward to details view** instead of throwing an error. This means that the ticket was added to the database, which you can confirm by going to Supabase Studio and opening the tickets table, where you'll see the newly created row:

Figure 8.19: The new row in the tickets table

Now, instead of `alert("Forward to details view")`, let's actually forward to the matching Ticket Details page. At the top of the `CreateTicket` function, make sure to instantiate the Next.js Router with `const router = useRouter();` (remember to import from `next/navigation`, not `next/router`). Then, adapt the `if` statement to move to `tickets/details/id`, but make sure, just as everywhere else, to use the `urlPath()` function to maintain full flexibility if you later want to change the system to be path-based again:

```
.then(({ error, data }) => {
  if (error) {
    ...
  } else {
    router.push(urlPath(`/tickets/details/${data.id}`, tenant));
  }
});
```

Now, when you create a new ticket with the UI, you'll be forwarded to the Ticket Details page with the correct ID. This will allow you to implement the Ticket Details page with real data soon. Awesome!

At this point, in your browser network panel, you may notice that the Supabase response, especially locally, is immediate, but the Ticket Details page shows after a few moments, making it seem slow. This is not due to Supabase but to the render mechanisms in Next.js. Let's improve that now.

Improving loading behavior after adding a ticket

If Next.js recognizes that it must re-render a new page because it's not in its cache, it obviously must make a new request. This can sometimes take a short but noticeable while and make the page feel like it's a bit stale. Fortunately, there's a simple solution to that: the `loading.js` file (`https://nextjs.org/docs/app/building-your-application/routing/loading-ui-and-streaming`).

At each level of the Next.js structure, we can add `loading.js` files – these are loading components that show when the page is navigated to but not yet loaded. You can add one anywhere you need but let's just add one inside of the `[tenant]/tickets/details` directory, such that when we create a ticket in the UI and Supabase has inserted the row, we immediately see at least the loading component.

So, create `[tenant]/tickets/details/loading.js` and export the following code:

```
export default function Loading() {
  return (
    <article aria-busy="true">
      <strong>Loading ticket details...</strong>
    </article>
  );
}
```

Having that loading UI, we also want to tell Next.js to prefetch it such that Next.js has the loader already available when navigating to the Ticket Details page. You can do so by adding the following code as part of your `CreateTicket` component:

```
const router = ...
useEffect(() => {
  router.prefetch(urlPath(`/tickets/details/[id]`));
}, [router]);
```

This way, on production, Next.js will know that you're about to open a details page soon and will prefetch it. Doing that provides a much more fluid user experience even if Next.js takes a second to render the dynamic details page.

Having improved the UX, we now want to harden the `tickets` table with constraints before we move on to implementing the Ticket Details page with actual data.

Enforcing checks on the database columns

We now want to bring the same check we added in the frontend (that the title must be bigger than four characters and the description bigger than nine characters) to the `tickets` table itself such that, even if the frontend would not have the validation, a title such as `Hi` or a description such as `test` would fail at the database level.

Postgres allows us to add checks to single columns of a table. Checks are expressions that, like RLS policies, must return true, or else the row data is rejected. In contrast to RLS policy expressions, checks can only do simple checks within the row data itself and cannot do sub-expressions on foreign tables.

What we want is to add a check for string length. Go to the `tickets` table, find the `title` column, and select **Edit column**:

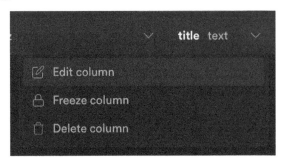

Figure 8.20: The column options

In the opened mask, scroll to **Constraints** and find the field labeled **CHECK Constraint**. There, you can simply enter any valid SQL expression that evaluates to a Boolean. As `length(some_string)` is a valid SQL function, you can check for the length of the title by filling the field with `length(title) > 4` like so:

Figure 8.21: The CHECK Constraint field

Click on **Save** and do the same for the `description` column by adding `length(description) > 9` to its **CHECK constraint** field.

Let's test whether the constraint works. You could adapt your frontend code to allow smaller titles and descriptions to prove that it will fail on the database, or instead, you could just test it right here in the Table Editor in one of the existing rows that you created with the UI. Just double-click the title field of a row to make it editable, enter something smaller than five characters (e.g., `abc`), and click outside the field to save it. It will reject and go back to the old value, as well as show an error telling you that the check constraint failed:

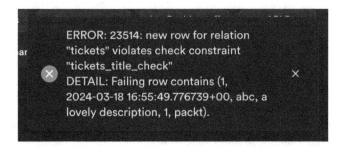

Figure 8.22: The error after trying to save abc as a title

Awesome, the added constraints work and the basic ticket creation flow is now complete. You're now able to create tickets that are automatically given your user ID with the help of triggers and database functions. Next, let's make the details view load the data matching the ticket ID.

Viewing the ticket details

We now want to display all ticket information from the database on the Ticket Details page. Right now, `TicketDetailsPage` is a server component – that's okay, as it allows us to fetch the ticket information with the page request and show it. So, let's start doing that.

Firstly, we require a Supabase client that respects RLS and is bound to the user. It's crucial to emphasize the significance of using a user-bound client here again. While we could instantiate an admin client within a server component, doing so would disregard all RLS policies and permissions. Therefore, let's ensure that we fetch the data using an RLS-respecting client within the `tickets/details/[id]/page.js` file:

```
export default async function TicketDetailsPage({ params }) {
  const supabase = getSupabaseCookiesUtilClient();
  const id = Number(params.id);
  const { data: ticket, error } = await supabase
    .from("tickets")
    .select("*")
    .eq("id", id)
    .single();

  return ....
}
```

> **Note**
>
> In the given code, I didn't use `params.id` but instead created an id variable that I enforced to be a number (if not, it will be NaN). Why? When we select data and do things such as `eq("id", id)`, if the ID is `123` for example, it does not matter whether it's `eq("id", 123)` or `eq("id", "123")`. The selection of data is always an API GET request and the parameters will be sent to Supabase as query parameters of an URL anyway – then, they're internally mapped to their correct type. However, inside the database, it will be an actual integer value. So, even if there's no immediate issue with having a string number instead of a real number, it's always handy to have the real type.

After making our `select` request, let's make sure that we show an error page when the ticket can't be fetched. We can do this like so:

```
if (error) return notFound();
```

Next, we want to render the ticket data on the page. You can simply deconstruct the `ticket` object for ease of use:

```
if (error) ...;

const {
  created_at,
  title,
  description,
  created_by,
  status
} = ticket;
```

Then let's put the variables in the right places, ending up with the following code (I replaced the `className` occurrences with . . . for better readability):

```
return (
  <article ...>
    <header>
      <strong>#{params.id}</strong> -{" "}
      <strong ...>{status}</strong>
      <br />
      <small ...>
        Created by
        <strong>{created_by}</strong> at <time>{created_at}</time>
      </small>
      <h2>{title}</h2>
    </header>
```

```
    <section>{description}</section>

    <TicketComments />
  </article>
);
```

Now go ahead and create a new ticket and you'll be forwarded to the details page with real data. However, you'll notice that it doesn't look complete:

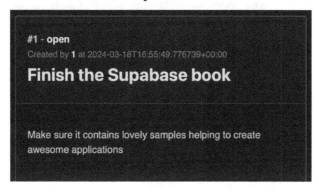

Figure 8.23: The ticket details view with data from the database

The status is in all lowercase and the date is a weird-looking ISO date string, but most strikingly, the author is a number. This makes sense because it is a number in the table row – the number is the service user ID of the author (so, your service user ID, in this case), but we want the *name* of the author instead.

If you tried to use the Supabase client to fetch the author's name by fetching the row from the service_users table and getting the full_name value from it, you'd be lucky because per RLS policies, you have access to your own service user row. However, if the ticket you're viewing was created from another account, the RLS policies would deny you access to that row.

You could do something different though. Knowing that you successfully fetched the ticket means that the user had access to that ticket. That means that if you now instantiate an admin client to load the related author name, you could do this:

```
const { data: { full_name} } = adminSupabase.from('service_users').
eq('id, created_by).single();
```

That would be safe and would work – but it would be an additional request after the first, which both have to be waited for. Can't we do that in one request? If you're well aware of database relations, you might say that joining the tickets table with the service_users table would provide the data from both tables. Although joining tables in Supabase is feasible and we will use joining mechanisms in *Chapter 10*, it naturally respects RLS policies so it wouldn't work for fetching the name of a different service user. So, what's the solution? Well, it's triggers. However, this time, we will use triggers for caching the author's name. Let me show you how.

Caching the author's name with a trigger

What we want to do is add some redundancy in the database, which adds resiliency and speed to our ticket data. Let me elaborate.

Earlier, we created a trigger that automatically set the `created_by` value to match the correct service user ID of the creator of the ticket. We don't want to change that. However, we want to add an additional text column to the `tickets` table that contains the user's full name. Yes, you heard me, we will be putting a copy of the name of the author in each ticket row. Your first thought might be: but David, isn't that bad, as we would have to sync the names with the new ones in the `service_users` table if they change? The answer is no. That's what makes the data resilient.

A user's real name doesn't change. When it does, it's likely due to a special event such as marriage. Now let's say that *David Lorenz* creates a ticket with the 333 ID in 2024. He then gets married in 2025 and takes his spouse's last name, so he is now *David Marx*. Who created the 333 ticket? It's the same person, but from a historical point of view, David *Lorenz* created the ticket, not David *Marx*. However, newly created tickets will be created by David Marx. So, by keeping a name copy in each ticket, you can ensure historical correctness. It's also super fast to get the name because it's literally part of the ticket row.

Even if you argue that you don't want historical correctness and, if a user name changes, you also want all old tickets of the user to have that name change, it'd be easy to do so because the name change would happen as part of your code logic. So, you could easily identify all old tickets of that user by the author's service user ID (`created_by`) and update the name accordingly.

First, make sure to delete existing tickets to have a clean setup to work with. You can do so by selecting the rows with the checkbox and then clicking on the **Delete x rows** button, where **x** will be substituted with the number of rows (*Figure 8.24*), or by executing TRUNCATE TABLE tickets; in the SQL editor of Studio.

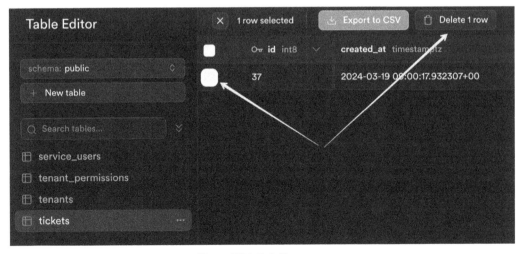

Figure 8.24: Deleting rows

When your `tickets` table is cleared, let's add a new column to it called `author_name`, setting the type as **text** and unchecking **Is Nullable**.

Next, let's implement setting the author's name with a trigger. You already know that before creating a trigger, you need a function that shall be triggered. Create one by moving to **Database | Functions**, choose the **set_ticket_author_name** name, keep the default schema as **public**, and make sure to select **trigger** as **Return type**:

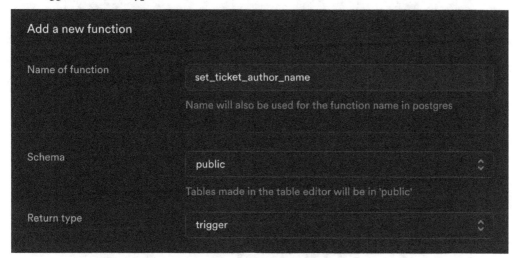

Figure 8.25: Creating the set_ticket_author_name function

In the **Definition** section, we can copy most of what we've done in `set_created_by_value`, with the difference being that this time, we want to set the `full_name` value of the matching `service_users` row to the `author_name` value of our ticket row instead of setting the ID. Hence, the definition should look like this:

```
BEGIN
  NEW.author_name = (SELECT full_name FROM service_users WHERE
    supabase_user = auth.uid());
  RETURN NEW;
END;
```

As we are not creating tickets on behalf of someone else (we will soon implement assigning tickets to others, but we will not let anyone create tickets for someone else), we are always only accessing our own service user row. Therefore, everything will work as expected permission-wise.

Save the trigger function and navigate to **Database | Triggers**. Add a new trigger, enter the `tr_tickets_autoset_author_name` name, choose **tickets** as the **Table**, choose **Insert** as the event, leave **Before the event** as is, and make sure to select **Row** in the **Orientation** section. Finally, choose your **set_ticket_author_name** function to be executed.

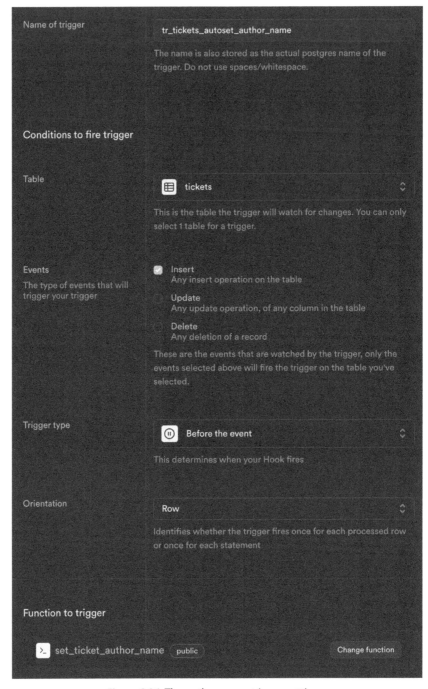

Figure 8.26: The author name trigger settings

Save that trigger. Now, you have two of them. Which is executed first? Well, for us, that doesn't matter. Our two triggers work independently from each other, so regardless of whether the ID or the name is set first, it will work. However, for your information, as per Postgres documentation (`https://www.postgresql.org/docs/9.0/trigger-definition.html`), multiple triggers on the same table are executed in alphabetical order. Hence, if you ever need to set the order of triggers, you can give them similar starting names with ascending numbers such as `trg_tickets_1_action` and `trg_tickets_2_other_action`.

> **Note**
>
> Instead of creating a second trigger, we could've edited the existing `set_created_by_value` function to execute both SQL expressions in one function. However, I wanted to separate them for the sake of atomicity.

Having the correct triggers and assuming that they'll work seamlessly together, we now need to make sure to reflect the `author_name` value in our template so we can see it in action. Go to the `TicketDetailsPage` component again and make sure to destructure `author_name`:

```
const {
  created_at,
  // ...
  author_name
} = ticket;
```

Now, use `author_name` instead of `created_by` in the rendered template and save it. Next, go to your application, sign in, and create a new ticket. After the ticket creation, you're forwarded to the ticket details, where your name will appear (you can jump ahead and see this in *Figure 8.27*).

Wonderful. At this point, really acknowledge that you've learned far more than just setting a matching name – you've learned how to enforce and cache values inside of the database while respecting RLS permissions. Since every user with access to the same tenant will be able to read the same tickets, users will now also see the author's name without hitting RLS restrictions and without any further fetch requests.

Next, let's make sure that the date isn't cryptic anymore and that the status, whatever it might be (`in_progress`, `open`, etc.), will be shown properly.

Improving the date and status view

The shown date is obviously the raw timestamp from the database, shown as an ISO string. With modern JavaScript, it's extremely easy to convert such a string into a human-readable date, even with correct language without you needing to translate anything. All you need to do is to create a new `Date` instance providing the ISO timestamp and then call `.toLocaleDateString(...)` to make it a human-readable, localized date string.

So, in your details page component, before the rendering return statement, create the date string as follows and use it accordingly, instead of `created_at`, in the `const dateString = new Date(created_at).toLocaleString("en-US");` rendering. Now your date will show in the correct locale format. In our case, this is the date from the United States.

Then let's make the ticket status more readable. For this, create a `src/utils/constants.js` file and export an object with the proper mapping of the database key inside:

```
export const TICKET_STATUS = {
  open: "Open",
  in_progress: "In progress",
  information_missing: "Information missing",
  canceled: "Canceled",
  done: "Done",
};
```

Now import it into the Ticket Details page and use it inside the rendering part:

```
<strong ...>
  {TICKET_STATUS[status]}
</strong>
```

The final detail view now only has human-readable values as shown in *Figure 8.27*.

Figure 8.27: The final details view

Having the ticket creation as well as the view details ready, we can now move on to listing all the tickets on the overview page at `/tickets`.

Listing and filtering tickets

Now we want to make sure that we list all tickets of a tenant when signed in, showing the open ones at the top, ordered by date. Not just that; we also want to allow users to filter tickets by title.

However, first, let's make sure that the current mocked ticket list is replaced with real data. To do this, make sure to create at least 15 tickets with two different user accounts so that we actually can filter different tickets and see that our list works. Once done, you're ready to continue.

> **Remember**
>
> If you're on `http://packt.local:3000/register`, you can enter just any email ending with **@packt.local** and then confirm it with the link sent to your Inbucket inbox.

To implement the ticket list loading, open `[tenant]/tickets/page.js` in your editor and remove the code that passes `tickets` to the `TicketList` component (`<TicketList tickets=…/>`). Instead, we want the `TicketList` component to be responsible for data fetching itself, so you'll end up with the following:

```
import { TicketList } from "./TicketList";

export default function TicketListPage({ params }) {
  return (
    <>
      <h2>Ticket List</h2>
      <TicketList tenant={params.tenant} />
    </>
  );
}
```

Then, move to the `[tenant]/tickets/TicketList.js` file, make it an `async` function to be able to fetch data on page load, instantiate a Supabase client, and fetch all tickets for this tenant:

```
export async function TicketList({ tenant }) {
  const supabase = getSupabaseCookiesUtilClient();
  const { data: tickets, error } = await supabase
    .from("tickets")
    .select()
    .eq("tenant", tenant);
...
```

Also, like we did in the details view, make sure to show the mapped status by replacing `<td>{ticket.status}</td>` with `<td>{TICKET_STATUS[ticket.status]}</td>`.

Without further additions, this will already work. However, it will load all tickets, even if there are 5,000 tickets (which, admittedly, would not really happen because Supabase limits the result of rows per request to 1,000 by default to avoid performance leaks (`https://supabase.com/docs/reference/javascript/select`)). So, let's make sure to just show a portion of tickets and allow you to load more tickets on demand.

Enabling paging

Let's now add the functionality to show a limited amount of tickets by enabling paging. However, first, I'd like to elaborate on why we're using paging and not the infinite loading approach.

Paging versus infinite loading

Nowadays, there are different best practices for loading follow-up data. Some use infinite loading (e.g., social media posts) by loading new data when the user has reached the end of the list. Some add a load more button to achieve the same thing.

A common problem with this approach, however, is that it's much harder to create deep links with such infinitely appended content. Say, for example, that someone scrolls through 5,000 posts (that's easier done than imagined) and then shares a link to the page they're on. Should the user receiving that link be taken to a page where 5,000 posts are loaded? Or should they be taken to a page where only 20 are loaded, starting from where the sharing user left off? Loading 5,000 posts at once is obviously ridiculous, but loading 20 from where the user left off means that these two users see two different states of a page with the same link.

That's not just a semantic problem but also a technical one. In hybrid applications such as ours, the frontend would have to prevent the backend from refreshing the page data when we infinitely load more data because otherwise, it would throw away the existing data.

Long story short, all of that can be solved in our ticket management application. Paging is not just the easier but presumably also the better option to navigate through a list of tickets as it's easy to implement, predictable, and allows for viewing the exact same view for different users sending links to each other (the well-known Jira system uses paging for navigating between tickets as well).

Implementing the paging logic

If you had 10,000 tickets, you wouldn't want to load or list all of them. Instead, let's only show six and, if there are more, add a button that moves to another page loading six more rows.

To do that, we must first count all available tickets such that we always know whether there are more tickets available to load. This is done by calling the same `select` command that exists in our `TicketList` component but with an additional option object where we tell it to only return the number of tickets, not the actual tickets:

```
const { data: tickets, error } = await ...
const { count } = await supabase
    .from("tickets")
    .select("*", { count: "exact", head: true })
    .eq("tenant", tenant);
```

Having that total count value, we can now determine whether we need to show more by subtracting the number of shown tickets from the total count:

```
const { data: tickets, error } = await ...
const { count } = await ...
const moreRows = count - tickets.length > 0;
```

Now, when `moreRows` is truthy, we add a button below the table that leads to the next page, which must contain a query parameter for us to know that we are on a different page:

```
return (
  <>
    <table>
    ...
    </table>

    <div>
      {moreRows && (
        <Link role="button" href={{ query: { page: 2 } }}>
          Next page
        </Link>
      )}
    </div>
  </>
);
```

Clicking the **Next page** button would lead to the same page but with `?page=2` attached to its URL.

Now, make sure that you actually limit the result of shown tickets to six by chaining your ticket selection code with `limit(6)`:

```
const { data: tickets, error } = await supabase
    .from("tickets")
    .select()
    .eq("tenant", tenant)
    .limit(6);
```

Next, we need to identify whether we need to load the first six rows (page 1) or the next six rows (page 2). That's easy. In Next.js, the URL query parameters are passed to the current page component as `searchParams`, so if `searchParams.page` is provided and is a valid, positive integer, we can tell Supabase to skip a specific number of rows.

However, to be able to do that, we need to make sure to pass `searchParams` from the page to our `TicketList` component:

```
export default function TicketListPage({ params, searchParams }) {
    ...
```

```
  <TicketList tenant={params.tenant} searchParams={searchParams} />
  ...
}
```

Then we can use it within our `TicketList` component, read the `page` parameter from it, and check whether the given string contains a positive integer. If so, we'll use that value as the page value; otherwise, we'll set it to be the first page:

```
export async function TicketList({ tenant, searchParams }) {
  let page = 1;
  if (
    Number.isInteger(Number(searchParams.page)) &&
    Number(searchParams.page) > 0
  ) {
    page = Number(searchParams.page);
  }
```

Now you always have the page number value within your `TicketList` component. That means that you can adapt your **Next page** button to always link to the next page, as well as show a **Previous page** button whenever the current page is bigger than 1:

```
<div style={{ display: "flex" }}>
  {page > 1 && (
    <Link role="button" href={{ query: { page: page - 1 } }}>
      Previous page
    </Link>
  )}

  {moreRows && (
    <Link
      style={{ marginLeft: "auto" }}
      role="button"
      href={{ query: { page: page + 1 } }}
    >
      Next page
    </Link>
  )}
</div>
```

However, what's still missing is actually showing the correct page data. This is called `OFFSET` in SQL, but in Supabase, you'll use the `range(startingPoint, end)` function for skipping rows where the end always has to be `startingPoint + numberOfRowsToFetch`. Hence, instead of using `limit(6)`, you can simply use the following:

```
const startingPoint = (page - 1) * 6;
const { data: tickets, error } = await supabase
```

```
.from("tickets")
.select()
.eq("tenant", tenant)
.range(startingPoint, startingPoint + 5);
```

Let me explain two things in the given code block:

- We use (page - 1) * 6 instead of page * 6 because otherwise, it would start skipping six rows from page 1, 12 rows from page 2, and so on

- As startingPoint itself is included, you need to add five additional rows for it to be six in total

Feel free to style the paging buttons a bit by making them smaller with CSS to not appear too bold. However, now, when saved, the ticket selection will be dependent on whatever page you're on. Go ahead and try it! In *Figure 8.28*, you see my view on page 2 of the packt tenant where I created 14 tickets earlier.

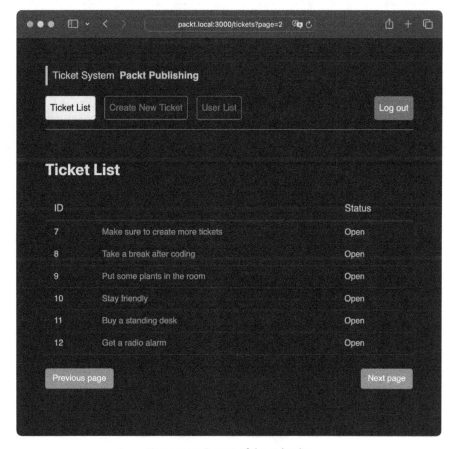

Figure 8.28: Page 2 of the ticket list

Now the paging is working but there's a little problem: `moreRows` isn't considering the current page yet, and that's why we always see the **Next page** button even though there are no more pages. So, what do we need to check? Well, we have more rows if the currently skipped/loaded rows are still smaller than the total count. Hence, `const moreRows = count - tickets.length > 0;` must become `const moreRows = count - page * 6 > 0;`.

Enforcing fresh data

The paging now works gracefully – what a success! – but I still see room for improvement. First of all, I'd like to indicate to Next.js to never cache that page on the server as we'll always want recent data. You can prevent page caching by going to `[teant]/tickets/page.js` and writing `export const dynamic = "force-dynamic";` after all of the `import` statements.

> **Note**
>
> When using `cookies()` or `headers()` in a Next.js page, which we do by using the `TicketList` component that instantiates a cookie-based Supabase client, Next.js automatically opts out of caching the page on the server. Despite that, it still makes sense to mark a page as dynamic if you know that it must always contain recent data. Doing so will ensure that it is considered dynamic even if you did not call `cookies()` or `headers()`.

However, this isn't all we need to do to allow for immediately fresh data. Next.js has a client-side cache, the **Router Cache**, which, in its current state, cannot be opted out to allow for "instant navigation" as described here: `https://nextjs.org/docs/app/building-your-application/caching#opting-out-3`.

> **Note**
>
> In Next.js 15, this has been overhauled and client navigation (Router Cache) is not cached by default anymore; see: `https://nextjs.org/blog/next-15-rc`.

This means that each page, even if it is dynamic, is held in the client cache for 30 seconds before it gets refreshed. To not have old data but immediately fresh data while navigating back and forth with the page buttons, we really want to opt out of this.

Luckily, you can do so with a little trick by adding a random value to the page's query parameters. This lets Next.js think it's always a new page and forces it to refetch it. So, for example, for the **Next page** button, `href` becomes `href={{ query: { page: page - 1, r: Math.random() } }}`. Having added `r: Math.random()` on both page buttons, they will now lead to the page being freshly fetched every time.

With that, there's one last bit to improve. When navigating between the pages there is, again, sometimes a short but noticeable rendering delay. Earlier, we dealt with this using the `loading.js` loader, but that is used for the whole page. Instead, we can show loaders for server components that are refreshed within a page. This is super easy.

If you wrap a server component with React's **Suspense** component, you can show a fallback while it's waiting for the server component to finish rendering. To make sure that it updates with updated query parameters, it's best to add a key (`https://react.dev/reference/react/Suspense#resetting-suspense-boundaries-on-navigation`) that changes when the query parameters change. Our `TicketListPage` component becomes this:

```
export default function TicketListPage({ params, searchParams }) {
  return (
    <>
    <h2>Ticket List</h2>
    <Suspense
      fallback={<div aria-busy="true">Loading tickets...</div>}
      key={JSON.stringify(searchParams)}
    >
     <TicketList
        tenant={params.tenant} searchParams={searchParams} />
    </Suspense>
   </>
  );
}
```

Now, if it takes a moment while navigating between pages, you'll see the loader fallback:

Figure 8.29: The Suspense fallback loader

With all of that working, we now want to make sure that more recent and work-relevant tickets are shown first. Let's look at how to do that.

Sorting tickets

Sorting rows with Supabase is extremely easy. You can simply use `.order(columnName, orderOptions)` (`https://supabase.com/docs/reference/javascript/order`) and chain it multiple times to order by multiple columns.

Let's make sure that the open tickets are shown first and that they are sorted by the newest date (so, in descending order based on the `created_at` value). The resulting fetch is as follows (new code is highlighted):

```
const { data: tickets, error } = await supabase
  .from("tickets")
  .select()
  .eq("tenant", tenant)
  .order("status", { ascending: true })
  .order("created_at", { ascending: false })
  .range(startingPoint, startingPoint + 5);
```

While the `created_at` part seems intuitive, you might ask yourself how the `status` part now sorts the `open` ones first. How would it know? Sorting on an enum value is done as per the order in which you defined it (*Figure 8.30*), which is only changeable on creation, unfortunately. The circle on the figure that follows indicates the drag handle with which you can change the order when creating an enum.

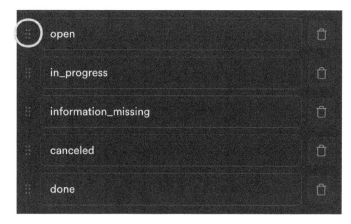

Figure 8.30: Sorting an enumerated type when creating it

This means that if you want to change the order of an enum type afterward, the best approach would be to create a new column (for example, `status_new`), set the new enum there, and once done, delete the old column.

Incredible! You've gained some pretty massive insights into dealing with the Next.js cache and have paging fully working with fresh data and properly sorted. Let's add a filter bar to search through the title or description of the ticket.

Creating a ticket filter

I want to show you how to create a simple filter bar for our tickets in which we can search for tickets by their content (title and description). I've prepared a component template with a search field and a button for you, including some minimal styling. Just create the `[tenant]/tickets/TicketFilters.js` file and put the following template inside it:

```
"use client";

import { useRef } from "react";

export function TicketFilters({ tenant }) {
  const searchInputRef = useRef(null);
  const onSubmit = (event) => {
    event.preventDefault();
    const search = searchInputRef.current.value;

    alert("Search tickets containing " + search);
  };

  return (
    <form onSubmit={onSubmit}>
      <div
        style={{
          alignContent: "center",
          display: "flex",
          gap: "15px",
        }}
      >
        <input
          type="search"
          ref={searchInputRef}
          id="search"
          name="search"
          placeholder="Search tickets..."
          required
          style={{ margin: 0, maxWidth: "350px" }}
        />

        <button type="submit" role="button"
          style={{ width: "auto" }}>
          Search
        </button>
      </div>
```

```
    </form>
  );
}
```

As you can see, there's nothing new in this template. It's just a form with an input field that triggers the `onSubmit` function showing the content of the input field in the alert.

Let's import this filter component in the `tickets/page.js` file and put it directly above the `TicketList` component's Suspense:

```
<h2>Ticket List</h2>
<TicketFilters tenant={params.tenant} />
<Suspense...
```

After that, opening your ticket management system should show a lovely `search` input field:

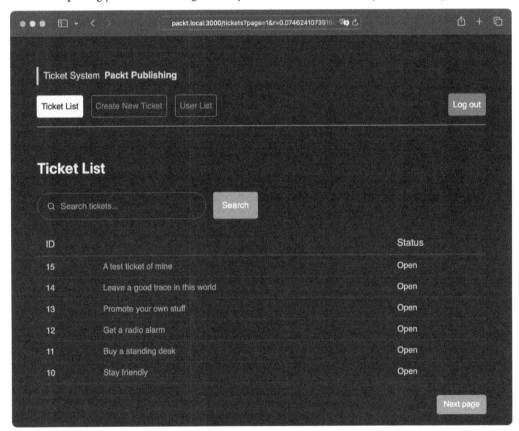

Figure 8.31: The Ticket List UI with the filter component

What we want now is to use that input value and search the tickets' titles and descriptions – no exact matches are required. So, if you enter `te`, it should at minimum include **A test ticket of mine** and **Promote your own stuff**, as both contain the **te** string.

As fetching our tickets happens in the `TicketList` component, we need to provide `TicketList` with the search value parameter, which we can do by passing it in the URL as a query parameter. Hence, our `TicketFilter` component must set the input value as a query parameter in the URL. To do that, we simply want to take the current URL, change the query parameters, and then update the URL. However, for any future implementations, we won't want to throw away other query parameters if there are any. This means that we want to get the current pathname, the current URL query parameters, and the router itself to push the new path. So, you need to import `useRouter`, `usePathname`, and `useSearchParams` from `'next/navigation'` and instantiate them, like so:

```
const searchInputRef = ...
const router = useRouter();
const pathname = usePathname();
const searchParams = useSearchParams();
```

> **Note**
>
> Before the App Router, the query parameters (`searchParams`) and the pathname of the URL were returned with `useRouter()`. In the newest architecture of Next.js, those have been separated, which is why you need to import three different functions for that. Also, please don't confuse yourself about the term "search parameters" – it's the same as query parameters, but search parameters is the more standard term.

Now, when the `TicketFilters` form is submitted, we want to update the URL and set the filter value as a query parameter (instead of `alert`). So, what we need to do in the `onSubmit` block is something like `router.push(pathname + '?' + updatedQueryParams)`. We can implement this by taking over the existing parameters, setting our new values, and then appending them as proposed. The returned object from `useSearchParams` uses the read-only, official `URLSearchParams` JavaScript standard. That's helpful because you can make a writeable copy by passing the existing one to a newly instantiated one. That means we can update our query parameters properly as follows:

```
const updatedParams = new URLSearchParams(searchParams);
updatedParams.set("search", search);
updatedParams.set("page", 1);
updatedParams.set("r", Math.random());
router.push(pathname + "?" + updatedParams.toString());
```

If you're wondering why I also set the page to the first page, imagine that you had previously navigated to page 8. Triggering a search is a whole new request and doesn't have anything in common with your previous paging, so we would start at the beginning with every new search request. The random parameter is again used for forcing a new request without a cache even when the input value didn't change (if I want to trigger the search over and over again with the same value, that's doable with that approach).

When you trigger a search, you'll see the value in the URL now. What's left is filtering the fetched tickets by that search input value. This is what we will do now. In the `TicketList` component, we have access to `searchParams`, which we currently use for reading the `page` value. We also want to read the `search` value. You already have one filter on your queries, which is an equality filter: `eq("tenant", tenant)`. If and only if the `searchParams.search` value is set and not empty, we want to filter by either title or description containing that value. That means that we need to avoid immediately executing (awaiting) the `supabase.from()` commands and instead prepare them.

So, instead of awaiting the result, you can create mutable (`let`) statements, which we will execute (`await`) after the `if` check. In the following code, I've not changed anything in the logic of our code other than being able to chain more functions in between:

```
let countStatement = supabase
    .from("tickets")
    .select("*", { count: "exact", head: true })
    .eq("tenant", tenant);

const startingPoint = (page - 1) * 6;
let ticketsStatement = supabase
    .from("tickets")
    .select()
    .eq("tenant", tenant);

const searchValue = searchParams.search?.trim();
if (searchValue) {
  // filter by searchValue as well
}

// continue chaining order and range
ticketsStatement = ticketsStatement
    .order("status", { ascending: true })
    .order("created_at", { ascending: false })
    .range(startingPoint, startingPoint + 5);

const { count } = await countStatement;
const { data: tickets } = await ticketsStatement;
```

Inside of the added `if (searchValue)` block, we now need to chain both requests with a text filter. If you only wanted to filter one column (such as `title`), that would be as easy as doing the following:

```
if (searchValue) {
  countStatement =
     countStatement.textSearch("title", searchValue);
  ticketsStatement =
     ticketsStatement.textSearch("title", searchValue);
}
```

That works for one column but if you added another (`.textSearch("title", searchValue).textSearch('description', searchValue)`), it would require both columns to contain the searched value and not just one column (chained filters are combined with AND, not OR).

> **Note**
>
> There is a solution to search multiple columns at the same time by querying not the columns themselves but a function that concatenates the columns together. If you want, you can try that as well: `https://github.com/supabase/postgrest-js/issues/289#issuecomment-1469967210`.

We now want to add an OR filter to our requests. This is done by chaining `.or(filterString)`, where `filterString` is a PostgREST-syntax filter string (`https://postgrest.org/en/v12/references/api/tables_views.html#operators`). The Supabase client has no convenient solution for what we want to do yet, such as `or({ title: { textSearch: searchValue}, description: { textSearch: searchValue }})`. So, instead, you have to construct the filter string on your own. It's not hard, but it's important to understand that the existing convenient filter functions such as `eq()`, `neq()` (not equal), `gte()` (greater or equal), and so on are all just simplified, convenient abstractions of actual PostgREST filter strings.

> **Recommendation**
>
> Always use a convenient wrapper function for filtering like `eq`, `gte`, etc. If none exist that solves your problem, only then should you create a custom filter string as we are doing now.

We want to have all results where the given search value is either inside of the title of a row or in the description. Let me first show you the solution and then I'll explain it:

```
if (searchValue) {
    const cleanSearchString = searchValue
      .replaceAll('"', "")
      .replaceAll("\\", "")
      .replaceAll("%", "");
```

```
const postgrestSearchValue = '"%' + cleanSearchString + '%"';
const postgrestFilterString =
  `title.ilike.${postgrestSearchValue}` +
  `, description.ilike.${postgrestSearchValue}`;

countStatement = countStatement.or(postgrestFilterString);
ticketsStatement = ticketsStatement.or(postgrestFilterString);
}
```

At first, I'm removing any potential double quotes, backslashes, and percentage signs from the user-provided input value. These symbols are barely used by users for actual searches and could be harmful as we're creating a filter string where these characters have technical meanings.

This clean search string is then explicitly surrounded as part of `postgrestSearchValue` with double quotes, which tells PostgREST where the search term starts and where it ends (that's why we removed them from the given input such that we're in control). Inside the quotes, we surround the string with the percentage signs, which is the wildcard in SQL for saying that there can be anything in front or after the searched string.

Then we use the `columnName.filterName.searchValue` filter syntax. In our case, that is `title.ilike."%WHATEVER_INPUT_VALUE_GIVEN%"`. However, what's `ilike`? Here, `ilike` is simply a case-insensitive search that allows to use wildcards (`%`).

Since we want to filter two columns, we will write that same filter once with the title column and once with the description column, and then connect them, separated by a comma, as one string. This way, the `or()` function will connect the comma-separated filters with an `OR` inside of SQL.

Save that and test your filter in the application. You should be able to search through the content and get proper results now – and thanks to the fact that we have added the filter to the count statement, it should show the **Next page** button if there are more results for your search.

There's one last thing to fix. If you click on **Next page** in a filtered result, you'll simply go to page 2 without the search because your current links don't pass on the search value. Let's change that by adding the search value to the query object of the links:

```
href={{
  query: {
    ...,
    search: searchParams.search
  },
}}
```

What a journey. You can not only create tickets and list them but also filter them easily. By implementing the paging and the filter, you've learned how you can select a range of results, combine multiple filters, and construct filters with the PostgREST syntax.

In the next section, you'll implement a delete button in the ticket details view.

Deleting tickets

If you can create a ticket, you should also be allowed to delete it. We'll now implement such a functionality in the details view and improve our RLS once more by doing so.

Go to `tickets/details/[id]/page.js`. Here, we first want to identify whether we want to show a **Delete ticket** button or not. For that, we want to check whether the service user ID of the ticket is the same as ours. That's why we need to fetch our own service user ID, which we do right below `if (error)`:

```
const supabase_user_id =
  (await supabase.auth.getUser()).data.user.id;
const { data: serviceUser } =
  await supabase
    .from("service_users")
    .select("id")
    .eq("supabase_user", supabase_user_id)
    .single();
```

We can safely assume, having previously fetched the ticket data and checked for an error, that this request will succeed. Hence, I won't do any further error checking here.

Next, we will define a simple `isAuthor` value if the ticket author and the service user ids match:

```
const isAuthor = serviceUser.id === ticket.created_by;
```

With that variable, we now conditionally render a **Delete ticket** button. For that, in the `<header>` part of the details page, I moved `id` and `status` into a `div` tag and moved this `div` tag into `className="grid"` together with the new button (with `Pico.css`, this aligns the elements at the top side by side):

```
<div className="grid">
  <div>
    <strong>#{params.id}</strong> -{" "}
    <strong className={classes.ticketStatusGreen}>
      {TICKET_STATUS[status]}
    </strong>
  </div>
</div>
```

```
  {isAuthor && (
    <button role="button" className="little-danger">
      Delete ticket
    </button>
  )}
</div>
```

You can see that the button also has a `little-danger` CSS class, which I added to `global.css` so it can be reused anywhere:

```
button.little-danger {
  font-size: 0.9rem;
  width: auto;
  justify-self: end;
  border-color: var(--del-color);
  background: transparent;
}
```

If you open a ticket you created with the currently signed-in user, you should now see the new button at the top right of the ticket box:

Figure 8.32: The Delete ticket button

For the ticket deletion, I don't want to make a server request, but I also don't want to make our existing page a client component because that would mean that all our lovely awaited requests would have to be done on the frontend too. Instead, let's create a new client component called `TicketDetails` (`details/[id]/TicketDetails.js`) to which we pass the data from the `page.js` file. Hence, we move the `return(..)` part from the page to the `TicketDetails` component and our `page.js` file will effectively return this:

```
return (
  <TicketDetails
    tenant={ticket.tenant}
    id={id}
    title={title}
    description={description}
```

```
      status={status}
      author_name={author_name}
      dateString={dateString}
      isAuthor={isAuthor}
    />
  );
```

In your `TicketDetails.js` file, make sure to mark the file as a client component with `"use client"`; at the top, import the required imports you previously had in `page.js`, and make sure you destructure the properties properly (check the repository for reference).

Within this new component, it's now super simple to trigger a deletion request. Instantiate a Supabase browser client at the top of the component, as well as the Next.js router (`useRouter`), then add the following `onClick` function to the delete button:

```
onClick={() => {
  supabase
    .from("tickets")
    .delete()
    .eq("id", id)
    .then(() => {
      router.push(urlPath("/tickets", tenant));
    });
}}
```

Clicking the delete button will redirect you to the ticket overview, but the ticket will not be deleted. A deletion in Supabase is silent; we have no way to confirm whether it was actually deleted besides trying to fetch the ticket again and expecting nothing in return (you can try this yourself but we won't here). The only reason why it didn't get deleted is because there are RLS policies for creating and reading tickets, but not for deleting. Let's add a policy that allows you to delete your own tickets.

Go to the Table Editor, move to the `tickets` table, click on **Auth policies**, and then click **New policy**. Enter `allow deletion of own tickets` as the name, select the DELETE operation, choose the **authenticated** role, and then create the policy expression. In the other policies for tickets, we gave permission to the tenant if the user has permission. This time, we simply want to check whether the signed-in user is the owner. That means that we want to check whether a service user row exists where the id value matches with the `created_by` value of the ticket, as well as where the service user matches the signed-in Supabase user:

```
EXISTS (
  SELECT FROM service_users
  WHERE
    service_users.id=tickets.created_by
    AND service_users.supabase_user = auth.uid()
)
```

Save that policy, move to your application, and open any ticket you have created. Now, clicking the deletion button will delete the ticket from the database. That was easy!

Summary

Congratulations on completing this chapter, where we ventured deep into the core of our application to implement the dynamic ticket management system.

You made your application user-centric by allowing interactions such as the creation of tickets as well as showing, listing, filtering, and deleting them.

Your effort didn't stop at merely manipulating data; you gained a deeper understanding of enums, tapped into the power of caching with database functions, and discovered the benefits of triggers. These skills are crucial for crafting an efficient and intuitive application.

With this newly added vitality, your application now stands ready to offer a functional and secure ticket management experience.

Looking ahead, in the next chapter, we will implement the user list and allow the ability to set assignees to tickets.

Creating a User List with RPCs and Setting Ticket Assignees

You've built a secure app with dynamic ticket functionality, but there's room to grow – the user list UI still relies on test data, and currently, there's no option of setting an assignee on a ticket.

This chapter aims to close these gaps by introducing a key feature of Supabase that will take your project to the next level: **remote procedure calls** (**RPCs**). We'll dive into creating a user list fueled by real data, exploring why RPCs are the optimal choice for this task, their versatility, and how to maintain their security. By doing so, you'll learn how the work we put into the user list streamlines the process of setting and changing ticket assignees.

This part of the journey will naturally lead you into the nuances and importance of UPDATE RLS policies, offering you a deeper insight into the layers of security that Supabase provides and how to leverage them effectively in real-world applications.

So, in this chapter, you'll tackle the following:

- Adding a user list with an RPC
- Allowing the setting and editing of an assignee to a ticket

Technical requirements

There are no additional technical requirements as opposed to the previous chapter. However, I want to point out that in SQL, you can use both lowercase and uppercase to write code. That means, for example, if you see USING or SELECT written in uppercase in one place but lowercase in another, there's no difference. In their newest templates, Supabase tends to use lowercase. I love using uppercase for differentiating built-in code from my custom code, but this is just my personal preference.

You'll find the related code of this chapter here: https://github.com/PacktPublishing/ Building-Production-Grade-Web-Applications-with-Supabase/tree/Ch09_ UserList_and_Assignees.

You can always run npm run reset to get a fresh Supabase setup matching the chapter's branch.

Adding a user list with an RPC

We have a user list page, which is supposed to show the users of the same tenant. Now, I'll show you how you can use a so-called RPC to achieve that aligned with your existing RLS permissions. In Supabase, an RPC is the process of calling a database function outside of the database. Similar to calling .select() with the Supabase client, calling .rpc() will allow you to run custom database functions and process their return. We will use this to return the list of users.

Ensuring there are enough users to test

To get started, you should make sure that you have at least two users per tenant to prove that everything is working as expected. Ultimately, if just your name shows up in the list, or you only create users in one specific tenant, then you cannot prove that you're seeing what you're supposed to see. So, make sure you have at least four accounts across two different tenants (so, two per tenant).

Once you have your accounts, let's adapt the table to reflect the structure of the UI.

Enhancing the table structure

Let's take a look at the user list UI of our application so far:

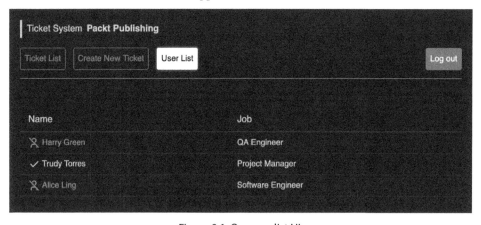

Figure 9.1: Our user list UI

As you can see, the UI shows each user's job title and availability (if they are unavailable, this could be due to sickness or vacation). We don't have these columns in our `service_users` table yet, so edit the table with the Table Editor and add two additional columns:

- `is_available`: This is of the `bool` type, with the default value set to `true`, and the **Nullable** box unchecked
- `job_title`: This is of the `text` type, with no default value and the **Nullable** box checked

Once you've added the two columns, just change the job titles of the existing users in the Table Editor's `service_users` rows directly to whatever you want, and set half of the users' `is_available` columns to FALSE so we have some alternating data to display in the UI. In my case, after adding the columns and values, the table looks like this:

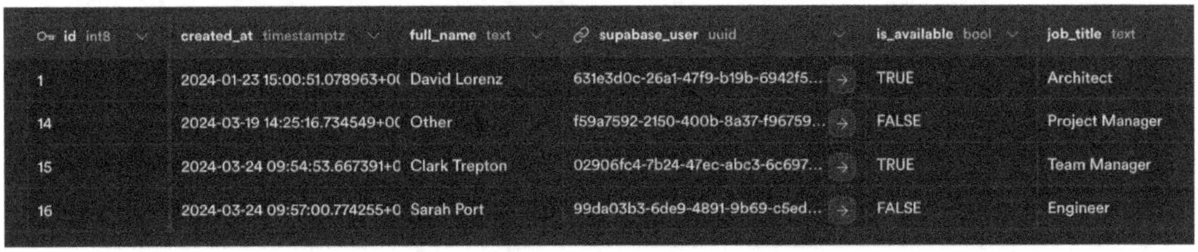

O▾ id int8	created_at timestamptz	full_name text	🔗 supabase_user uuid	is_available bool	job_title text
1	2024-01-23 15:00:51.078963+0(	David Lorenz	631e3d0c-26a1-47f9-b19b-6942f5... →	TRUE	Architect
14	2024-03-19 14:25:16.734549+0(	Other	f59a7592-2150-400b-8a37-f96759... →	FALSE	Project Manager
15	2024-03-24 09:54:53.667391+0	Clark Trepton	02906fc4-7b24-47ec-abc3-6c697... →	TRUE	Team Manager
16	2024-03-24 09:57:00.774255+0	Sarah Port	99da03b3-6de9-4891-9b69-c5ed... →	FALSE	Engineer

Figure 9.2: Service users with their availability and job title

We won't implement editing the job title or the availability in the UI, but feel free to implement that on your own after this chapter.

Now, it's all about how to get that list of users into the UI, even though, by the RLS policies, every user only has access to its own service user row. Let's learn how to achieve that next.

Fetching the users with an RPC

I will now show you how to fetch a list of users with an RPC.

Our user list at `tickets/users/page.js` is a server component, which means we could make use of the admin client to retrieve a list of the users of the same tenant. That's one option but I'd like to remind you that `middleware.js` doesn't shield users from reaching this page if there's a valid structured session (in *Chapter 4*, I explained that session data can be manipulated). Hence, security-wise, you still need to first confirm that the current user has rightful access to the current tenant before loading sensitive data with the admin client.

So, in theory, you could instantiate a normal RLS client, select the tenant from the database, and, if the returned tenant data isn't empty, you would be assured the current user has rightful access. Then, you can use the admin client to select the other users to display them.

But I want to implement a cleaner option without the need to call two clients or even the admin client at all. We will do so by creating an RPC function. You already created a database function for your triggers in the last chapter. When you created those functions, you set the return type to `trigger`. That made those functions only accessible to be used by triggers.

However, if you create a function in the `public` schema that returns anything else but a `trigger` type, it can be called explicitly from the Supabase client.

Inside the function that we are about to create, we will return the result of a SELECT statement, which gives us all service users for a specific tenant. Before I show you the definition of the function, let's pin down the SQL expression that will provide us with what we need for our UI.

Writing a SQL expression for getting all users of the same tenant

We want to return a list of the values of the `service_users` table including id, `full_name`, `job_title`, and `is_available` for a specific tenant. For this, we will now create a SQL expression to be used in our function later.

Given a specific `tenant_id`, we can select all users from the `service_users` table where the user also has a defined permission on the given tenant. It would look like this:

```
SELECT s.id, s.full_name, s.job_title, s.is_available FROM service_
users s
    WHERE  EXISTS (
      SELECT FROM
        tenant_permissions
        WHERE tenant = tenant_id AND service_user = s.id
    );
```

This would work to get all service user records on the given tenant. By using FROM `service_users` s, we're aliasing the table with the s character for even simpler access, so we can write s.id instead of `service_users.id`.

However, although the given SQL will work just fine, if you're very familiar with SQL, you'll know that there are always a lot of different options to achieve the same set of results. I prefer joining multiple tables instead of writing a sub-select in the WHERE clause. See the following alternative SQL statement with the same result:

```
SELECT s.id, s.full_name, s.job_title, s.is_available
    FROM service_users s, tenant_permissions p
    WHERE s.id = p.service_user AND p.tenant = tenant_id;
```

If you're not well aware of the different SQL joining methods, let me explain quickly. If you didn't have a `WHERE` clause here, it would be like having one big table where each row of the first table is combined once with each row of the second table. So, if you had three service user rows and three permission rows, you'd have nine rows. In other words, selecting from two tables at the same time simply results in all possible combinations between the rows of the first and second tables.

With the `WHERE` clause condition, `s.id = p.service_user`, we're saying that we only want to return rows where the service user ID (`s.id`) is equal to the permissions service user (`p.service_user`), hence only rows where for each user the user matches on both tables. With `p.tenant = tenant_id`, we additionally limit the returned rows to return only those where the permission is for the tenant we want (`p.tenant = tenant_id`).

If the second SQL expression confuses you, I recommend sticking with the first one (using `EXISTS`) for now. But whatever you choose is fine; they are both internally optimized by the database and have no different performance implications.

Having cleared that up, let's move on to defining the database function.

Defining the database function

To define the database function, in Supabase Studio, go to **Database | Functions** and click **Create new function**. Name it `get_tenant_userlist`, keep the schema public, and choose `json` as the type, as we want to return a JSON that is an array of users of the tenant. The other types are atomic types and should only be used if you write a function that expects a simple return – such as counting users and returning `int8` or checking whether a user exists and returning `bool`.

Now, very importantly, we have to add a parameter to that function by clicking on **Add a new argument**. This parameter will let us know the tenant needed to fetch the user list.

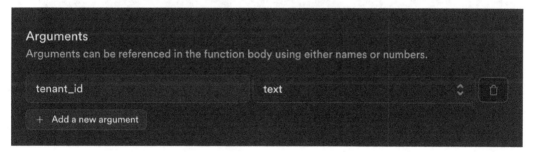

Figure 9.3: Adding arguments in the function creation mask

Simply name the argument `tenant_id` and choose the `text` type (as our tenant ID is a text). In the function **Definition** section, this `tenant_id` argument will be available to use as a variable.

Now, we want to use one of the previously defined SQL expressions to get the rows, but the problem is, no matter which one you previously chose, they both return database rows, not a JSON array. In Postgres, there is a `jsonb_agg(row)` function, which, when used as part of a SELECT statement, will aggregate all rows into a JSON array.

In the given SQL expressions, you could write SELECT `jsonb_agg(s)`, which would work but would include all values from service users – not just specific columns. However, as every SELECT can be used as a sub-select, we can do the following:

```
SELECT jsonb_agg(sub) FROM (
    SELECT s.id, s.full_name, s.job_title, s.is_available
    FROM service_users s, tenant_permissions p
    WHERE s.id = p.service_user AND p.tenant = tenant_id) sub;
```

This will transform the rows from the given SELECT to a proper JSON array with JSON objects. You can test this expression by running it in your SQL Editor, replacing `p.tenant = tenant_id` with whatever tenant you want to test. When I run it with `p.tenant = 'packt'`, I see the following:

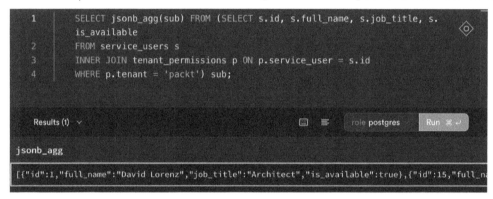

Figure 9.4: Creating a JSON array from a selection

The SQL expression is now ready to be used in our function definition. Add the following into the function definition field to let the function return the result of the given expression (but don't save it just yet; we're not done):

```
BEGIN
  RETURN (
    SELECT jsonb_agg(sub) FROM (
    SELECT s.id, s.full_name, s.job_title, s.is_available
    FROM service_users s, tenant_permissions p
    WHERE s.id = p.service_user AND p.tenant = tenant_id) sub
  );
END;
```

This would make the function return a JSON array of service users matching the tenant that was provided as the function's argument. It looks like the function works as we want it to, but as this function is bound to RLS, it would only return the signed-in user itself inside of an array, as the signed-in user does not have access to the other users' rows.

The first step to getting foreign user rows is to make the function run with admin rights. To achieve that, click on **Show advanced settings** (*Figure 9.5*). This will open additional settings in the mask where you'll find the **Type of security** section. The security type is preselected to be **SECURITY INVOKER**, which means that it respects the rights of the user that is invoking it, hence RLS-bound. But to make this function run unrestricted, you need to select **SECURITY DEFINER**.

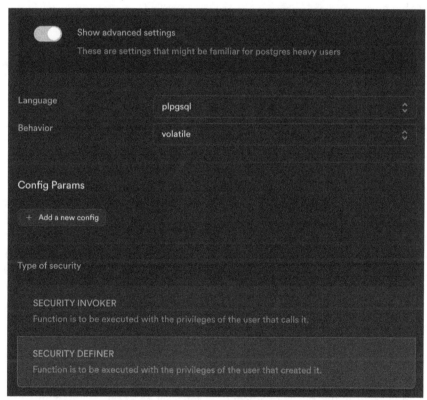

Figure 9.5: Advanced settings of database function creation

The problem now is that this function doesn't respect your defined RLS anymore; anyone can execute this function with the exposed Anonymous Key and get the list of users for each tenant. So, what do we do?

Well, what is always bound to the current user, no matter whether the function runs with admin rights or not, is the `auth.uid()` function. Knowing this, we can add our own security check inside our function definition and simply throw a database error to avoid further executions if permissions aren't given:

```
DECLARE
has_access bool;

BEGIN
   has_access = ( EXISTS (
     SELECT FROM service_users s, tenant_permissions p
     WHERE
      s.supabase_user=auth.uid()
      AND s.id=p.service_user
      AND p.tenant = tenant_id
   ));

   IF (has_access != true) THEN
     RAISE EXCEPTION 'no access to the data';
   END IF;

   RETURN (...);
END;
```

In the given code, you can see that before the `BEGIN` keyword, we declare a local Postgres variable. Inside the function, we have an `EXISTS (SELECT (..))` statement where we check whether the current user has a matching permissions row defined in the database. We then assign the result of that `EXISTS(..)` query to `has_access`. Finally, we check whether `has_access` is not `true` and, if so, throw an exception, which will cause Postgres to abort further execution. If the `if` block does not trigger, it will just continue fetching the user list with admin rights. This way, you ensure that only users with access to the tenant will get the user list of that tenant.

> **Note**
> You can add as many local variables to functions as you want between `DECLARE` and `BEGIN`.

After that, here's the full function definition, which you can finally save:

```plpgsql
1    DECLARE
2    has_access bool;
3
4    BEGIN
5      has_access = (EXISTS(SELECT FROM service_users s,
       tenant_permissions p WHERE s.supabase_user=auth.uid() AND s.id=p.
       service_user AND p.tenant = tenant_id));
6
7      IF (has_access != true) THEN
8        RAISE EXCEPTION 'no access to the data';
9      END IF;
10
11     RETURN (
12       SELECT jsonb_agg(sub) FROM (
13       SELECT s.id, s.full_name, s.job_title, s.is_available
14       FROM service_users s, tenant_permissions p
15       WHERE s.id = p.service_user AND p.tenant = tenant_id) sub
16     );
17   END;
18
```

Figure 9.6: Function definition in the editor

At this point, your `get_tenant_userlist(tenant_id)` function is now ready to be called. But before I show you how to use it as an RPC in your application, I'd like you to test it in your Supabase Studio SQL Editor first. You can run it, using the `packt` tenant, for example, by writing `SELECT public.get_tenant_userlist('packt');`. You'll notice that it will throw an error:

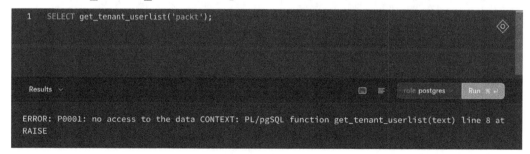

```
1    SELECT get_tenant_userlist('packt');




Results ⌄                                            🖺  ☰   role postgres    Run ⌘ ↵

ERROR: P0001: no access to the data CONTEXT: PL/pgSQL function get_tenant_userlist(text) line 8 at
RAISE
```

Figure 9.7: Getting an error running the new function with admin rights

Do you know why we're getting this error, even though it's running with the `postgres` role with admin rights? Shouldn't we be able to execute it? In fact, no, we shouldn't be able to execute it just because we're admin.

To answer the questions, everything is exactly working as expected as the thrown error is RAISE EXCEPTION, which we've added to our get_tenant_userlist function. This is triggered because it checks for the existence of a permission row with auth.uid(), and auth.uid() is NULL for our postgres user so it cannot find a matching row.

But you already know that you can impersonate other users using the **role** dropdown. So, if you select a user that has access to that tenant (other@packt.local, in my case), you will get a successful result:

Figure 9.8: Getting data with the impersonated user

Now, let's use this function in our application.

Using the function with an RPC

To run the created function and process the results, open the user list page (tickets/users/page.js) and delete the mock array first (const users = ...) as we'll be using real data now. Then, make the UserList function async so we can await data from Supabase and add the following code:

```
export default async function UserList({ params }) {
  const supabase = getSupabaseCookiesUtilClient();
  const { data: users, error } =
   await supabase.rpc("get_tenant_userlist", {
     tenant_id: params.tenant
   });

  return ...
```

With rpc(function_name, function_arguments), you can execute any function that you defined in the public schema and get its returned value within data.

As we get `id`, `full_name`, `is_available`, and `job_title` per each user object from the database, you need to adapt your render template with the following highlighted code:

```
{users.map((user) => (
  <tr key={user.id}>
    <td style={{ color: !user.is_available ? "red" : undefined }}>
      {user.is_available ? <IconCheck /> : <IconUserOff />}
      {user.full_name}
    </td>
    <td>{user.job_title}</td>
  </tr>
))}
```

With that done, we can now see the user list, like so:

Figure 9.9: The user list with data from RPC

Here, you just learned how you can use admin rights to fetch more data than what is allowed per RLS, while still keeping the boundaries of tenant access per user. With this wonderful bag of knowledge, let's move on to another RPC quest by allowing you to set a ticket assignee at ticket creation.

> **Important security notes**
>
> If you need to define functions that aren't supposed to be called from the outside via `.rpc()`, then you should create them in another custom schema (e.g., `my_protected_schema`), as only the public schema is exposed by default to have anonymously callable functions with the Supabase client.

Allowing the setting and editing of an assignee to a ticket

For any ticket, we want to allow the setting of an assignee. The `assignee` column is very much identical to the `created_by` column, with the only difference being that it shall also cache the name if the ticket was ever edited and updated to another assignee. Let's first adapt the database table accordingly.

Adding assignee columns in the tickets table

To add more columns to the tickets table, click on **Edit Table**:

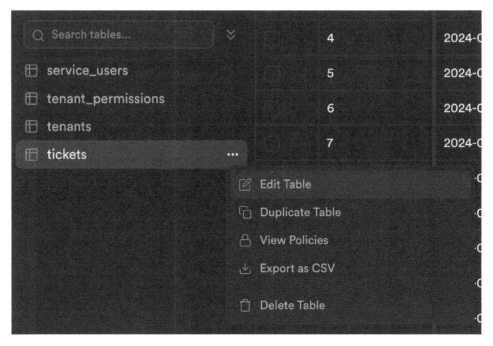

Figure 9.10: Edit tickets table

Then, add two new columns:

- assignee: Like the created_by column, this must have a relation to the service_users table

- assignee_name: This is similar to author_name, being of the text type

We don't enforce assignees so both can be nullable.

Figure 9.11: The assignee columns with relation added

Once done, save the changes, and let's make sure the name is cached.

Creating the trigger function to cache the name

Similar to `set_ticket_author_name`, you need a function that caches the name of the assignee. Go to **Database | Functions**, click **Create new function**, name it `set_ticket_assignee_name`, and choose `trigger` as the return value.

In the function **Definition** section, we need to consider two things that are different from setting the author's name:

- The assignee could be null and hence we cannot derive the name
- If the assignee is not the user itself, it won't have access to the other user's name

So, the definition of the function will look like this:

```
BEGIN
  IF (NEW.assignee IS NULL) THEN
    NEW.assignee_name = NULL;
  ELSE
    NEW.assignee_name = (
      SELECT full_name FROM service_users WHERE id = NEW.assignee
    );
  END IF;

  RETURN NEW;
END;
```

This code should be intuitive enough to understand, but with **SECURITY INVOKER**, this function wouldn't get access to other users' names. To solve it, you could either make use of the previously created function (`get_tenant_userlist`) and, from its results, filter out the correct user. Or, you simply decide to run this function with admin rights by choosing **SECURITY DEFINER**.

But if we leave it as is, then it would also be able to retrieve a name for a user ID that isn't assigned to the tenant of the ticket. Let's fix that:

```
BEGIN
  IF (NEW.assignee IS NULL) THEN
    NEW.assignee_name = NULL;
  ELSE
    NEW.assignee_name = (
      SELECT full_name FROM service_users
        WHERE id = NEW.assignee
        AND EXISTS (
         SELECT FROM tenant_permissions p WHERE
         p.tenant = NEW.tenant AND p.service_user=NEW.assignee
        )
    );

    IF (NEW.assignee_name IS NULL) THEN
     NEW.assignee = NULL;
    END IF;
  END IF;

  RETURN NEW;
END;
```

This function even sets the given assignee ID to NULL if there is no name found.

Next, let's make sure that the trigger function is executed. Go to **Database** | **Triggers** and click **Create new trigger**. For the name, enter `tr_tickets_autoset_assignee_name`, then select the appropriate table (`tickets`). As opposed to setting the author's name in a ticket, the assignee could change over time, so let's select two trigger events at which to call the function, **Insert** and **Update**:

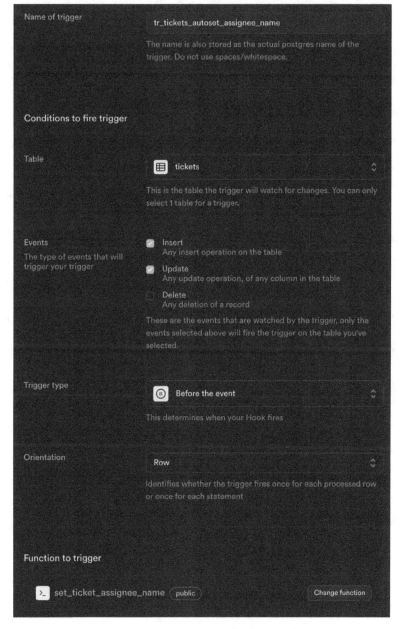

Figure 9.12: Trigger options for the assignee

Set **Trigger type** to **Before the event** and choose **Row** for **Orientation**. Finally, choose the set_ ticket_assignee_name function to be triggered and save the trigger.

Now, let's move on to adding an assignee at ticket creation.

Adding an assignee at ticket creation

Thanks to our previously created get_tenant_userlist function, we can use the same RPC to load the users of the tenant and put them into a <select> element of our ticket creation form. Let's create a reusable assignee component that will call get_tenant_userlist on its own.

Creating the assignee select component

We've always co-located components with their related pages but the select assignee shall be a component that could be used for different use cases across the application; we will at least use it in the ticket creation page as well as the Ticket Details page. Create a new folder called src/components and within that, create a file called AssigneeSelect.js with the following content to be our AssigneeSelect component:

```
"use client";
import { getSupabaseBrowserClient } from "@/supabase-utils/
browserClient";
import { useEffect, useState } from "react";

export function AssigneeSelect({ tenant, onValueChanged }) {
  const [users, setUsers] = useState(null);
  const supabase = getSupabaseBrowserClient();

  useEffect(() => {
    supabase
      .rpc("get_tenant_userlist", {
        tenant_id: tenant,
      })
      .then(({ data }) => {
        setUsers(data ?? []);
      });
  }, []);

  return (
    <select
      name="assignee"
      disabled={users === null}
      onChange={(e) => {
        const v = e.target.value;
        onValueChanged(v === '' ? null : v);
```

```
      }}
    >
      <option value="">
        {users === null ? "Loading..." : "No assignee"}
      </option>
      {users &&
        users.map((user) => {
          return (
            <option key={user.id} value={user.id}>
              {user.full_name}
            </option>
          );
        })}
    </select>
  );
}
```

As you can see, the component triggers the RPC function client-side when the component is mounted, and then stores the returned users in the state. setUsers(data ?? []) ensures that even if the returned value is falsy, it will set at least an empty array so we know it has finished loading (as it's null initially). Then, in the rendering part, as long as it's loading, there's just a disabled select showing with the **Loading...** text.

Once loaded, the first option item in select will be an empty value displaying **No assignee**. For the users we fetched, we create select options displaying the name of the user and the value being the user ID. When the selection is changed by the user, we pass the newly selected value to an onValueChange(newValue) callback function. So, wherever we use this component, we only need to pass a callback function.

Let's use this component in our ticket creation form.

Adding the assignee selection to the ticket creation

Now, inside new/page.js, we can import this new component and use it as follows:

```
export default function CreateTicket({params}) {
  ...
  const [assignee, setAssignee] = useState(null);
  ...

  return (
    ...
    <input .../>
    <AssigneeSelect
    tenant={params.tenant}
```

```
        onValueChanged={(v) => setAssignee(v)} />
      <textarea .../>
    ...
  )
}
```

Once added, on my `packt` tenant, I can see the following list of options:

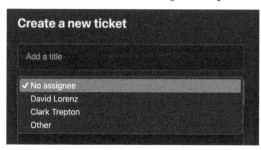

Figure 9.13: The users of the tenant in the assignee select

The only thing left to do is to make sure the chosen assignee gets inserted along with the other values when the form is submitted:

```
.insert({
  title,
  description,
  tenant,
  assignee,
})
```

That's all! But even though this is expected to work just fine, you won't be able to see the assignee in the UI of our application after creating a ticket, as we need to add it to the ticket details view as well. Let's do that.

Showing the assignee in the details

We will not just display the assignee as text in the ticket details; we will display it using the same kind of `select` component to allow changing the assignee to someone else. Doing that, you'll not just learn how to edit values but you'll also be able to confirm that our trigger is working just fine.

So, go to the `TicketDetails.js` file and add the possibility of passing the assignee ID to it (we don't need to pass the assignee name, and you'll soon see why):

```
export function TicketDetails({
  tenant,
  id,
```

```
    status,
    title,
    description,
    author_name,
    dateString,
    isAuthor,
    assignee
}) { ...
```

Then, on the `tickets/details/[id]/page.js` file, make sure to pass the assignee value to the `TicketDetails` component.

Back in your `TicketDetails` component, import the `AssigneeSelect` component and pass the required properties to it. Put it wherever you want – I'll put it in one `div` together with the **Delete ticket** button:

```
<div className="grid">
    ...
    <div
      style={{
        display: "flex",
        flexDirection: "column",
        alignItems: "end",
      }}
    >
      <AssigneeSelect
        tenant={tenant}
        onValueChanged={(v) => {
          // todo
        }}
        initialValue={assignee}
      />
      {isAuthor && <button... />}
    </div>
</div>
```

As you can see, I'm also passing the assignee value as an `initialValue` property, which will help us set the current value, if provided.

In the `AssigneeSelect` component, make sure to destructure the `initialValue` property:

```
export function AssigneeSelect({
    tenant,
    onValueChanged,
    initialValue
```

Then, in the `AssigneeSelect` component, you can set the most recent value as an attribute (or fall back to the default being an empty string):

```
<select value={initialValue === null ? "" : initialValue} ...
```

Now, if you create a new ticket with an assignee, the current assignee will be shown in the ticket details:

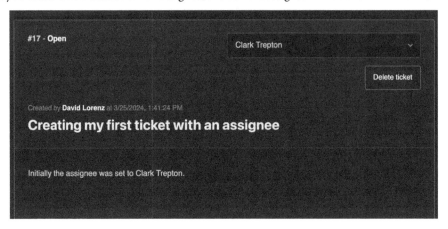

Figure 9.14: The assignee in details view

As a last step, we want to allow for dynamically changing the assignee.

Updating the assignee

Inside `TicketDetails.js`, we want to immediately update the assignee after the selection has changed. This is very easy using `.update(..)` and the existing `onValueChanged` callback:

```
<AssigneeSelect
  tenant={tenant}
  onValueChanged={(v) => {
    supabase
      .from("tickets")
      .update({
        assignee: v,
      })
      .eq("id", id)
      .then(() => router.refresh());
  }}
  initialValue={assignee}
/>
```

This will indeed update the value, but only if you have the proper RLS policies set up, which you don't. You have policies for SELECT, INSERT, and DELETE but none for UPDATE. If you try to change the value, you'll notice it will fall back to the original value again as it will refresh the page after trying to update the value. So, let's add the UPDATE policy now.

You've created enough policies by now to know where to create an additional policy for the tickets table. But the UPDATE policy is a little bit special and requires deeper understanding; so, let's have a closer look.

Go to the **Auth Policies** area of the tickets and open the policy creator as usual. Choose **UPDATE** and you'll see that you can activate the **Use check expression** checkbox. Click it and you'll see two fields that you can edit, one starting with using and one starting with with check:

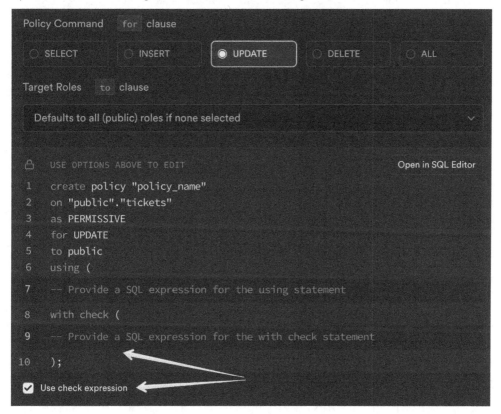

Figure 9.15: The UPDATE RLS policy mask

Now, what you might not have noticed when creating the other RLS policies is the fact that for SELECT, it said USING, and for INSERT, it said WITH CHECK, and for DELETE it also said USING.

An update is kind of a mix of both, which is why there are both fields. To understand it, you must burn this into your memory: *the RLS policies we write act as constraints to which rows can be "touched" and which cannot.* So, for SELECT (USING), it means that an RLS policy defines which rows one can read. For INSERT (WITH CHECK), it means that an RLS policy is a write constraint where the policy logically constrains the data you're allowed to insert. To perform UPDATE, both must be fulfilled – you must be able to select the row you want to update to then be able to perform writes on it.

So, let's just put true in the WITH CHECK expression for the UPDATE, no? That would mean that a user could only update tickets from a tenant to where the user has access, but the user could set new values that don't meet this same requirement (for example, the user could maliciously set a foreign tenant ID). In turn, that would mean that the USING expression is evaluated against the existing values in the table, and the WITH CHECK expression is evaluated against the new values to prevent certain values from being set.

But what do we want and need to set for the UPDATE policy? In both, we will set the same expression that we have defined for SELECT. That means a user can update any ticket where the user has tenant permissions. Let's do that now. Give your policy a proper name, and set both fields to the following:

```
COALESCE(
  (auth.jwt() -> 'app_metadata' -> 'tenants') ? tickets.tenant,
  false
)
```

Once saved, you can confirm that updating the assignee on a ticket details view will now work as it will show the changed value.

To close, let me give you a little bit more information about the implications of UPDATE policies.

Understanding the limitations of UPDATE policies

Most often, you'll not find a difference in the UPDATE policy between the USING and WITH CHECK expressions, instead providing the same expression to both. But in our case, that means that a user can theoretically update not just all of the tickets where tenant access is given, but also all of the columns of such a ticket, even the author or created_by value.

This is extremely important to understand. Usually, valid users of a tenant won't have malicious intents, but still, it's important to know the limitations of RLS – it is *row*-level security, not *column*-level security.

If you need to constrain updates to certain columns, there are ways of doing that, as shown in *Chapter 12*. There, you'll learn that such capabilities don't replace RLS but complement them.

With all that, you've learned about setting and updating values with Supabase and using a trigger that works on multiple events.

Summary

Congratulations on completing this chapter! You've implemented a user list and ticket assignees, both using RPCs. You not only learned a new approach to fetching data but also ensured that our operations remained secure within a multi-tenant context. With the implementation of assignees, you also gained an understanding of the delicate distinctions between assignees and authors in tickets even though they reference the same table and seemed to be of the same requirements initially.

Your exploration into UPDATE RLS policies unveiled important insights regarding its security implications within our multi-tenant system. These advancements mark a significant milestone in our journey, setting the stage for further enhancements in real-time data interactions in the upcoming chapter.

10

Enhancing Interactivity with Realtime Comments

In the last chapter, you learned how to conquer **remote procedure calls** (**RPCs**) and edit row data, and gave your application a first touch of post-ticket-creation interactivity by allowing the user the ability to change assignees.

But what if your application could become more dynamic, interactive, and in tune with the real-time web? Imagine a scenario where updates happen live, without the need to hit the refresh button. This isn't just a convenience – it's becoming an expectation in modern web applications.

Welcome to *Chapter 10*, where we leap into the integration of the Realtime service of Supabase. Along this path, you'll be designing a comments table that automates more values than any other table we've created so far, and learn yet another method to optimize RLS by using custom functions.

After having implemented the creation and view of comment data, you'll also learn how to complement the existing data with delightful real-time functionality to receive instant UI updates reflecting any changes in the comments.

By the end of this chapter, you'll know the extent of the Realtime features and their potential pitfalls.

So, in this chapter, we will cover the following topics:

- Creating the comments table
- Adding a trigger to set the tenant automatically
- Adding and optimizing RLS policies
- Implementing comment creation
- Listing existing comments from the server
- Implementing Realtime comments
- Embracing additional Realtime insights and learning about common pitfalls

Technical requirements

There are no additional technical requirements as opposed to the previous chapter. You'll find the related code of this chapter here: `https://github.com/PacktPublishing/Building-Production-Grade-Web-Applications-with-Supabase/tree/Ch10_Realtime`.

However, there's one important thing to note: if you check out this branch, or any upcoming branch, and run `npx supabase start` and `npx supabase db reset` to get the exact same setup as in the branch, you'll notice that Realtime is always disabled (you'll see how to activate that in the *Implementing Realtime comments* section). That's because, at the time of writing this book, the Supabase CLI doesn't support dumping the on/off state of Realtime in tables – yet. Hence, if Realtime isn't working, you most likely have to activate it.

Creating the comments table

First and foremost, we need a table to store our comments data. We want to keep the data structure of a comment as simple as possible – in our case, the comment is a piece of written text with a specific creation date and an author.

With the primary key, every comment will get a unique ID. Also, a comment will point to the ticket it belongs to (a relation). Adding a tenant-comment relation to the `comments` table isn't a necessity as the related ticket (which has a tenant relation) would imply who the tenant is. But, to be able to use simpler RLS expressions, we'll also add the tenant reference to each comment.

So, go to the Table Editor and create a `comments` table. It will consist of the following columns: `id`, `comment_text`, `created_by` (the author), `created_at`, `ticket` (the reference), `tenant`, and `author_lname` (to cache the author). I will spare you from the explanation of how to create the table, but here's a screenshot of the result, with all columns set to **Not Nullable**:

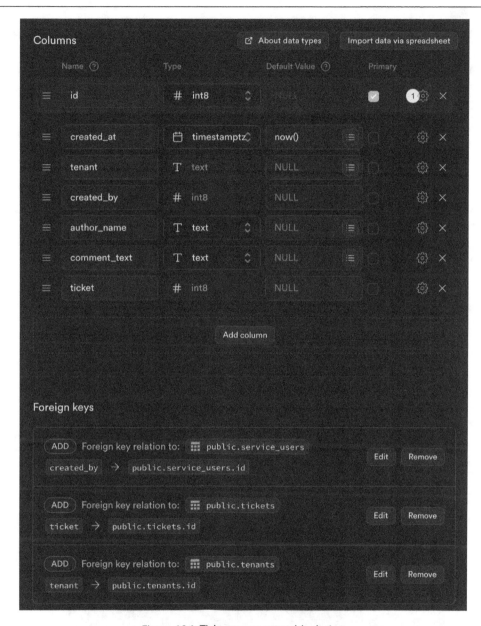

Figure 10.1: Ticket comments table design

Awesome. Now, let's ensure that there is a trigger that will set the author automatically and cache the author_name value. The best thing is, since we've already used the exact same column names as in the tickets table, we can reuse the existing trigger functions for this table, too.

Hence, let's not create a new trigger function but just a new trigger by going to **Database | Triggers**, and clicking **Create new trigger**. Name the trigger `tr_comments_autoset_author_name`, select the `comments` table, and set the **Events** type as **Insert**. Then, set the **Trigger** type as **Before the event** and **Orientation** as **Row**. Finally, select the `set_ticket_author_name` function to be triggered and save it.

Figure 10.2: Author name trigger for comments table

The only thing that can be slightly bothersome is the fact that we now use a function called `set_ticket_author_name` for the comments. So, let's change this function name now to the generic name, `autoset_author_name`. To do this, go to **Database | Functions**, edit the function, and change the name. As the database keeps an internal reference to this function, everywhere where the function is used, the new name is now automatically updated.

Next, we also need to ensure that not just the cached name is set in the comment but the actual service user ID as well, so make sure that you create another trigger called `tr_comments_autoset_created_by` with the same settings, but setting the function to trigger as `set_created_by_value`.

Once you've added both triggers, let's continue by ensuring the `tenant` column value is automatically derived from the related ticket.

Adding a trigger to set the tenant automatically

It makes sense to keep the tenant ID with the comment to be able to use simpler RLS policies, but it doesn't make sense to allow the user to set or change it – the tenant ID should always be derived from the related ticket.

For that, let's create a new trigger function named `derive_tenant_from_ticket`. We don't need any special rights for this function – whoever has the rights to add or update a specific comment row will also have access to its related ticket – so simply put the following code in the function definition:

```
BEGIN
  NEW.tenant = (
    SELECT t.tenant FROM tickets t WHERE t.id = NEW.ticket
  );
  RETURN NEW;
END;
```

This code, when executed, sets the `tenant` value on the newly added row, derived from the given ticket ID (`NEW.ticket`).

You can see the complete setup of the `derive_tenant_from_ticket` trigger function here:

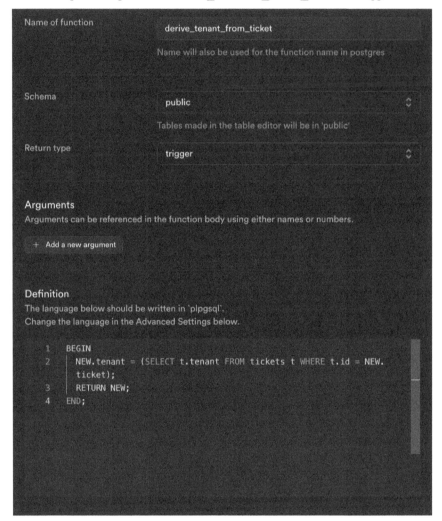

Figure 10.3: Creating the tenant ID trigger function

Once you have saved the trigger function, we need a trigger to trigger the function. So, go to **Database |
Triggers**, click **Create new trigger**, select the `comments` table, and choose the same settings as we did
for `tr_comments_autoset_author_name`, but named `tr_comments_set_tenant_id`.
Here's screenshot of the trigger creation:

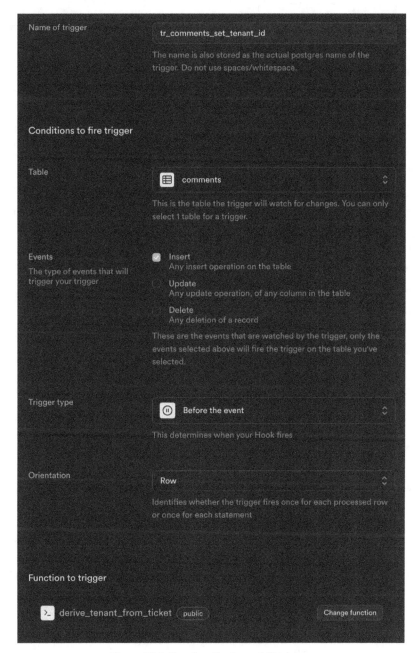

Figure 10.4: Creating the tenant ID trigger

Now that we have the comment's tenant ID as well as the author name being stored automatically, let's create RLS policies that allow us to fetch, insert, and delete comments.

Adding and optimizing RLS policies

As I said, we won't be implementing editing comments in this book, but I'll still create a policy for it so I can test updates with active policies directly in the database (with the impersonation feature you already know from *Chapter 6*). UI-wise, I want to be able to fetch and insert comments within a tenant to which I have access. Policy-wise, I want the updating and deletion of comments to only be allowed for its creators (`created_by`).

If we were to implement the `DELETE` or `UPDATE` policy for the author of a comment, it would look like this (however, don't add this yet):

```
EXISTS (SELECT FROM service_users s WHERE s.id = created_by AND
s.supabase_user = auth.uid())
```

You've already seen a similar statement for the deletion policy of the ticket.

For inserting and fetching comments, again, we would have to use the well-known expression that we've used in other policies (don't add this one yet either):

```
COALESCE(
  (auth.jwt() -> 'app_metadata' -> 'tenants') ? tickets.tenant,
  false
)
```

But it is starting to become clear that we're essentially reusing the same policies over and over again:

- A policy for checking whether the signed-in user is the author – or more specifically, checking whether the signed-in user matches with a service user ID of a column
- A policy for checking whether the user has access to the tenant that is defined in the row

Copy and pasting the policies is starting to become annoying, so let's change that and make smaller reusable policies by using database functions.

Creating RLS helper functions

RLS expressions can execute functions – that much is obvious, as we've used functions from the `auth` schema such as `auth.uid()` or `auth.jwt()`. That means that you can also create your own function to abstract away some logic, which we'll do now.

We want two functions:

- A function named `is_same_user(given_service_user_id)` to check whether the signed-in user matches the one provided as a parameter; we use that to to check whether the user is the creator of a row
- A function named `has_tenant_access(tenant_id)` to check whether the signed-in user has access to the provided tenant, which will shorten most of our expressions

So, let's create these RLS helper functions. It would be completely okay to add these functions in the public schema as they will both be RLS-bound and not run with admin rights, but it would also mean that they could be called as an RPC. There's no issue in doing so, but also, there's no value in doing so either.

As I don't want them to be exposed to be used as an RPC, we will create those helper functions in a custom schema. Let's create a schema called `rls_helpers`. To create that schema, go to the Table Editor, open the **schema** dropdown, and click **Create a new schema**.

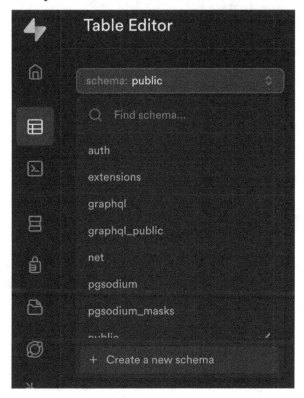

Figure 10.5: Creating a new schema button

A mask opens with the only field being the schema name, so enter `rls_helpers` and save it. That's all that we need to be able to create functions inside of it.

Now, navigate to **Database | Functions**, click **Create new function**, and make sure to select the `rls_helpers` schema in the mask that appears. Fill in `is_same_user` as the function name, mark **Return type** as **bool**, and add a new `service_user_id` argument of the `int8` type.

For the function definition, we want to check whether the provided service user ID matches the one from the currently signed-in user. So, enter this in the **Definition** section:

```
BEGIN
  RETURN (
    EXISTS (
      SELECT FROM service_users
      WHERE
        id=service_user_id AND
        supabase_user = auth.uid()
    )
  );
END;
```

In the function code, we search the existence of the user with the provided service user ID and add a condition to match with the signed-in user on the `supabase_user` field. Save that function but keep the default **SECURITY INVOKER** – there's no need to run this with admin rights as a user will anyway only be able to access its own data. The first task is now solved.

> **Note**
>
> In the last chapter, we will discuss a few performance tricks with regard to improving RLS performance. Since the `rls_helpers.is_same_user` function checks the signed-in user with `auth.uid()`, you could also run it with **SECURITY DEFINER** rights and potentially get some performance gains if your table has a lot of data. However, at this stage, I don't want to discuss the benefits and downsides of that. You'll read more about it in *Chapter 13*.

Let's now move on to creating the `has_tenant_access` function in the `rls_helpers` schema. You already know the flow – open the mask, create a function with the name `has_tenant_access` that returns a Boolean value, and make sure it takes `tenant_id` (a text type) as an `app_metadata` argument of the signed-in user or else fall back to `false`. You already know how to do this from our RLS policies:

```
BEGIN
  RETURN (COALESCE(
    (auth.jwt() -> 'app_metadata' -> 'tenants') ? tenant_id,
    false
  ));
END;
```

Here's the complete second function definition (by not setting anything else in the advanced settings, we again use the default **SECURITY INVOKER** rights):

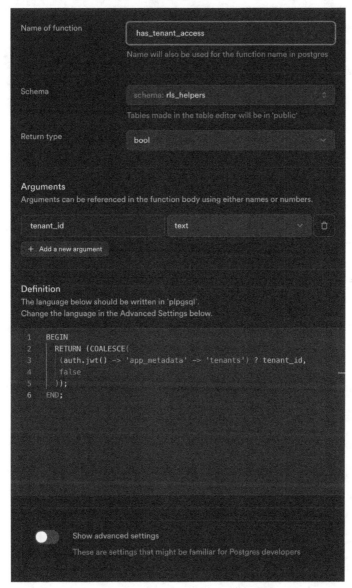

Figure 10.6: The has_tenant_access function creation

To see your newly created functions in **Database | Functions**, you need to switch the schema to rls_helpers in the dropdown:

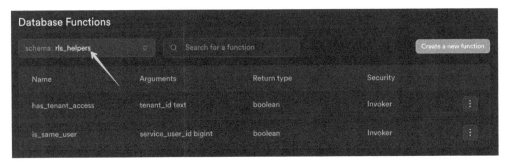

Figure 10.7: The rls_helpers functions list

Awesome! Now, let's create our comments RLS policies with the help of these new functions.

Creating the policies

Let's create the **DELETE** policy for the comments first. Open the policy creation mask for the comments table, choose **DELETE** for **Policy Command**, and, as always, choose **authenticated** for **Target Roles**.

Then, enter the following code in the **USING expression** box:

```
rls_helpers.is_same_user(comments.created_by)
```

And that's the full RLS expression – it will execute the is_same_user function, passing the comment author (created_by), and then only returning true if that's the same user as the one who is signed in. Awesome, right?

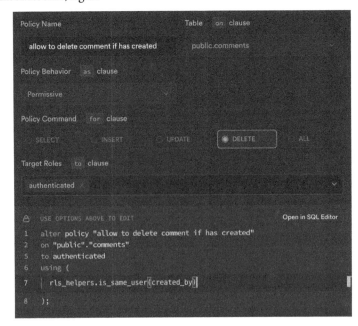

Figure 10.8: Adding the DELETE policy for comments

After saving the **DELETE** policy, create an **UPDATE** policy with the same expression as the **DELETE** policy (but only if you want to allow updating comments).

Next, let's create the **SELECT** and **INSERT** policies. Can you guess the expression to use? It's simply the following:

```
rls_helpers.has_tenant_access(comments.tenant)
```

The following screenshot shows the implementation of the **SELECT** policy specifically:

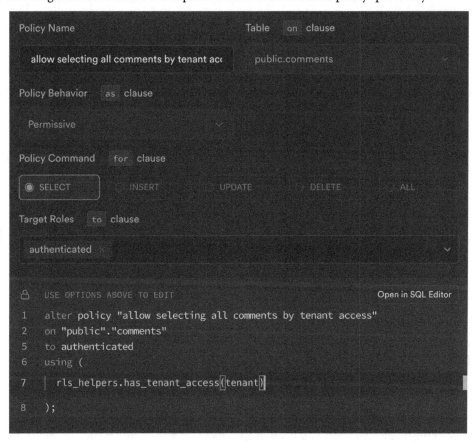

Figure 10.9: Adding the SELECT policy for comments

Now, those RLS expressions feel like fresh water after sunbathing compared to the previous longer expressions in other RLS policies, don't they?

After adding all the policies for **SELECT**, **INSERT**, and **DELETE**, you're done with setting up the comments table. Not only do you now know how to optimize RLS policy complexity with custom functions and how to create them in a custom schema but, by doing so, you can also change the policy logic in one place rather than adapting every single policy. Say, for example, you could later on decide to allow special admin users tenant access even if they're not assigned to a tenant. This can then be done by simply adapting the `has_tenant_access` function instead of changing any RLS policies.

> **Note**
> Now that you have these new, shiny helper functions, feel free to also update all existing policies in the other tables to simplify your RLS expressions. You'll find the updated expressions in this GitHub repository branch: `Ch10_Realtime`.

Let's move on to implementing comment functionality in the application.

Implementing comment creation

In your existing UI, in the Ticket Details page, there's already a `TicketComments` component used that has the components we need. So, the only thing we have to do is to process the comment and send it to the Supabase server for insertion into the database.

To do this, open up the `TicketComments.js` file and instantiate a Supabase client. Also, we need to make sure we add a property to `TicketComments` such that we can pass the ticket ID, such as `<TicketComments ticket={ticket.id} />`:

```
export function TicketComments( {ticket} ) {
  const commentRef = useRef(null);
  const supabase = getSupabaseBrowserClient();
...
```

Then, we need to ensure to pass that ticket ID from `TicketDetails` to the `TicketComments` component. In the `TicketDetails.js` file, pass it like so:

```
<TicketComments ticket={id} />
```

Next, back in the `TicketComments` component, change the `onSubmit` code of the `<form>` element using the following code to add a new comment in Supabase when the form is submitted:

```
onSubmit={(event) => {
  event.preventDefault();
  const comment_text = commentRef.current.value.trim();

  if (!comment_text) return alert("Please enter a comment");

  commentRef.disabled = true;
```

```
supabase
  .from("comments")
  .insert({
    ticket,
    comment_text,
  })
  .then(() => {
    commentRef.current.value = "";
    commentRef.disabled = false;
  });
}}
```

The code doesn't contain any surprises or anything that you do not know yet. When the form is submitted, it reads the value directly from the `textarea` element (`commentRef.current`) and checks that it's not empty. If it's not empty, it sets the element to `disabled` explicitly (I could've used a state for that but it also works this way) and then empties the `text` field and enables it again once Supabase is done inserting the row into the database.

What you have now is code that inserts a comment. What you don't have is the comment showing up in the UI. For freshly added comments, we will use Realtime to add them to the UI the moment they're added to the database, immediately visible to anyone.

But before we do, let's confirm first that our existing code is working. Go to any ticket, submit a comment, and check whether it's properly stored in the database. I've added two comments in one ticket on the **activenode** tenant, and another comment in a ticket of the **packt** tenant. In the following screenshot, I can see that they were properly added to the database:

Ow id int8	created_at timestamptz	created_by int8	comment_text text	ticket int8	tenant text	author_name text
3	2024-03-26 12:21:49.088216+0(	16	A lovely comment section!	18	activenode	Sarah Port
4	2024-03-26 12:23:28.305001+0	16	Another comment	18	activenode	Sarah Port
5	2024-03-26 12:36:43.718334+0(	1	A wonderful comment in the p:	17	packt	David Lorenz

Figure 10.10: Comments table entries

That's lovely. I want to emphasize the beauty of the work we've done – here, we've only provided two values in our `insert` statement and everything else was set automatically.

As a next step, let's ensure to list real existing comments below the ticket.

Listing existing comments from the server

Right now, we only have test comments below each ticket. Let's change that. In this section, we want to ensure that existing comments of a ticket are loaded at page load, and hence, are shown immediately with the ticket data.

> **Note**
>
> Since we add new comments to the database without any page refresh, for newly added comments, we want to use a real-time callback in the upcoming section. In this section, we make sure that all existing comments that are available when loading the page will be visible at page load.

Inside our client `TicketComments` component, we could use `useEffect(() => {}, [])`, and inside of it, use the Supabase client to load all data from the `comments` table with `supabase.from('comments').eq('ticket', ticket)`. That's a completely valid option. However, this would be a lazy-loading approach as the ticket data would immediately be available from the server and, when the `TicketComments` component mounts, it would load the comments.

A better way for our project is to get all of the ticket comments with one request alongside loading the ticket data, and that's what I want to show you now.

To do that, we have to go to the place where the ticket data is fetched: `tickets/details/[id]/page.js`. There, you currently have the following code, which fetches all the ticket data:

```
const { data: ticket, error } = await supabase
    .from("tickets")
    .select("*")
    .eq("id", id)
    .single();
```

Let's change that with just a tiny adjustment to request it to load the related comment data as well:

```
const { data: ticket, error } = await supabase
    .from("tickets")
    .select("*, comments (*)")
    .eq("id", id)
    .single();
```

Okay, this code probably needs some explaining as it's not so intuitive why and how it works – especially since we use `from("tickets")` and not something like `from("tickets, comments")`.

We learned in *Chapter 1* that Supabase uses PostgREST to build an automatic API layer between you and the database. PostgREST doesn't just forward requests to the database – you already know that it sets the current user if one is signed in so it can be used with the `auth.*` functions in the database – but it also reflects the database and its relations. That means PostgREST knows the details of which table has references to which other tables.

Now, what happens when you write `select("*, comments (*)")`? PostgREST will detect that special `comments(*)` syntax and know that it's not just a simple column of the table we're fetching from (`tickets`). Hence, with this syntax, PostgREST will look up references between those tables and check whether there is a matching relation between the `comments` table and the table we're selecting from. Then, it will automatically join them together and load all related comments as a sub-object of the usual object.

This is super-easy to understand if you look at the following sample output of `console.log(ticket)` from a ticket on the **activenode** tenant where I added two comments:

```
{
  id: 18,
  created_at: '2024-03-26T11:06:52.091522+00:00',
  title: 'Just a ticket on activenode.learn',
  description: 'Hello there!',
  created_by: 16,
  tenant: 'activenode',
  status: 'open',
  author_name: 'Sarah Port',
  assignee: null,
  assignee_name: null,
  comments: [
    {
      id: 3,
      tenant: 'activenode',
      ticket: 18,
      created_at: '2024-03-26T12:21:49.088216+00:00',
      created_by: 16,
      author_name: 'Sarah Port',
      comment_text: 'A lovely comment section!'
    },
    {
      id: 4,
      tenant: 'activenode',
      ticket: 18,
      created_at: '2024-03-26T12:23:28.305001+00:00',
      created_by: 16,
      author_name: 'Sarah Port',
      comment_text: 'Another comment'
    }
  ]
}
```

Figure 10.11: console.log(ticket)

As you can see, the `comments` property is an array of comment rows. If there is no related comment, the array will just be empty (`[]`). You can even enforce a specific order of comments by chaining an `order()` call as follows:

```
select("*, comments (*)")
 .order("created_at", { ascending: true, foreignTable: "comments" })
```

This tells Supabase to sort the related comment rows in ascending order (sorting is applied within the `comments` table, based on the `created_at` column, oldest first).

Now, we can deconstruct the comment rows alongside the other ticket values within `details/[id]/page.js` and forward them to `TicketDetails` as a property, using the following highlighted code:

```
const {
  ...,
  comments
} = ticket;
...
<TicketDetails
  ...
  initialComments={comments}
/>
```

Then, inside the `TicketDetails` component, you need to forward them once more to the `TicketComments` component so that we can render them there:

```
export function TicketDetails({
  ...,
  initialComments
}) {
  ...

  ...
    <TicketComments ticket={id} initialComments={initialComments} />
  ...
```

Next, in the `TicketComments` component, we want to read `initialComments` from the properties and get rid of the existing `const comments = ...` mock data:

```
export function TicketComments({ ticket, initialComments }) { ...
```

Now, create a new variable inside the component and assign `comments` to `initialComments`:

```
const comments = initialComments || [];
```

You might ask yourself why you'd have to do this when you could also just use `initialComments`. You're right. But the reason is that, in the upcoming section on Realtime, we will make this a state to implement Realtime comments merged with `initialComments`, so this way, we are preparing for that.

Now, we have to adapt our render template of the `TicketComments` component because our previous mock data had slightly differently named properties (`.author` instead of `.author_name`, etc.). So, let's adapt it to this to use the proper database properties:

```
{comments.map((comment) => (
  <article key={comment.id} className={classes.comment}>
    <strong>{comment.author_name} </strong>
    <time>
      {new Date(comment.created_at).toLocaleString("en-US")}
    </time>
    <p>{comment.comment_text}</p>
  </article>
))}
```

Let's also improve it just a little bit more so that it also shows the comments count as part of the `Comments` header element:

```
<h4>Comments ({comments.length})</h4>
```

Now, if you visit any Ticket Details page with existing comments, you'll see them wonderfully listed, like so:

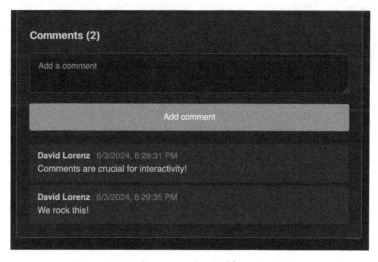

Figure 10.12: Comments loaded from the server

Perfect! Let's make this thing show new tickets with light speed; let's get into Realtime.

Implementing Realtime comments

In this section, we will enhance the `TicketComments` component to immediately show the new comment list if there's a new comment added – without any delay. It's time for Realtime!

Before we use the Realtime feature, there's a bit of prework to be done in the component to be able to update the existing comments in the UI with new comments coming from the Realtime service. We need to be able to change the `comments` value of our `TicketComments` component. And what's better to change an existing value than a state?

So, change your `TicketComments` component to use a state for `comments` by passing `initialComments` as its initial value:

```
const [comments, setComments] = useState(initialComments || []);
```

Wonderful. Now, we're able to change the `comments` array and, with it, update the UI, whenever needed. Next, I'll show you how to enable Realtime and subscribe to the updates of the `comments` table.

Enabling Realtime and subscribing to it

To be able to receive real-time updates, we must activate the Realtime feature on that table. You can activate Realtime either at table creation by checking the **Enable Realtime** checkbox or, as we'll do now, by going to the table view and clicking the **Realtime off** button:

Figure 10.13: Realtime button

Make sure to confirm that you want to enable Realtime when the system asks.

Then, to subscribe to Realtime insert events, we can use the Supabase client to listen to open a Realtime channel and tell Supabase to deliver changes in the Postgres database to this channel. We do this at the top inside our `TicketComments` component within `useEffect` to make sure it is only executed once at startup:

```
useEffect(() => {
  const listener = (payload) => {
    console.log("Realtime event received!", payload);
  };
```

```
  const subscription = supabase
    .channel("my-channel")
    .on(
      "postgres_changes",
      {
        event: "*",
        schema: "public",
      },
      listener,
    )
    .subscribe();

  return () => subscription.unsubscribe();
}, []);
```

What this code does is create a Realtime channel with your Supabase instance with an arbitrary name; here, it is my-channel. This channel name allows you to separate concerns but it's free to choose.

Next, we chain .on(listener_type, listener_options, listener_function), where we use the postgres_changes listener type to listen to database changes. The second parameter contains the listener options where we state that we want to get updates from tables from the public schema. The event value takes one of UPDATE, INSERT, DELETE, or *, where the latter means all. The third parameter is the listener function, which receives a so-called payload for each event that is happening.

But if we leave it like that, it would also fire the listener function for any other table where Realtime was potentially enabled since we are not constraining it to a specific table in the options. Also, we not only want to constrain it to changes on the comments table but we also only want comment updates for this specific ticket that is currently opened in the UI. Let's fine-tune the listener_options object as follows (the highlighted code is the new code):

```
.on(
  "postgres_changes",
  {
    event: "*",
    schema: "public",
    table: "comments",
    filter: `ticket=eq.${ticket}`
  },
  listener,
)
```

Now, we tell the Realtime service that we only want updates from the `comments` table and that we also only want updates where the ticket column value is equal (`eq.`) to the ticket ID that we have.

The filter takes the same kind of syntax as the PostgREST filters we talked about in *Chapter 8*, and you can find the full list of available Realtime filters for database changes here: `https://supabase.com/docs/guides/realtime/postgres-changes#available-filters`.

Let's test this with `console.log` just to see what we're getting. Go to any ticket page and open the browser console.

Now, to be able to confirm that the event is triggered even for other users, not just for yourself, make sure to open two different browsers (e.g., one in a private window), log in with two different accounts on the same tenant, and open the same ticket.

Then, when you create a new comment with either of the accounts, you'll see the payload and its structure being logged in the inspector – in all the windows in which you've opened the ticket:

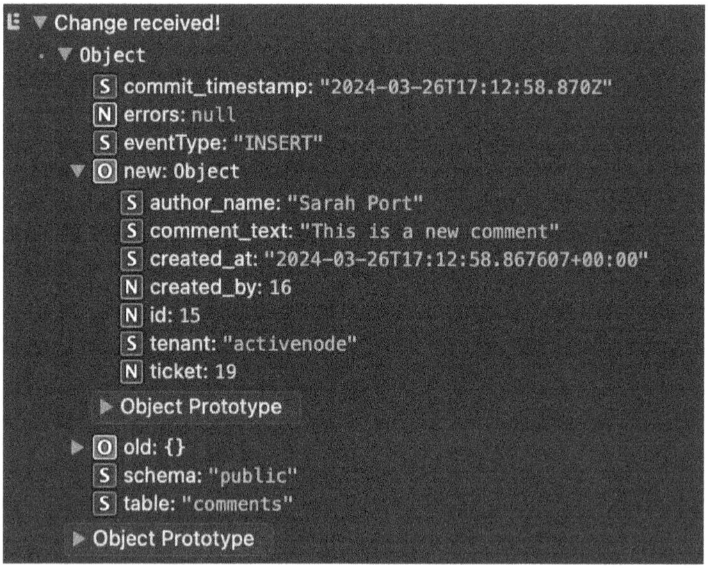

Figure 10.14: Logged Realtime object after inserting a comment

The object that is logged (*Figure 10.14*) for a newly inserted row is structured as follows:

```
{
  commit_timestamp,
  errors,
  eventType,
  new,
  old,
```

```
    schema,
    table
}
```

Let's have a look at each property:

- `commit_timestamp` is just the time of the event (but most of the time, you don't need it).

- You don't have to necessarily care about the `errors` value because if you're getting a proper `eventType`, you can be sure that there will be proper data – and if not, you don't process the event. But sure, if you're interested in logging an error value, this is your property to use.

- `eventType` is a database operation. It can be `INSERT`, `UPDATE`, or `DELETE`.

- The new and `old` properties are very interesting. When an `UPDATE` or `INSERT` event happens, the `new` property will contain the new database row. However, the `old` property will only be set for the `UPDATE` and `DELETE` event returning an object with only the primary key (which usually is `id`). So, if you update a comment with the ID 9100, `old` will be `{ id: 9100 }` but `new` will contain the complete row with the new values.

- Finally, there are the `schema` and `table` properties, which are self-explanatory.

With that knowledge, we can now use the payload to make real-time changes to our `comments` array.

Updating the UI with Realtime data

When we get a payload that contains the `eventType` called `INSERT`, we want to append that new comment to our existing comments. Doing so is rather easy, by changing the existing `listener` function:

```
const listener = (payload) => {
  const eventType = payload.eventType;

  if (eventType === "INSERT") {
    setComments((prevComments) => [...prevComments, payload.new]);
  }
};
```

Here, we check `eventType` and, if a new row is given, we simply append `payload.new` (which is a comment row object) to the value of the `comments` array. Now, if you add a new comment, it will immediately appear in all open windows. Isn't that amazing?

Now, even though we haven't implemented the ability to edit or delete comments in the UI, we can still test it with Supabase Studio. But first, we have to account for these events in our `listener` function. This is the final listener processing all of the types:

```
const listener = (payload) => {
  const eventType = payload.eventType;

  if (eventType === "INSERT") {
    setComments((prevComments) => [...prevComments, payload.new]);
  } else if (eventType === "DELETE") {
    setComments((prevComments) =>
      prevComments
        .filter((comment) => comment.id !== payload.old.id)
    );
  } else if (eventType === "UPDATE") {
    setComments((prevComments) =>
      prevComments.map((comment) =>
        comment.id === payload.new.id ? payload.new : comment
      )
    );
  }
};
```

For `DELETE`, we take the existing `comments` array and filter out the one that matches an existing `payload.old.id`. For `UPDATE`, we map each item to itself, except when it matches the ID of the one from the `payload` object; in that case, we replace the existing object with `payload.new`, so the new row overrides the old data.

At present, you have a fully working real-time comments system that can add, delete, and update comments in the state without delay. Now, let me show you how to test the deletion and update functionality by using Supabase Studio.

Triggering impersonated real-time updates with the Table Editor

Realtime isn't just firing when you make changes with the Supabase client such as `.insert()`. It fires due to the changes inside of Postgres. That means you can just use the Table Editor to change row data and it will trigger a real-time update in the UI wherever you subscribe to it.

First, open a ticket in your browser and add a few comments to it. In the following screenshot showing the **activenode** tenant, user **Sarah** has added one comment, and user **Copple** has added another two comments:

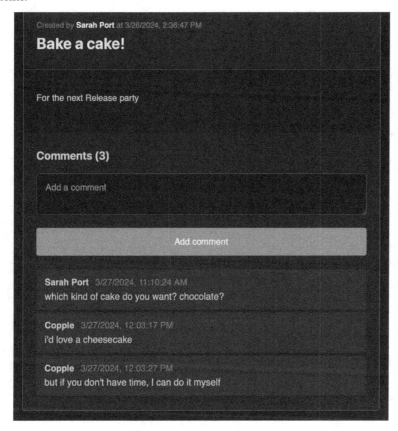

Figure 10.15: Comments on a ticket

To update the first comment as **Sarah**, go to the Table Editor, then the comments table. From here, we can impersonate that account by clicking on the small **role** selection dropdown (*Figure 10.16*) where I chose **authenticated role**. Then, next to **sarah@activenode.learn**, click **Impersonate**. You already used the impersonation feature in *Chapter 6* with the SQL editor.

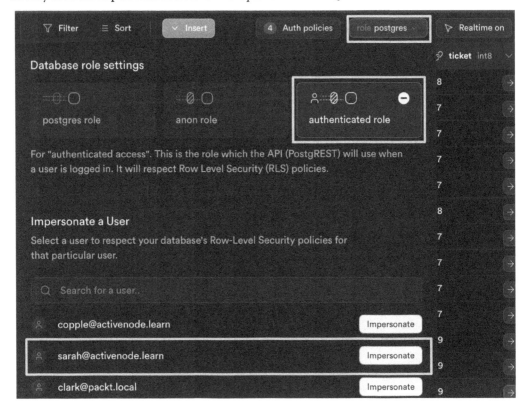

Figure 10.16: The Impersonate button

Now, we'll see all the ticket comments that Sarah has access to (my **Sarah** account only has access to the **activenode** tenant). Now, make sure to keep a browser window open where you have your ticket management system with the ticket details open of the ticket that you're about to edit in the database so you can observe what happens with what you're doing next.

Within that ticket comments list, double-click on the comment_text column of the comment you want to edit:

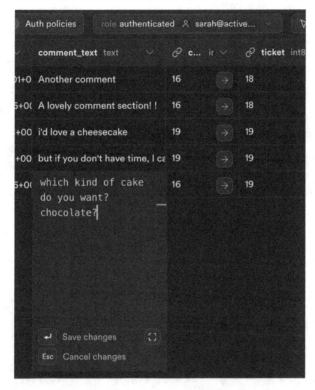

Figure 10.17: Double-clicking on a ticket column

Enter a new value and press *Enter* to save it:

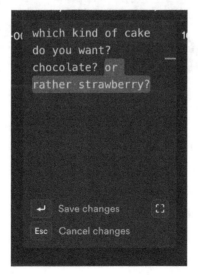

Figure 10.18: Updating comment_text with additional text

As soon as we press *Enter*, it's immediately updated in the UI:

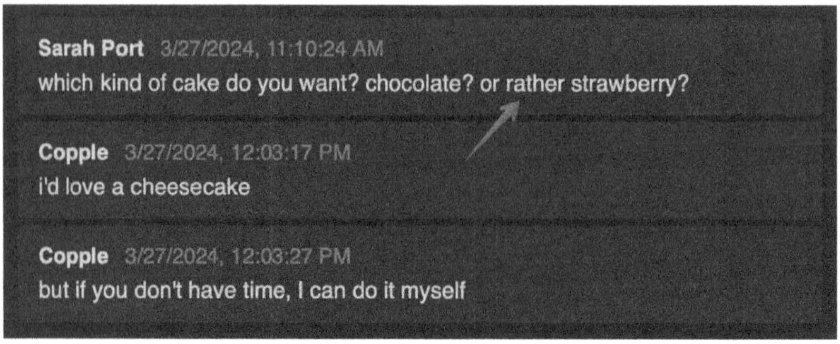

Figure 10.19: Updated comment

Now, let's get rid of the last comment from **Copple**. For that, we can impersonate **Copple**, select the comment in the list, and click **Delete 1 row**:

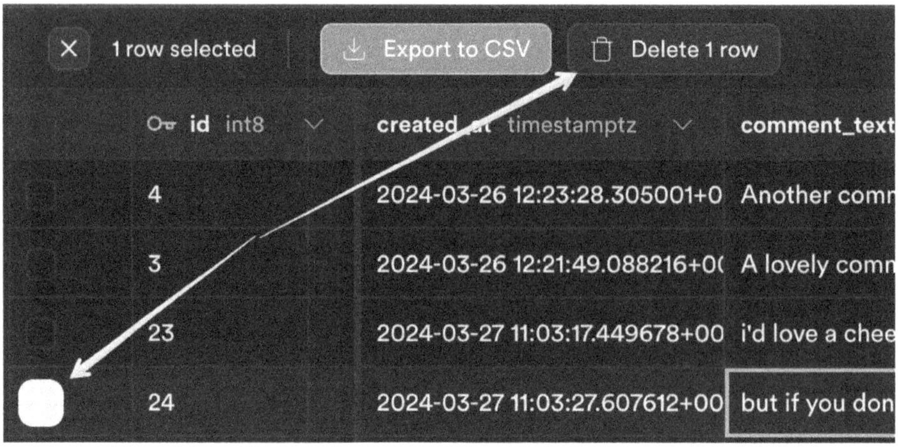

Figure 10.20: Deleting a comment

Upon row deletion, the comment seamlessly disappears from the comment list on the Ticket Details page. If Realtime doesn't impress you, I don't know what does!

Congratulations! You've acquired all the essential knowledge needed to fetch and trigger real-time updates from the database. To cap it off, let's explore some additional insights about the Realtime service in the final section of this chapter.

Embracing additional Realtime insights and learning about potential pitfalls

Although you've learned the most important part of Realtime, it's important to know that it can do more than just listen to the database. In this section, I won't go into too much detail, but I'll give you a few starting points to expand your own thoughts:

- Just like any other connection, a real-time connection can break (for example, because your Wi-Fi disconnects). In .subscribe(), you can pass a function to receive updates about the connection itself, like so:

```
.subscribe((status) => console.log('connection status', status))
```

Herewith, if you're notified with a status that isn't SUBSCRIBED, you might want to consider resubscribing:

```
export enum REALTIME_SUBSCRIBE_STATES {
    SUBSCRIBED = 'SUBSCRIBED',
    TIMED_OUT = 'TIMED_OUT',
    CLOSED = 'CLOSED',
    CHANNEL_ERROR = 'CHANNEL_ERROR',
}
```

Figure 10.21: Realtime status values

- Deletions in the database ignore the given filter property due to technical constraints (https://supabase.com/docs/guides/realtime/postgres-changes#delete-events-are-not-filterable). To put that in simpler terms, enabling Realtime in the comments table means that everyone, no matter the provided RLS, will be informed when any comment row is deleted. Functionality-wise, we couldn't care less as we only remove comments from the UI if the provided ID is in the list, so it doesn't disturb the UI.

However, security-wise, it's important to be aware of that. Imagine that Person X has access to a comment with id=123 but Person Y doesn't. If Person X deletes the comment, Person Y – the one who doesn't have access to the comment data – can still receive the event that { id: 123 } was deleted. In our ticket management application, this is no real security risk – no important data is exposed other than a number – but this knowledge becomes extremely important with the next bullet point.

- When a row is deleted, Realtime will not return the full data of that row, just the primary key as part of the `old` property of the payload – for example, `{ ..., old: { id: 123 } }`. You can change it to return all of the row's data (which is explained here: `https://supabase.com/docs/guides/realtime/postgres-changes#receiving-old-records`); however, I urge you to *never* do that as it comes with a huge security pitfall. As described in the previous point, the `DELETION` event is the only event that is not protected by RLS within Realtime. So, if you do enable the ability to return all of the data, you'll expose all of the data to everyone at the moment of its deletion.

- You can use the Realtime service to send arbitrary messages using arbitrary channels. You can open a channel by calling `supabase.channel('your-channel-name')` (this works both on the server and client). Then, you need to call `.subscribe()` on that channel, as you've seen before. Once subscribed to a channel, you can broadcast messages to it to everyone in the same channel (`https://supabase.com/docs/guides/realtime/broadcast`). Here's a sample of how to simply subscribe to a channel, without listening to its events, and sending a message once it's ready:

```
const c = supabase.channel("cool-channel-name");
c.subscribe((status) => {
  // Wait for successful connection
  if (status !== 'SUBSCRIBED') {
    return null
  }
  c.send({
    type: 'broadcast',
    event: "test",
    payload: { message: "hello, world" }
  })
})
```

Here, `event` can be a freely chosen string to differentiate (and filter) broadcasted messages, and `payload` can be an arbitrary object you want to broadcast.

To receive such broadcast messages, you can use the known `on(type, options, listener)` function following `subscribe()`:

```
const c = supabase.channel("cool-channel-name");
c.on("broadcast", {event: "*"}, (payload) => {
  console.log("Received", payload)
}).subscribe();
```

- You must make yourself aware that broadcasted messages within a channel are not protected by default. That means the `broadcasting` listener type behaves differently from the `postgres_changes` listener type; the latter respects RLS and only sends database events, as was previously explained. Broadcasted messages, however, will be readable by *anyone* who joins the channel. The first question that should pop up in your mind is, *Can someone list all available channels?* The answer to that is yes, but only if that someone has admin rights (service role); then, channels can be listed with `supabase.getChannels()`. This also means that if you choose complicated, arbitrary channel names, it would be hard for other people to guess them. Still, I don't recommend sending sensitive data via `broadcasting` but go for database changes (`postgres_changes`) instead.

> **Note**
>
> As stated, broadcasted messages are publicly accessible if someone knows the channel name. However, there's a bigger change coming allowing for more fine-grained access control even on broadcasted channels. Realtime messages, in the future of Supabase, will be directly mappable to RLS policies. I cannot explain how because this is still a work in progress. Until then, you need to rely on `postgres_changes`, which is a stable solution even with the changes coming. Information about the upcoming solution can be found here: `https://github.com/orgs/supabase/discussions/22484`. The new docs are found here: `https://supabase.com/docs/guides/realtime/authorization`.

- Broadcasting and receiving messages don't tell anything about the number of users using that channel. You can get that information using the Presence API, which you can learn more about here: `https://supabase.com/docs/guides/realtime/presence`. This is useful if, for example, you want to implement a whiteboard or a game where it's helpful or necessary to detect users leaving and joining the channel.

Now, you have everything at hand for working with the Realtime feature in Supabase. Let's wrap everything up!

Summary

That's a wrap on *Chapter 10*! We've really amped up our application by bringing it into the real-time era, moving far beyond the functionalities we started with. Stepping into real-time functionality felt like opening a whole new door for us.

Creating the `comments` table wasn't just about defining another table; it was about making something that works smarter, not harder. It was about automation and simplification (deriving the ticket ID) as well as using custom functions in the respective RLS policies (`rls_helpers`), hence, making them much more readable and maintainable.

But the actual game-changer? Live updates in the comments. There's something pretty satisfying about seeing user interactions unfold in real time. It takes the user experience from "good" to "wow." We navigated through the potentials and pitfalls of Realtime features, learning valuable lessons along the way.

Closing this chapter, our application has transformed. It's not just functional—it's interactive and engaging, keeping users hooked with instant updates. We've laid a solid foundation in Realtime features, and now, the possibilities for expanding our app seem endless.

The next chapter will give a lovely finish to all of this by adding the possibility of uploading files and learning about how to use them conveniently and securely within the application.

11

Adding, Securing, and Serving File Uploads with Supabase Storage

In the previous chapter, you made your app more interactive by implementing Realtime comments. What's missing is the possibility of uploading files in a ticket to allow users to communicate better.

In this chapter, you'll become friends with the Supabase Storage feature to implement file uploads within ticket comments. You will see how to upload and delete files, secure them from unwanted access with special RLS policies, and bind them logically to comments and consequently to users.

Besides learning how to enable and control file uploads, specifically for image files, you'll learn how to serve them visually right in your application in combination with the Supabase Image Transformations. Also, I'll show you how you could build your own image-serving backend.

After having learned about setting RLS policies for Storage, at the end of the chapter, I'll give you additional guidance on understanding what it means to write RLS policies directly on the storage tables. On top of that, I'll give you guidance on how you can create more complex storage restrictions based on tiers or user roles.

So, in this chapter, we will cover the following topics:

- Creating and understanding Storage buckets

- Enabling the addition of comments with file attachments

- Serving image attachments directly in the UI

- Writing RLS policies directly on buckets and objects

- Diving into advanced storage restrictions

Technical requirements

In this chapter, there are no new, specific technical requirements. However, you should generally know what a **content delivery network** (CDN) does – if you don't, there's a note box in the first section to help you out.

It can be also helpful if you have some knowledge of bucket-based Storage (S3 API compatible object storage); if not, I simply recommend going back to *Chapter 1* where the Supabase Storage is explained.

The code for this chapter is found here: `https://github.com/PacktPublishing/Building-Production-Grade-Web-Applications-with-Supabase/tree/Ch11_Storage`.

Creating and understanding Storage buckets

The Supabase Storage service will power any of your Supabase-based applications with whatever file-based feature you can imagine (for example, creating your own Dropbox clone). In our case, we want to use the Storage service to allow attachments in ticket comments for our ticket system.

Before we start with this implementation, I should introduce you to the heart of Supabase storage, **buckets**. A bucket exists to hold files. Hence, you need at least one bucket to store files.

You can create as many custom-named buckets as you want to have a good separation of concerns. For example, let's take a production application I worked on in the past. The application was a municipality manager with specific users, internal Dropbox-like file access, and the possibility to post news internally allowing for comments. In this case, I used three buckets – one named `avatars` for profile pictures, one named `intranet` for internal file uploads, and one for comment attachments named `comment-attachments`).

Not only does having multiple buckets clarify which bucket holds which files, but as you'll see, you can have different RLS policies for each bucket (you'll learn more about storage-bound RLS throughout this chapter).

Now, let's open up Supabase Studio and get to know buckets by actually creating one. Navigate to Supabase **Storage** in your instance and you'll see no existing buckets:

Figure 11.1: The Storage sidebar with no buckets

Click **New bucket** and you'll face a bucket creation mask. Here, you can choose the bucket name, however, it is not changeable after creation, so choose wisely. Let's name it `comment-attachments`.

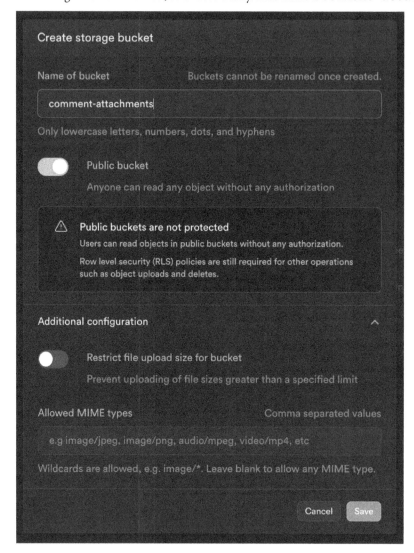

Figure 11.2: Bucket creation mask

In the mask, you should also choose whether your new bucket is **Public** or **Private** (default); we explore the differences in the next subsections, but for now, choose **Public**.

Next is an **Additional configuration** dropdown that allows you to define file type limitations and upload size limitations. If you wanted to have a bucket that will solely hold images and videos of whatever type but with a maximum size of 50 megabytes, you'd choose this file size limit and type image/*, video/* into the **Allowed MIME types** field. For our ticket system, we don't want to set any limitations, so, let's leave the **Additional configuration** fields empty.

Finalize the bucket creation by clicking **Save**. Your bucket will then appear in the sidebar with a clear **Public** indicator, including a File Explorer on the right (which is empty right now).

Figure 11.3: The comment-attachments bucket view

Now, let's first explore what it means to have a public bucket, how to manage files within it, and afterward, how it differentiates from a private bucket.

Examining public buckets

A **public bucket** is a bucket where a file can be accessed without any restrictions. However, this doesn't mean that a person can explore all of the files inside the bucket. Let's understand this by seeing it in action.

In the File Explorer of your bucket, click on **Create folder** and name the new folder my-folder. Then, click on the folder to enter it, and select **Upload files** to upload whatever file you want; I'll upload an image named coffee.jpg. After the upload, the File Explorer will show you the uploaded image and you can preview the image by clicking on it:

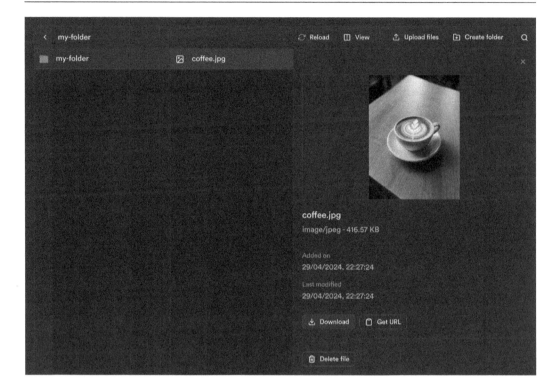

Figure 11.4: coffee.jpg in Supabase Storage

As you can see, the bucket acts just like a hard drive. You can create folders and files within it however you like.

Now, when we click the **Get URL** button, we'll get the direct URL to that file. On my local instance, that is `http://localhost:54321/storage/v1/object/public/comment-attachments/my-folder/coffee.jpg`. As you can see, the URL includes the bucket name and file path.

More generally, the URL is composited by the following logic: `API_URL + /storage/v1/object/public/ + BUCKET_NAME + PATH_TO_FILE`. On an instance on `supabase.com`, the same file would just have a different `API_URL`.

This URL isn't restricted and cannot have restricted access. Whoever knows the URL will be able to see that file. That's why it's called a public bucket.

The public bucket comes with an important upside: it can serve files very efficiently. The accessed files in a public bucket can be cached by the CDN server that lives between the user and the Storage server. If one person accesses a file in a public bucket, the CDN server will receive this request and go and get this file; then, it will deliver the file to the user. Next time, when another person accesses that file, it will be delivered even faster as the CDN server will act and deliver the file directly without asking the Storage server, as it's cached by the CDN.

> **Note**
>
> The simplest definition you can get of a CDN is this: it's a server that will deliver files for you (i.e., content delivery).
>
> However, usually when we talk about CDNs, we talk about more than simply delivering files. A CDN is usually a whole infrastructure that is reliable (safe from outages through fallbacks) and extremely fast. The speed comes from the fact that a CDN replicates files in multiple international servers, detects the location of the one who requests a file, and will then, with a so-called Load Balancer, decide which server will answer (this happens in basically "no time"). This ensures that files are delivered as fast as possible.
>
> However, it's not just about the speed. Because a CDN will deliver files for you and not your own server, you take a load of load away from your server and hence your application isn't affected by file delivery performance-wise.

Given that a public bucket can provide faster access and cached access to files, due to missing authorization checks, it's a good choice for non-sensitive data such as profile pictures, blog content pictures, or any other similar kind of potentially public media that can be delivered easily and fast by using URLs (e.g., directly in an `<img />` tag).

While you cannot restrict access to a specific known file path, you can limit file exploration within a bucket, such as preventing the listing of existing files. In the following sections, I'll show you how you can use the `list()` function and how it behaves with regard to permissions.

Exploring files within a bucket programmatically

By default, exploring files in a Supabase storage bucket is restricted to users with the Service Role Key (admin) and can be extended to other roles with RLS policies on buckets (as specifically shown in *Understanding RLS policies directly on buckets and objects table*).

With the appropriate permissions, you can use the `list()` method to explore files within a Storage bucket using the Supabase client. Here's an example:

```
supabase.storage
    .from("comment-attachments")
    .list()
    .then(({ data, error }) => {
      console.log("List of files in bucket", data);
    });
```

By default, `list()` lists files and folders in the root of the bucket but you can also pass it a specific path (`.list(path)`) to only get files within that path. There are more options in the official docs for this method: `https://supabase.com/docs/reference/javascript/storage-from-list`.

If you call this `.list()` with a normal, non-admin Supabase client, `data` will contain an empty array (`[]`) because the storage information is stored in the database and protected by RLS (and you already know, no existing RLS policy means no access).

But how and where can you create an RLS policy for storage? As mentioned, we will learn more about RLS in relation to storage later in this chapter, but let me give you a snack sample for now.

Learning how a basic RLS policy can be added to your bucket

Let's say you want everyone to be able to browse all of the files in the `comment-attachments` bucket. That means you want `supabase.storage.from(bucketName).list(folderName)` to return a list of files in the bucket for everyone including non-authenticated users. For that, within the **Storage** section, you'll find the **Configuration | Policies** button. Once selected, you'll see all the storage-related RLS policies:

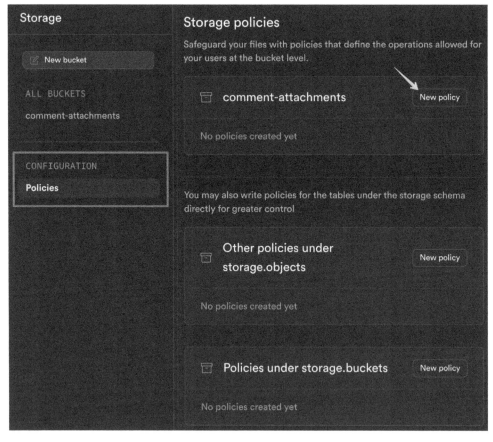

Figure 11.5: Storage policies

Within `comment-attachments`, click **New Policy** and the storage RLS editor will open, where you can give the policy a descriptive name, such as `allow everyone to list all files`:

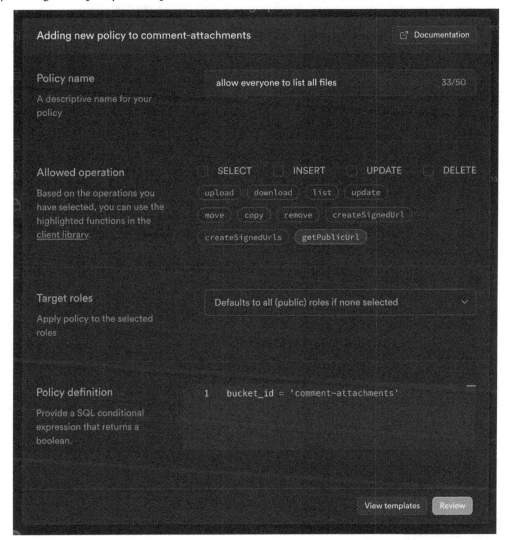

Figure 11.6: Policy creation mask

The `bucket_id = 'comment-attachments'` policy definition is prefilled by default. This is the most important requirement as you want to create a rule for that specific bucket (otherwise the rule would apply to all buckets).

Within the **Allowed operation** section, you can see that getPublicUrl is highlighted already. This is the simple functionality Supabase provides to get the full URL to a known file. This indicates once more that it's a public bucket and this permission cannot be revoked (it's also not really related to your policy but it's there for completeness' sake).

If you now check the **SELECT** checkbox in the **Allowed operation** section, you'll see that a few more permissions are highlighted:

Figure 11.7: Highlighted permissions after checking SELECT

The permission feature we are interested in is list, which allows the listing of files concerning our policy expression. As our policy expression only states bucket_id = 'comment-attachments', we allow everyone to list all files within this bucket, hence everyone can browse the bucket data freely.

The other highlighted permission features we can ignore – the download permission is always enabled on a public bucket (as you can access the file directly), and creating a signed URL will only be of interest for private buckets as you'll learn later.

At this point, you could click on **Review** and save that policy, and you'd have a policy created that allows browsing the bucket, however, I won't save it as I don't want anyone to browse the files.

I'd assume that understanding a public bucket and adding a simple bucket policy is rather straightforward. However, the question remains whether it is a good choice for us to go with a public bucket or whether we should switch to a private bucket for our comment attachments. Let's learn about private buckets in the next section and see where it leads us.

Understanding private buckets and revising our bucket choice

A **private bucket** is one where the file information is protected by RLS, just as in a public bucket, but also where the file, even if the path is known, cannot be accessed. The only way to access a file within a private bucket is either by triggering a special request to the file URL or by creating a signed URL.

While you've learned that a public bucket can benefit significantly from CDN caching, a private bucket faces a different scenario. Due to the need to check permissions, the cache hit rate (how many times the CDN is hit without hitting the actual Storage) is much lower for private buckets, as the request is different for each user.

How else does a private bucket differentiate from a public bucket? A private bucket doesn't allow direct access with the known URL schema, like so:

```
API_URL + /storage/v1/object/public/ + BUCKET_NAME + PATH_TO_FILE
```

Instead, there is a different composition for file URLs within private buckets, as follows:

```
API_URL + /storage/v1/object/authenticated/ + BUCKET_NAME + PATH_TO_
FILE
```

You can see that the difference is that the public bucket uses public within the URL and the private bucket uses authenticated as part of the URL.

However, now it gets interesting. Most people assume that given proper RLS you can use it the same way as with the public bucket, for example:

```
<img src="PRIVATE_BUCKET_FILE_IMAGE_URL" />
```

The assumption is that an authenticated user with proper permissions will be able to see the image and everyone else won't. *But this assumption is wrong.* No one will be able to see this image. Even if a signed-in user has the right to access that file, you cannot use it as a direct link as given in the example.

Let me elaborate. A direct link within an image (as given in the example line of code) is a GET request. However, that doesn't include the Authorization HTTP header, which Supabase requires to trigger a permission check. So, the request will be rejected. You could use the JavaScript-native fetch(), for example, to fetch a file (PRIVATE_BUCKET_FILE_URL) and add the Authorization: Bearer YOUR_SUPABASE_ANON_KEY HTTP header to the request. This will allow Supabase to check the permissions and return the contents of the file to you. But you cannot use a private bucket file URL as a direct link.

So, constructing this logical URL schema and sending the Authorization header along is an option you can use to retrieve a file – but it is not an option to send a link to someone else (as a link obviously wouldn't have an Authorization header).

This is why Supabase provides functionality to simplify getting a file from a private bucket, which creates **signed URLs** of files. Let's learn about those now.

Learning about signed URLs

A **signed URL** is a generated URL to an existing file within a bucket. It has time-limited validity and can be created by whomever has SELECT rights as per RLS policy. Once a signed URL is created, the URL will skip checking access rights for that specific file – that's what differentiates it from fetching the file with an Authorization header – a signed URL is self-sufficient and doesn't need that header. But since the signed URL is unguessable and has a limited lifetime, usually signed URLs are your best bet when working with private buckets so you have a lot of explicit control.

Next, you'll see how to create such a signed URL programmatically but if you want, you can also create a signed URL of a private bucket within Studio by clicking the **Get URL** button, just as we did in the public bucket. However, as opposed to a public bucket, within a private bucket, the **Get URL** button requires us to define the expiry period of the URL:

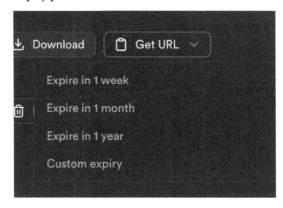

Figure 11.8: Get URL within a private bucket

With the Supabase client, you can create such URLs programmatically as follows:

```
const { data: { signedUrl }, error } =
  await supabase.storage
    .from("comment-attachments")
    .createSignedUrl("my-folder/coffee.jpg", 60);
```

In the given code sample, 60 refers to a 60-second expiry time.

With all the information we have about private buckets now, can you guess as to which is a better fit for our goal of saving file attachments of ticket comments? Lean back and think about it a bit before jumping to the next section, where I'll give you the answer.

Choosing a private or a public bucket?

So, what to do now? Should we make comment-attachments a private bucket or leave it public? As we're creating a multi-tenant application publicly facing the internet, with data uploaded to tickets that might be highly confidential, I would never rely on security through obscurity, no matter how unlikely one would be in guessing a very complex path structure. Even if no one would ever guess any path, using a public bucket would mean that once a person is removed from a tenant, they'd still be able to access the files within the tenant if they had saved the URLs to any files.

Hence, let's make our bucket *private* to provide a secure application. To do that, simply click on **Edit bucket** in the sidebar:

Figure 11.9: Edit the bucket

Then, toggle the **Public** option off:

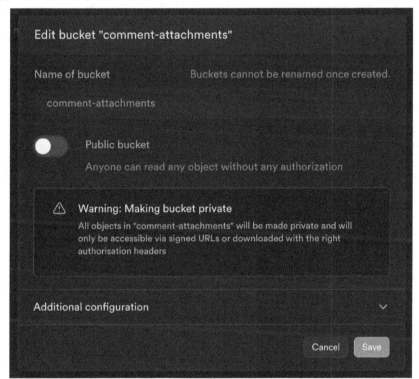

Figure 11.10: Making a bucket private

Once you have saved the new private bucket, let's recap what we've covered. So far, you've learned what the Storage service is good for, that it behaves like a hard disk, and that you can either have a fully public bucket that profits from the CDN or a private bucket that gives you controlled security. You've also learned that even a public bucket is not explorable (e.g., using `.list()`) by default if you don't specify explicit RLS to allow it.

Now, let's continue with implementing file attachments as part of comments in tickets of our application.

Enabling the addition of comments with file attachments

In this section, we want to allow to add as many additional file attachments to a comment as the user wants. We start by enhancing our `TicketComments` component UI first by adding some logic to be able to choose files within a comment. Then, we will continue with the Supabase-related flow, which will look as follows:

1. We upload the files to the storage, if there are any, and wait for the upload to finish. This will require us to set up proper Storage RLS policies.

2. Then, we insert the comment into the database.

3. After that, we connect the entry in the comments table with the uploaded file for them to be related technically.

4. Finally, we ensure that the uploaded files are shown in the UI along with the comment text.

Let's get started.

Preparing the UI with file upload possibility

Open `TicketComment.js` and find `<textarea />`. Then, right below the `textarea` field, add a new `input` field for file uploads:

```
<label htmlFor="file">
  <input
    type="file"
    id="file"
    name="file"
    multiple
    ref={fileInputRef} />
</label>
```

The `multiple` attribute will make it simple to select multiple files in the system's file browser when clicked. Then, make sure that `fileInputRef` is created at the top of the component so we can access this `input` field when the comment is submitted.

Right below the existing `commentRef`, we also create a reference for the `input` field for easy access to the files:

```
const fileInputRef = useRef(null);
```

We will upload these files when the form is submitted but we need to keep a reference of the selected files, so, we need to add an `onChange` listener to `<input type="file" />` combined with a state where we store the file references. To do this, first, create the React state below the `comments` state:

```
const [comments, ...
const [fileList, setFileList] = useState([]);
```

Then, to get a better autocompletion within the editor, we will hint the state type to be of what the browser gives us when adding files to an input: `FileList`. We will hint by providing a little JSDoc comment, which is supported in most editors (it needs the additional redundant parentheses, or else the editor won't match). I would like to point out that this is just a helpful comment, not a necessity (with TypeScript, you could just use `useState<FileList>([])`):

```
const [fileList, setFileList]
    = useState(/** @type {FileList} */ ([]));
```

Next, we must ensure that this state is going to be updated whenever `onChange` on the file input is triggered:

```
<input
  type="file"
  id="file"
  name="file"
  multiple
  onChange={(e) => {
    setFileList(e.target.files);
  }}
/>
```

With that set, we have a reference for the selected files, and when the form is submitted, we can upload the files to Supabase storage. Let's upload the files now.

Uploading files to storage

To upload the selected files, we can use the Supabase instance we already have in the `TicketComments` component.

But first, within `onSubmit`, we want to check whether there are files attached. If so, we want to trigger the upload and wait for the upload success, then afterward, we insert the comment into the database. We can do so by using promises as follows (the new code is highlighted):

```
onSubmit={(event) => {
  ...
    commentRef.disabled = true;
  let uploadPromise = Promise.resolve();
```

```
    if (fileList.length) {
      // upload files here and replace the promise accordingly
    }

    uploadPromise.then(() => {
      supabase.from("comments").insert(...)...
    });
  }}
```

As you can see, the comment insertion command has moved inside the `uploadPromise.then()` callback. Now let's replace the `// upload files ...` code comment with actual upload logic. But where exactly do we upload the file inside the bucket? Which folder should we upload it into? Do we even need folders? If so, do we need to create them first?

First and foremost, when we upload files to folders that do not exist, Supabase will create them for us, no matter how deeply nested they are, so there's no need to worry about creating folders; Supabase will only need the file and the path where it shall put it.

But what's a good path for our comment attachments? You want files to be structured to resemble a similar logic as you have in your database. Let me tell you what this means. Here's the path that we choose for comment attachments within the Storage bucket:

```
tenant_id/ticket_id/random_hash/original_filename
```

This means that if we are on the `packt` tenant and the current ticket ID is `753`, we would upload a file named `cat.jpg` (within the `comment-attachments` bucket) to a path like this:

```
packt/753/80119f9a045d68/cat.jpg
```

This is a logical structure and you know immediately which tenant this file belongs to and to which ticket ID it was added. But why the random directory? Well, for each upload, we want to maintain the original information, but we also want to avoid potential collisions. So, if multiple users are uploading the same filename within a ticket, we want to allow that, so we simply add a random value in between.

Getting such a short, random string value can be done with the following function, which will be defined within a `utils/helpers.js` file as follows:

```
export function getRandomHexString() {
  return Math.random().toString(16).slice(2);
}
```

Make sure to add this function to your project. This function generates a random value, converts it to a hexadecimal string, and cuts off `0.` at the front, which gives you a value such as `c02e54b6b99058`, for example.

Alright, you know the file structure we want to use and you have the `util` function to generate a random string. Let's move on to the upload logic.

Before we can implement the upload logic, we must ensure we have access to the tenant ID within the `TicketComments` component. So, let's pass it inside the `TicketDetails` component onward to `TicketComments`:

```
<TicketComments
  ticket={id}
  initialComments={initialComments}
  tenant={tenant}
/>
```

Next, make sure to destructure the tenant ID in the `TicketComments` component accordingly:

```
export function TicketComments({ ticket, tenant, initialComments })
{...
```

Then, within `onSubmit` of `TicketComments`, our upload logic will look like this:

```
let uploadPromise = Promise.resolve();
if (fileList.length) {
  uploadPromise = Promise.all(
    Array.from(fileList).map((file) =>
      supabase.storage
        .from("comment-attachments")
        .upload(
          [
            tenant,
            ticket,
            getRandomHexString(),
            file.name,
          ].join("/"),
          file
        )
    )
  );
}
```

Here, we use `Promise.all` to wait for all of the provided promises to be completed. The provided promises are upload promises from Supabase. We created those upload promises by mapping each entry in `fileList` to a promise from Supabase using the `.upload(path, file)` functionality. The path of the file is just a string as `tenant`, `ticket`, `getRandomHexString()`, and `file.name` are concatenated with `/` via `.join("/")`.

Like database actions such as `select()` or `insert()`, an `upload()` call returns a promise, which will resolve to a `{data, error}` object. When successful, data will contain a `{ path: "your/ uploaded/file/path" }` object. This means, in our case, when `Promise.all([upload1, upload2, ...])` resolves, it will just be multiple of such `{data, error}` objects returned within one array.

The good news here is that the code for uploading the files is now already done. However, the bad news is this won't work yet. Can you guess why? Our bucket is private and we haven't set up any RLS policies for it. To be able to upload files, we need to create a policy with INSERT permissions. So, let's do that and create a tenant-bound RLS storage policy that enforces the given file path convention.

Creating an RLS storage policy for comment uploads

In the bucket, add a new policy (as shown earlier). For the policy name, enter tenant ticket comment uploads; for **Allowed operation**, choose INSERT and SELECT; and for **Target roles**, add authenticated. Then, you should see a view like this:

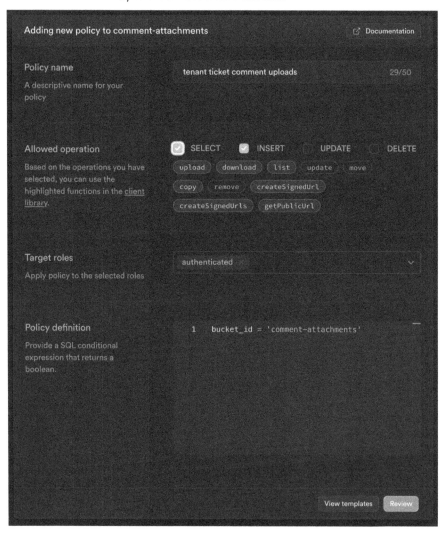

Figure 11.11: RLS creation mask baseline

Now, we want the policy to not just contain `bucket_id` but also contain the policy such that the first folder must be an existing tenant ID and the second folder must be an existing ticket ID.

So, how can we enforce this structure with an RLS policy? We need a way to split a given file path into its pieces (tenant and ticket ID) and then check whether the current user has access to such. If we cannot determine that the user is allowed to create a file, then either the file path was wrong or the user doesn't have the proper permissions. So far, implementing a path-based RLS is the plan.

It isn't immediately obvious but in the given mask you're writing a policy for the table `storage.objects` table. Hence, in this policy, you can refer to all columns that are available in that table (you can switch the schema to `storage` in the Table Editor to check out all existing columns of `objects`).

The `name` column of the `objects` table contains the full file path. But what we need is the single pieces/folders of the file path. This can be easily achieved with the `storage.foldername(path)` helper function. This function is best understood with an example: `(storage.foldername("foo/bar/123.jpg"))[2]` would be equal to `bar` as the function itself returns `["foo","bar","123.jpg"]` and we access the second element within that array. This can be seen in the following raw SQL execution within Supabase:

Figure 11.12: Using the storage.foldername() function

Enforcing the first folder to be a valid tenant and the second to be a valid ticket comes down to the following policy expression:

```
bucket_id = 'comment-attachments'
AND
EXISTS (
  SELECT FROM tenants
  WHERE tenants.id = (storage.foldername(objects.name))[1]
)
AND
EXISTS (
  SELECT FROM tickets
  WHERE tickets.id = (storage.foldername(objects.name))[2]::int8
)
```

Here, I've highlighted `objects.name` to make sure you use `objects.name` and not just `name`. Usually, when there's only one candidate for a column in an expression, you don't need to necessarily add the explicit table (`objects.`) but our `tenants` table also has a `name` column; that means `name` is ambiguous and you must make sure the correct column is referenced by specifically writing it out – you don't want the database to pick one for you!

So, within our `comment-attachments` bucket, we ensure that the first folder must meet with an existing tenant ID and the second folder must match with an existing ticket ID. As folder names are of the `text` type, we need to cast the ticket ID to an integer value with `::int8`.

The coolest part is this respects the existing policies. So, if someone does not have access to a tenant, one can also not create a file in the respective directory. Isn't that great?

Once you've saved your RLS policy, you'll see that Supabase creates two policies from it, one specifically for `SELECT` and one for `INSERT`. So, there are no combined policies for it but at least you can create them with just one mask.

Now, if you submit a comment with uploads, you'll see that they're being uploaded (you can spot them in the bucket in the Supabase File Explorer). However, there's no relation to the comments yet. This is why we're only seeing the comments without the attachments in the comment list. Let's fix that in the next section.

Connecting uploaded files with the written comment

If you just uploaded files for a comment like I did, you can get rid of them in the file explorer, as we will now implement the instant connection of such files with comments.

As one comment can have `0:n` attachments, it's time to go back to database design – we need to create a `comment_attachments` table that contains the comment ID to which it relates, as well as the path to the file as `file_path`. You already know how to create a table with relations, so do that now – you should end up with a table that has the default, untouched `id` and `created_at` columns, as well as the following two columns: `comment` with a relation to the `comments` table and `file_path` being just a string.

Once you're done creating the table, let's make sure to add the proper `comment_attachments` entries after files have been uploaded and the comment was inserted (the newly added code is highlighted):

```
uploadPromise.then(((fileUploads) => {
  supabase
    .from("comments")
    .insert({
      ticket,
      comment_text,
    })
    .select()
```

```
    .single()
  .then(({ error, data: commentData }) => {
    commentRef.current.value = "";
    commentRef.disabled = false;
    fileInputRef.current.value = "";
    setFileList([]);

    if (error) return alert('Error adding comment');

    if (fileUploads) {
      supabase
        .from("comment_attachments")
        .insert(
          fileUploads.map((file) => ({
            comment: commentData.id,
            file_path: file.data.path,
          }))
        ).then();
    }
  });
});
```

As you can see, we've moved the previously existing `.from("comments").insert()` into the callback of `uploadPromise.then(callback)`. Also, we added `.select().single()` to then get the data from the inserted comment immediately; once we have that comment data, we reset the form values and the form state and check whether `fileUploads` contains anything (it doesn't obviously if you haven't selected a file to upload).

If `fileUploads` is not empty, we trigger an insert for each uploaded file to the `comment_attachments` table such that each upload is linked to the comment. As we don't just pass a single object but an array of multiple objects to `.insert()`, Supabase will consider each of them to be one insertion – this is called a **bulk insert**.

Also, if you are wondering why `then()` is there without a function, we don't need to provide a function but we need `then()` to execute the insertion.

> **Hint**
>
> You can use triggers for inserting the attachment relation. You already learned about triggers in *Chapter 8*. Certainly, instead of adding `comment_attachments` with additional JavaScript code, you could use a trigger to do that. Given that we have a file path convention that includes the tenant and the ticket id, we cannot derive the comment it would match. However, if we also added a random hash to each created comment and used that hash as part of our file path structure, we could use a trigger to match those within the database as part of a trigger.

As always, the insertion into comment_attachments will only work when we have the proper RLS set up for the newly created table. So, let's make sure we have those RLS policies in place. Add the following INSERT RLS policy to comment_attachments to allow the owner of a comment to insert a related attachment to it:

```
rls_helpers.is_same_user(
  (SELECT c.created_by FROM comments c
  WHERE c.id = comment_attachments.comment)::int8
)
```

Here, we make use of our already existing rls_helpers.is_same_user function, which we created in the previous chapter. It allows us to check the signed-in user against the value provided, which, in this case, is the comment author (c.created_by). There's no need to cast the statement to an integer like I did using ::int8 but I like the added explicitness in this case.

The created policy is for inserting rows but to be able to read the attachments again, we also need a SELECT policy on the same table. We want to allow the user to read any attachment row where the comment access also exists:

```
EXISTS (
  SELECT FROM comments c WHERE c.id = comment_attachments.comment
)
```

This is straightforward (reminder: this policy works because it triggers the respective RLS policy within the comments table; if we don't have access to a comment, we cannot read the respective comment attachment data as we cannot even read the comment).

If you now go to any ticket and add a comment with selected files in the file upload, you'll still only see the comment's text in the UI. However, you'll find the uploaded files in the proper directories within the bucket, as shown in *Figure 11.13*:

Figure 11.13: Uploaded files in bucket

Then, in *Figure 11.14*, you can see related entries in the comment_attachments table. Go ahead and confirm that this works. Comparing *Figure 11.13* and *Figure 11.14*, you can see two uploaded file entries for the packt tenant on ticket ID 17 with comment ID 24.

Figure 11.14: comment_attachments rows

Now that you have solved uploading and referencing of comment attachments, you must also show these attachments alongside the comment, including the option to download them. This is what we do next.

Showing the connected files

We want to achieve three things now:

- Showing the attachments alongside the comment on page load
- Allowing the user to download the attachments
- Showing attachments for comments added by the Realtime listener (as it isn't the same as loading them on page load)

The easiest part will be listing the attached files for a comment, as we can fetch this data synchronously with the comment itself. In the previous chapter, you already learned how to retrieve related data from tables. You selected all data from the tickets table as well as all of its related comments by using a join in the select statement as follows:

```
.select("*, comments (*)")
```

The best thing is that this also works for related data of related data, so you can nest this. As we want potential `comment_attachments` data for each comment, we extend this selection string within the `tickets/details/[id]/page.js` file to become the following:

```
.select("*, comments (*, comment_attachments (*) )")
```

Now, it's party time because at this point we only have to check in a given comment object if there are entries within `commentObject.comment_attachments` and, if so, display it in the UI. So, go to your `TicketComments` component and navigate to the rendering logic (`{comments.map((comment) => ...`). There, right below `<p>`, add the following code:

```
<article ...>
  ...
  <p>{comment.comment_text}</p>

  {comment.comment_attachments?.length > 0 && (
```

```
    <>
      <small style={{ display: "block" }}>Attachments</small>
      {comment.comment_attachments.map((attachment) => (
        <button key={attachment.id} className="file-badge">
          {attachment.file_path}
        </button>
      ))}
    </>
  )}
</article>
```

With this, when attachments are given within a comment, we'll list them as buttons.

To make them look nice and small I have simply added the `file-badge` CSS class, which I added to `globals.scss` for simplicity:

```
button.file-badge {
  font-size: 0.9rem;
  display: inline-block;
  width: auto;
  margin-right: 0.5rem;
  padding: 0.26em 0.5em;
  border-radius: 4px;
  border: 0;
  background-color: rgba(255, 255, 255, 0.15);
  cursor: pointer;
}
```

Now, when I then visit the ticket where I added uploaded files to a comment, I see this:

Figure 11.15: Showing attachments of a comment

This is good, but we only want the file names to show in the UI, not the whole bucket path. Let's fix that by splitting the file path by the slashes and then choosing the last entry:

```
{attachment.file_path.split("/").pop()}
```

Now, the screen will look like this:

Figure 11.16: Showing only filenames of attachments

We also want to provide direct access to that file when the filename is clicked. This is easy. Add the following click listener to the `file-badge` button:

```
onClick={() => {
  supabase.storage
    .from("comment-attachments")
    .createSignedUrl(attachment.file_path, 60, {
      download: false,
    })
    .then(({ data, error }) => {
      window.open(data.signedUrl, "_blank");
    });
}}
```

When an attachment is clicked, we create a signed URL that is only valid for 60 seconds before it expires. Then, we open that link in a new browser tab. I've explicitly set the `download` option to `false` (that's the default), which means the browser will only trigger a file download if it cannot display the file (such as a `.exe` file). However, if it's a document/media type that the browser can view directly (such as a PDF), the browser will just show it.

Alternatively, if you want the browser to download a file instead of viewing it, simply put `download: true`.

Lovely! Now, when I click the `cheesecake.jpg` button (shown back in *Figure 11.16*), I see this:

Figure 11.17: The cheesecake.jpg image

Now, our files are attached to comments represented in the UI, but we aren't completely done yet, as the attachments are not shown when a new comment is added with Realtime. Adding a new comment isn't triggering a fresh page load but will be added to the UI with a Realtime subscription on the `comments` table. We get a newly added comment via Realtime from the database in the immediate moment it is created before the related `comment_attachments` is created. That means that even if we did a selection on the `comment_attachments` table after receiving a comment via Realtime, it could happen that (with a bad connection) the related `comment_attachments` table might not even be added – so, we're requesting something before it exists and we're getting nothing back.

To solve that problem, the easiest solution is to not just listen to the `comments` table's real-time changes but also to the `comment_attachments` table's real-time changes. So, go to the `comment_attachments` table and enable the Realtime feature (as we did in *Chapter 10* for the `comments` table). Then, we will reuse the Realtime subscription we have. Right now, there's a listener with the following configuration within the `TicketComments` component:

```
const subscription = supabase
  .channel("my-channel")
  .on(
    "postgres_changes",
    {
      event: "*",
      schema: "public",
```

```
      table: "comments",
    },
    listener
  )
  .subscribe();
```

We have two options to listen for `comment_attachments` changes:

- We could remove the `table` configuration option and get all real-time updates on the public schema, then decide what to do based on the table that fired an update provided within `payload.table`.

- We could chain the subscription with yet another `.on()`.

I love my code to be specific for extra clarity, so I will go with the second option, like so:

```
const subscription = supabase
  .channel("my-channel")
  .on(...)
  .on(
    "postgres_changes",
    {
      event: "*",
      schema: "public",
      table: "comment_attachments",
    },
    attachmentsListener
  )
  .subscribe();
```

Hence, in the first `on()`, you have your listener for the `comments` table changes, and the second `on()` is listening for `comment_attachments` changes.

Now, all we have to do is to build the `attachmentsListener` function, which you should create below the existing `listener` definition. This function will only listen to `INSERT` events and add a new `comment_attachments` object to the comment object that we have in the state:

```
const attachmentsListener = (payload) => {
  const eventType = payload.eventType;

  if (eventType === "INSERT") {
    setComments((prevComments) =>
      prevComments.map((comment) => {
        if (comment.id === payload.new.comment) {
          return {
            ...comment,
            comment_attachments: [
```

```
            ...(comment.comment_attachments || []),
            payload.new,
          ],
        };
      }

      return comment;
    })
  );
  }
};
```

Now, when a Realtime `INSERT` event for `comment_attachments` is fired, the comment's state is updated, and if a matching comment is found, it will add the new attachments to the comment object. When you add a new comment with files attached to it now, it will also immediately render the file attachments. How cool!

With that, we have implemented file attachments on comments. In the next section, I'll give you insights about how to instantly deliver image files.

Serving image attachments directly in the UI

We have already discussed how direct URL usage, such as using `<img src="BUCKET_FILE_URL" />`, is possible only with a public bucket. However, when using a private bucket, we still want to preview uploaded images within a comment in the UI. Additionally, even with a public bucket, where you can easily use the file link to display an image, there are times when you might want to show a lower-resolution version of the image instead of the original.

That's why, in this section, I want to expand on the possibility of using direct URLs to serve and view optimized images within an application, even with private buckets, and show you how to use Image Transformations to get magically converted versions of images.

Using Image Transformations

Even though a public bucket allows us to view an image without restrictions, it might not be the cleverest choice to simply put the original file path into an image. This is because images can be quite big, and you often don't need to load an original 50mb-sized picture, especially not for small-scale preview purposes such as avatars.

This is where the Supabase Image Proxy comes in, providing a tool called **Image Transformations**. This is an additional tool in the Supabase stack that connects directly to the Storage server. I won't go into detail as it's mostly self-explanatory and well-documented (`https://supabase.com/docs/guides/storage/serving/image-transformations`) but I want to give you an example to understand how it can be used and how it benefits you.

Let's say you have a public bucket with a /coffee.jpg image and you want it to be 500 x 500 pixels. You can use the following URL schema to let the image transformer of Supabase convert your image on the fly and return an adapted 500 x 500 pixels version like so:

```
<img src="API_URL/storage/v1/render/image/public/your-bucket-name/
coffee.jpg?width=500&height=500" />
```

As you can see, the URL is constructed with the API_URL/storage/v1/render/image/ public/your-bucket-name/your-image-name?TRANSFORMER_OPTIONS_HERE schema.

Image transformations can be done with images from private buckets as well but only as part of creating signed URLs, like so:

```
supabase
.storage
.from('private-bucket')
.createSignedUrl('coffee.jpg', 6000, {
  transform: {
    width: 500
    height: 500,
  },
})
```

With this knowledge about Image Transformations, you have a tool at hand that will take the pain off image conversion and manipulation. Go ahead and explore more options in the link provided a few paragraphs ago.

Next, I want to quickly guide you on how you can enable direct links, such as we have with public buckets, with private buckets by implementing a pseudo-CDN.

Building a pseudo-CDN for private buckets

Let's say that whenever someone uploads an actual image as a comment attachment, we want to show a tiny preview of it instead of just the filename. What we could do is issue a signed URL with Image Transformations for each attachment that is an image, for example, on page load, by checking the file extension.

But that would mean creating a lot of signed URLs all the time, and how would you choose the expiry dates? What if someone leaves the tab open for 10 minutes and then scrolls to a section where the image is loaded but our signed URL already expired after 1 minute? Should we extend the expiry dates to somewhat infinity (which is a security issue) or should we ensure to issue a signed URL only when the image is really in the viewport? Also, what if the same user refreshes the page; should we cache both the URL and expiry date to avoid creating another signed URL if the previous one is still valid? Shouldn't we take care of not issuing another signed URL?

So, as you can see, as great as signed URLs are, they might not be a good fit for mass delivery of images. Instead, what you can do is build your pseudo-CDN. It's quite easy: you can build a GET-based API endpoint that will take an image path and deliver the processed image only if the user has the proper rights. Let me guide you through a sample version of such an implementation.

First, create an app/[tenant]/cdn/route.js file, which contains a GET Route Handler:

```
export async function GET(req) {
}
```

Next, our goal is to be able to use an image, as follows:

```
<img src="http://packt.local:3000/cdn?image=the/image/path.jpg" />
```

Hence, our GET handler needs to read the image parameter that is sent. Let's implement that:

```
export async function GET(req) {
  const { searchParams } = new URL(req.url);
  const image = searchParams.get("image");
}
```

Now, it gets interesting – we want an RLS-based Supabase instance to not just check whether such an entry exists, but also implicitly confirm that the user requesting the entry has access to it. In our case, all we must do is check for a matching comment_attachments entry:

```
import { getSupabaseCookiesUtilClient } from "@/supabase-utils/
cookiesUtilClient";
export async function GET(req) {
  const { searchParams } = new URL(req.url);
  const image = searchParams.get("image");

  const supabase = getSupabaseCookiesUtilClient();

  const { data: cdnImage, error } = await supabase
    .from("comment_attachments")
    .select("file_path")
    .eq("file_path", image)
    .single();

  if (error) {
    return new Response("Error fetching image", { status: 500 });
  } else {
    // deliver the image
  }
}
```

You can see that all we do is fetch `file_path` from the `comment_attachments` table. That's why I said "implicitly" because if one doesn't have access per our defined RLS policies, one will also not get a result and `single()` (requesting exactly one row) will make sure to return an error then.

Alright, there's no further explanation needed for that, but how do we deliver the image? We download it on the server and send it back to the user! We only want our pseudo-CDN to deliver small-scale versions of images, so we will make use of Image Transformations to downscale them (this is also much less memory-intensive for our API). So, the solution for delivering the image is as follows:

```
...
// deliver the image
return supabase.storage
    .from("comment-attachments")
    .download(cdnImage.file_path, {
      transform: {
        quality: 70,
        width: 100,
        resize: 'contain'
      },
    })
    .then(({ data: imageBlob, error }) => {
      return new Response(imageBlob);
    });
```

We request a small, downscaled version of the image with a width of 100 pixels and a quality of 70 percent, and make sure that the image ratio is respected with `resize: 'contain'`. When the transformer is done, it returns the image as a blob image (`https://developer.mozilla.org/en-US/docs/Web/API/Blob`), which we can simply return as the server response.

Your simple pseudo-image CDN for the private bucket is now done! Next, let's make use of it to show previews for images within our comments.

Using the pseudo-CDN inside our UI

When a filename indicates it's an image, we want to add an `<img />` tag with the preview image. For this implementation, I'll only be checking for `.jpg` but feel free to add other image extensions.

Within the `TicketComments` component, right below the filename, add the following code:

```
{attachment.file_path.endsWith(".jpg") && (
  <img
    style={{ marginLeft: "10px" }}
    src={urlPath(
      `/cdn?image=${attachment.file_path}`,
      tenant
```

```
    )}
   />
 )}
```

That's all there is to do! We conditionally render an image element with the URL to the pseudo-CDN we created. Here is the result:

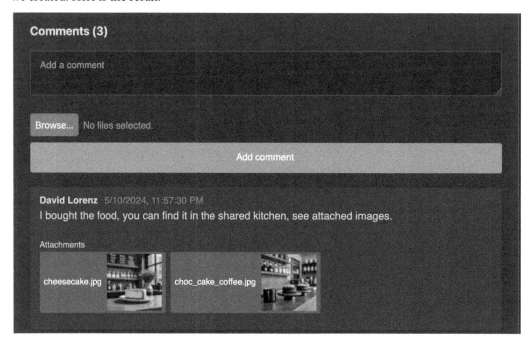

Figure 11.18: Preview images with pseudo-CDN

That's great. Before we wrap up the topic though, let me give you a little more insight into how to make this even better:

As we have not just used the RLS-based client for checking the database but also for triggering `.download()`, it will check permissions once more during the download request, and reject downloading the file if the user doesn't have the proper rights defined in the bucket.

In theory, this is unnecessary. For triggering `.download()`, you could use the admin client instead, which avoids checking permissions another time, given that you've checked the permissions with the RLS-based client beforehand. However, that's just an FYI for you on how you could optimize this.

Once you've downloaded such a file from the Image Proxy, you could then store it within a protected folder on your server for even faster access next time. This means that next time, you'd only check permissions and then not talk to the Image Proxy again.

The Image Proxy comes with potential additional costs (on `supabase.com`), depending on the number of transformations you do. With a pseudo-CDN, you can also use your own transformation libraries, such as `sharp` (`https://www.npmjs.com/package/sharp`). Be aware, though, that this can be quite resource-intensive so using it on hosting platforms such as Vercel or Netlify might easily hit the limits of the execution.

Now, you know more about delivering files from buckets and you've learned that you can make your application faster and more efficient by using Image Transformations. With that, the chapter is nearly complete, but I'd like to give you some more useful insights about Storage in the upcoming section. There, I'll tell you when and why to write policies directly on the `storage.buckets` or `storage.objects` table.

Writing RLS policies directly on buckets and objects table

So far, we've only written RLS policies on the `comment-attachments` bucket, but we can also write RLS policies directly for the `storage.objects` and `storage.buckets` table. When would we do that? Let's have a quick peek at what Supabase Studio shows me in the RLS section right now:

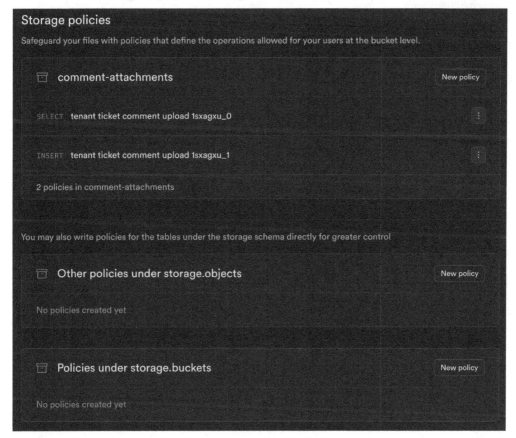

Figure 11.9: Storage RLS

We have two policies on `comment-attachments`, and no policies on `storage.objects` or `storage.buckets`.

However, this isn't quite correct. Both of our `comment-attachments` policies are inside of `storage.objects` as this table is where the file information is stored. It's just a visual helper from Supabase Studio that detects which buckets are targeted within the policies and separates them visually. This means that if there was a more general policy without `bucket_id = 'your-bucket-name'` in the RLS expression, it would show under **Other policies under storage.objects**. Further, this means that you could create policies that scope beyond just one bucket – if you want to.

But what about the `storage.buckets` policies then? The `buckets` table stores information about the available buckets. If one has writing permissions (`INSERT`), it would mean that one could programmatically create a bucket according to the policy (via `supabase.storage.createBucket()`). Likewise, if you write a policy that gives someone read rights (`SELECT`), it would mean that someone can list buckets via `supabase.storage.listBuckets()`.

With this, you gained more clarity about RLS policies in the storage schema. Next, I want to finalize this chapter by giving you a sample of how you can think of more advanced samples to implement storage restrictions with.

Diving into advanced storage restrictions

Let's say our ticket system has different tiers – for the sake of simplicity, the free tier would only allow one file upload per ticket (so only one comment with a file; other comments would be without files) while the pro tier would allow multiple uploads. How would we restrict uploads accordingly?

Inside your `comment-attachments` storage policy, you'd need to write an expression that checks for existing files for the same user and the same ticket. We have made the ticket ID part of the file path, so we can extract that from the file path, but not the user ID. You could obviously add that as part of the file path, but the Supabase user ID is also stored along with the file upload within `storage.objects.owner_id`.

So, if we wanted to restrict someone from uploading another file for the same ticket, we would adapt our existing RLS expression to the following:

```
bucket_id = 'comment-attachments'
AND
EXISTS(
  SELECT FROM tenants
  WHERE tenants.id = (storage.foldername(objects.name))[1]
)
AND
EXISTS (
  SELECT FROM tickets
```

```
    WHERE tickets.id = (storage.foldername(objects.name))[2]::int8
)
AND NOT EXISTS (
    SELECT FROM storage.objects o
    WHERE (storage.foldername(o.name))[2] = (storage.foldername(objects.
    name))[2]
)
```

This won't work just yet though! It will let you save that expression but it will fail when executed. That's because within your policy for `storage.objects`, you select something from `storage.objects`, which triggers the underlying RLS policy within `storage.objects`. Now, you might think, so, what? That should still work. It should but it doesn't. Postgres tries to help you avoid self-referencing connections like these to avoid potential infinite recursions, so it will simply fail to execute this (even if it could work):

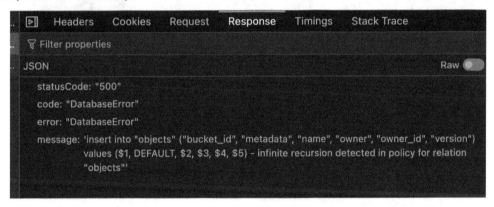

Figure 11.20: Error on self-referencing

Solving that is straightforward though. To check whether there is already an entry in storage with the same ticket ID, we can just create a `file_exists_on_ticket` helper function within our `rls_helpers` schema, which runs with admin rights (`SECURITY DEFINER`) and hence will skip RLS. Why does this help?

When we select data within an admin-powered function, checking RLS is skipped for that function. Hence, Postgres will not call RLS policies recursively and not throw an error for this potential infinite loop.

Let's create such a helper function by going to your database functions, selecting the `rls_helpers` schema, then adding a function called `file_exists_on_ticket`, with **Return type** set to `bool` and **Arguments** set to `ticket_id`:

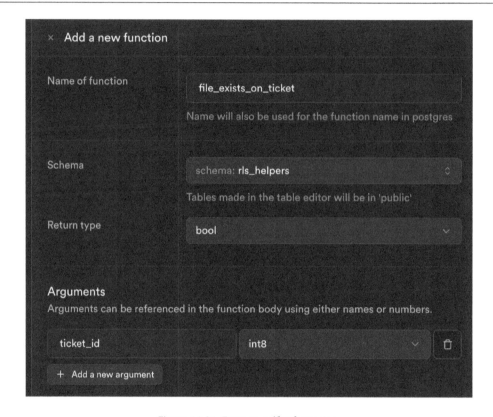

Figure 11.21: Error on self-referencing

Then, fill out the function with the following simple logic:

```
BEGIN
 RETURN EXISTS (SELECT FROM storage.objects o
   WHERE (storage.foldername(o.name))[2]::int8 = ticket_id);
END;
```

So, all this function does is check whether there is a file in the bucket for which `ticket_id` is the second-nested folder (remember that we use the following structure in our bucket: `tenant_id/ticket_id/random_hash/original_filename`).

Then – and this is very important – you must select `SECURITY DEFINER` in the advanced options to run with admin rights skipping self-referencing RLS checks.

Once saved, you can go back to edit the RLS expression like this:

```
bucket_id = 'comment-attachments'
AND
EXISTS(...)
```

```
AND
EXISTS (...)
AND rls_helpers.file_exists_on_ticket(
  (storage.foldername(objects.name))[2]::int8
) = false
```

Now, it will work like a charm and allow only one upload per ticket, no matter how many comments.

> **Note**
> I will not leave this RLS check in the policy because I don't want to activate it. However, you
> will find the `rls_helpers.file_exists_on_ticket` function within the repository
> branch to play around with it.

With that, you've seen how you can wrap your head around even more complex Storage restrictions.
You're ready to implement anything you can imagine storage-wise.

Summary

In this chapter, we explored file uploads with Supabase Storage, revolutionizing our application's capabilities.

We started by dissecting the differences between public and private buckets, delving into their caching
disparities and the implications for our application. Next, we implemented file uploads for comments
and learned about the Supabase client storage functionality and the required policies we must create.

Later, we explored the concept of pseudo-CDNs, unlocking the potential to use direct links for
private buckets and unlocking more potential for delivering files programmatically. We learned
about the convenient Image Proxy server, which allows us to transform images on the fly and dive
into advanced storage restrictions, demonstrating how to enforce access limitations based on existing
storage information.

With all of this, let me tell you that the major functionality of our application is complete, and with
everything you've learned up to this point, you're ready to kickstart projects!

However, let's not stop there. To help you become a real Supabase expert, in the next chapters, you'll
gain knowledge that will help make your application even more solid, fast, and secure. In the next
chapter specifically, we will think more about security.

Part 4: Diving Deeper into Security and Advanced Features

In the last part of this book, we dive into advanced techniques and security measures to fortify your Supabase application. You'll explore new features, optimize performance, and ensure your app is secure from potential threats. This part will empower you to push the boundaries of what your application can do, preparing you for any challenge. Let's supercharge your Supabase skills!

This part includes the following chapters:

- *Chapter 12, Avoiding Unwanted Data Manipulation and Undisclosed Exposures*
- *Chapter 13, Adding Supabase Superpowers and Reviewing Production Hardening Tips*

12

Avoiding Unwanted Data Manipulation and Undisclosed Exposures

Supabase is secure by default, but as with any other technology, the problem is when people cobble together a few copy-pasted parts from Google, StackOverflow, or ChatGPT without deeper knowledge of what it will imply. During the writing of this book, I found a massive security leak in a Supabase-powered application, leaking all the personal clients' data. The problem wasn't Supabase at all; it was sloppiness within certain areas of their application's SQL expressions. This is what you want to avoid.

You already know how to protect your table data with RLS, that you can use additional internal schemas, and that an SQL function can adhere to RLS or run with admin rights. This chapter will give you the last bits of security-related information that you need to know to avoid potential unwanted data exposure and strengthen your application's security.

This chapter is made up of mostly self-contained, but security-related, topics. Though they don't all follow on from one another, I recommend that you read them all.

In this chapter, we will cover the following topics:

- Understanding PostgREST's OpenAPI Schema exposure
- Being careful with `current_user` usage and understanding `auth.role()`
- Generating new Anonymous Keys, Service Role Keys, and database passwords
- Benefiting from Supabase Vault
- Utilizing silent resets to avoid data manipulation
- Enabling column-level security/working with roles

- Understanding security on views and manually created tables

- Changing the `max_rows` configuration

- Understanding safe-guarded API updates or deletion

- Adding middleware inside Postgres for each API request

- Using the Security Advisor

- Allowing the listing of IPs for database connections

- Enforcing SSL on direct database connections

Technical requirements

The code for this chapter can be found here: `https://github.com/PacktPublishing/Building-Production-Grade-Web-Applications-with-Supabase/tree/Ch12_Security`.

Understanding PostgREST's OpenAPI Schema exposure

In this section, I want to talk about schemas, their structures, how PostgREST exposes them, as well as how to avoid exposure.

You must be aware that your complete `public` schema structure is, in fact, publicly visible to everyone who has the Anonymous Key to your project. But you might wonder, why is that? And shouldn't we have discussed this in *Chapter 11*? The short answer is this – PostgREST follows the OpenAPI specs, and even though someone can see the schema, they can't see your data, so it's not a security issue per se. Let's dig into this further.

In *Chapter 1*, I explained how PostgREST introspects the database and provides an API on top of it. PostgREST hereby follows the OpenAPI specification and allows us to explore the possible API calls (`https://postgrest.org/en/v12/references/api/openapi.html`) and build upon them. This means that once someone has your API URL and Anonymous Key (which is exposed when you use Supabase with frontend clients), everybody knows what your schema looks like, which relationships between the tables exist, and which RPCs are available to be called.

There is an open-source Supabase Schema Visualizer (created by GitHub user @zernonia and available at `https://github.com/zernonia/supabase-schema`) that creates a visual representation of your schema. You can find it integrated within your project's Supabase Studio Dashboard (see *Figure 12.1*), but it is also independently usable and will create a schema visualization, with a provided API URL and an Anonymous Key.

Figure 12.1: A visualized schema with relations (don't worry about the specifics;
this is just an overview of the Supabase Schema Visualizer)

This means that a potential hacker could use the existing information as an easy-to-understand visualization of your structure. Also, it means that your maybe-super-precious architecture is partially exposed to the outside world.

PostgREST exposes this information at the base path of the Supabase REST API. You can just make a GET request to `YOUR_API_URL/rest/v1/` and add an HTTP header with the Anonymous Key (in curl, this is `-H 'Apikey: YOUR_ANON_KEY'`). The server will then respond with the full JSON output of the schema (which the visualizer uses to create a diagram):

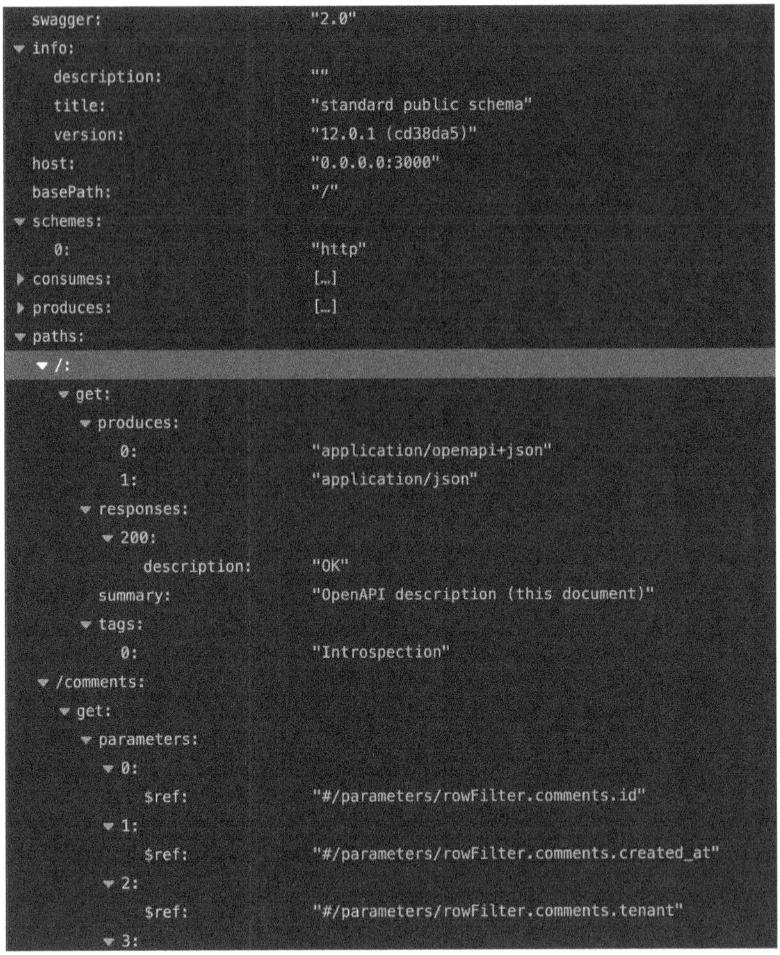

```
swagger:                         "2.0"
▼ info:
    description:                 ""
    title:                       "standard public schema"
    version:                     "12.0.1 (cd38da5)"
  host:                          "0.0.0.0:3000"
  basePath:                      "/"
▼ schemes:
    0:                           "http"
▶ consumes:                      [...]
▶ produces:                      [...]
▼ paths:
  ▼ /:
    ▼ get:
      ▼ produces:
          0:                     "application/openapi+json"
          1:                     "application/json"
      ▼ responses:
        ▼ 200:
            description:         "OK"
        summary:                 "OpenAPI description (this document)"
      ▼ tags:
          0:                     "Introspection"
  ▼ /comments:
    ▼ get:
      ▼ parameters:
        ▼ 0:
            $ref:                "#/parameters/rowFilter.comments.id"
        ▼ 1:
            $ref:                "#/parameters/rowFilter.comments.created_at"
        ▼ 2:
            $ref:                "#/parameters/rowFilter.comments.tenant"
        ▼ 3:
```

Figure 12.2: A small extract of the PostgREST OpenAPI introspection

> **Note**
> On localhost, don't use ?apikey, as it works without it and throws an error otherwise.

But what about other schemas, like the rls_helpers schema we created? Is that exposed as well? No. The default schemas that are exposed are public, storage, and graphql_public (graphql_public is a schema that reflects data from the public schema for the GraphQL endpoint of Supabase, which you can learn more about here: https://supabase.com/docs/guides/graphql).

Okay, but what if I don't want my schema structure to be visible/exposed? Let's discuss that now!

Preventing schema exposure

If you want to avoid people scanning through your schema with the PostgREST APIs (which are available at YOUR_API_URL/rest/v1/), there's a *simple* solution and a *correct* solution.

The *simple* solution is to completely disable OpenAPI exposure by running this SQL expression:

```
ALTER ROLE authenticator SET pgrst.openapi_mode TO 'disabled';
NOTIFY pgrst, 'reload config';
```

By running this, OpenAPI generation will be disabled, but this also means that Supabase's health check assumes that PostgREST is down, as it doesn't respond anymore on the base URL. So don't do this!

If you did, reverse your code to this:

```
ALTER ROLE authenticator RESET pgrst.openapi_mode;
NOTIFY pgrst, 'reload config';
```

Instead, let's look at the *correct* solution – this involves returning a mocked OpenAPI response with empty API data, as described at https://postgrest.org/en/v12/references/api/openapi.html#overriding-full-openapi-response. To do so, we need to define a mock function in our database that returns the most basic JSON as a valid OpenAPI response:

```
create or replace function mock_openapi() returns json as $_$
declare
openapi json = $$
  {
    "swagger": "2.0",
    "info":{
      "title":"Overridden",
      "description":"This is a my own API"
    }
  }
$$;
begin
  return openapi;
end
$_$ language plpgsql;
```

Once executed, we need to tell PostgREST to use this function to generate the schema response:

```
ALTER ROLE authenticator SET pgrst.db_root_spec
 TO 'public.mock_openapi';
NOTIFY pgrst, 'reload config';
```

> **Note**
>
> In Supabase Studio, there is an **API Docs** tab that usually shows you documentation specifically for each of your tables. As it is based on the returned OpenAPI response from PostgREST, which is now empty, this will not show auto-generated docs anymore – however, this should be your least concern, as your documentation is now this book!

Now you know how to hide away your schema while still being able to use PostgREST on the schema. If you want to undo this action, just use RESET:

```
ALTER ROLE authenticator RESET pgrst.db_root_spec ;
NOTIFY pgrst, 'reload config';
```

We can also remove schemas for use with PostgREST completely. Let's see that next.

Removing schemas from usage via API

If you really want to, you can go as far as saying that you don't want PostgREST to act on a schema at all. This also means disabling the usage with the Service Role Key. The only thing that you can then do to perform a query on the database is to connect directly to it.

With a local instance, you can control the API-exposed schemas in the `config.toml` file under the `[api]` section. There, you will find the `schemas = ...` configuration. On `supabase.com`, you'll find this in the **API | API settings** section, where you can add or remove schemas to be used by the API:

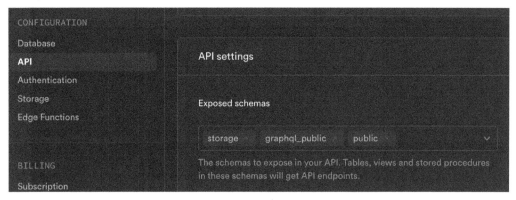

Figure 12.3: Setting the PostgREST-enabled schemas

That's all you need to do to disable API access to a schema. But what if you want the opposite – to specifically allow a custom schema to be used with PostgREST? Let's see that too.

Specifically exposing a schema to the API

Removing a schema to be able to be used by the API was straightforward; however, adding a custom one to be exposed isn't as intuitive as just creating it. Obviously, after creating a new schema, you must add it to the **Exposed schemas** section shown in *Figure 12.3*, but you also need to set a few permissions for the API to be able to read the tables properly. This is done with SQL commands granting the proper permission roles within Postgres.

You can find the set of commands to execute for an API-exposed schema on this page: `https://supabase.com/docs/guides/api/using-custom-schemas`. I will refrain from copying and pasting them into this book; note that you need to replace `myschema` with the custom schema you want to expose.

As an additional note, creating a Supabase client with a different schema than `public` can be done by passing `{ db: { schema: 'schema-name' } }` to the options when creating a client, like so:

```
createClient(url, key, { db: { schema: 'your-schema' } });
```

That's all there is to know about schema exposure. Let's hit the next topic.

Being careful with current_user usage and understanding auth.role()

`current_user` is a variable in Postgres that identifies the active user role within the current SQL execution context. We haven't discussed this variable yet, but I want to make you aware of its potential drawbacks. Some developers who have worked with Postgres databases might be familiar with using `current_user`. However, relying on this variable can lead to issues, and I want to ensure you understand its limitations to avoid potential pitfalls.

Let's create an example with `current_user`. When we make a request with PostgREST (by using the normal API/Supabase client) without authentication, the role is `anon`. This means that, in pure theory, you could make use of this `current_user` value – for example, within an RPC, as shown here, differentiating the access of admin users versus non-admin users:

```
CREATE OR REPLACE FUNCTION get_latest_user_data() RETURNS TEXT
SECURITY INVOKER
VOLATILE
LANGUAGE plpgsql
AS $$
BEGIN
  IF (current_user = 'service_role' OR current_user = 'postgres')
```

```
  THEN
    RETURN 'I will return the user data of the newest user';
  ELSE
    RETURN 'I will return the user data of the signed in user';
  END IF;
END
$$;
```

If we call this RPC with an admin Supabase client, we will see the first text, **I will return the user data of the newest user**. Otherwise, via the API, we get **I will return the user data of the signed in user**. That means it works. If I call it within the database (e.g., in Supabase Studio with the SQL Editor), I get **I will return the user data of the newest user**, as I execute SQL statements with `current_user` as `postgres` by default (if we don't explicitly change it with the impersonation feature below the SQL Editor). So that works as well. Everything seems just fine. However, I have now lured you into a potential pitfall.

The given function code, `get_latest_user_data`, only derives the proper role because it uses **SECURITY INVOKER**, and because it is called directly. However, if we call this function from within another function, things are becoming more difficult.

If you call a **SECURITY INVOKER** function such as `get_latest_user_data()` from a **SECURITY DEFINER** function, you open the gates to oblivion, as it will run with admin rights then. That's because the **SECURITY DEFINER** will run as `postgres`, and hence, the whole calling context of any function called within will be `current_role=postgres`.

So, you might ask, why would someone call a non-admin function within an admin context? As always, in programming, reusing logic can be useful. For example, if you want to avoid RLS checks for either performance reasons or to avoid recursion (as we did with `rls_helpers.file_exists_on_ticket`), then you would make that function a `SECURITY DEFINER` function and be quick to reuse an existing non-admin function, without thinking much about the implications. This is called human failure and happens every day.

So, what can we do about it? The solution lies within a function that will be bound to authentication and unaffected by the calling context. Let me show you a safe variant of the `get_latest_user_data` function and then explain why it's safe:

```
CREATE OR REPLACE FUNCTION get_latest_user_data() RETURNS TEXT
SECURITY INVOKER
VOLATILE
LANGUAGE plpgsql
AS $$
DECLARE
actual_role text;

BEGIN
```

```
  actual_role = coalesce(auth.role(), current_user);

  IF (actual_role = 'service_role' OR actual_role = 'postgres')
  THEN
    RETURN 'I will return the user data of the newest user';
  ELSE
    RETURN 'I will return the user data of the signed in user';
  END IF;
END
$$;
```

Here, we don't simply use the `current_user` value; we define a new variable, `actual_role`, which we set to `coalesce(auth.role(), current_user);`. Here, `auth.role()` immutably gives us the role of an API request (so whenever you use the Supabase client to retrieve data). `auth.role()` doesn't care if it's called within a `SECURITY DEFINER` function or not; it returns the currently claimed role of an API request.

But what is a **claimed role**? In *Chapter 6*, you learned about custom claims, where you used the `auth.jwt()` function to get the verified **JSON Web Token (JWT)**. With that JWT, containing also the user data of the signed-in user, you extracted `auth.jwt() -> 'app_metadata'` to get the custom-set claim (the tenant). Just as the `tenants` value is a custom claim (see *Chapter 6*), the role is also a claim (not a *custom* one but a built-in one from Supabase), and that claim can be accessed using `auth.role()`.

This claimed role is the role with which PostgREST executes the SQL commands. When you call `supabase.from('tablename').select('*')`, PostgREST will execute a `SELECT ...` command in Postgres with that claimed role.

Depending on how you connect with PostgREST (or, in our project, depending on how we connect with the Supabase client), the following roles are claimed:

- The anon role is claimed for using the Anonymous Key (normal RLS client) and no signed-in user.
- The `authenticated` role is claimed to use the Anonymous Key when a signed-in user makes a request. It then claims the role that is set in the `auth.users` table in the `role` column, which is `authenticated`.
- The `service_role` role is claimed to use the Service Role Key.

Returning to our code block, if we call `auth.role()` from within the SQL Editor, `auth.role()` will be `null` because there is no API request, just manual execution; in that case, with `coalesce`, we fall back to `current_user`.

This is a pretty safe approach, as whenever the Supabase client is used for fetching (i.e., a request with the API), it will not accidentally acquire admin rights.

> **Note**
>
> At this point, I'd like to emphasize once more that a *claim* is literally just a piece of information in the JWT (i.e., within the token that carries user data). Just like how we added a custom claim in *Chapter 6* and used it in our RLS policies, a JWT can have an infinite amount of claims.
>
> As long as the JWT is verified, a claim only means that it's a trustworthy piece of information. The claims that you get within `getSession()` are not trustworthy, as you learned in *Chapter 4*. However, for everything that happens between the Supabase services, rest assured that the JWT is verified or otherwise rejected.
>
> This blog post about JWTs is very helpful to dig deeper into the topic further: `https://stytch.com/blog/jwt-claims/`.
>
> This is also a good resource to understand how PostgREST generally deals with roles: `https://postgrest.org/en/v12/references/auth.html#overview-of-role-system`.

Generating new Anonymous Keys, Service Role Keys, and database passwords

Whenever you accidentally expose one of your secrets – for example, through screen sharing or another simple mistake – you need to know how to change this secret immediately. Also, it's a good idea to change them regularly anyway, just like changing regular passwords.

On your local development, there is no such option, but there's also no need to change the default secrets – that's because it's your local machine for local development, and it's not meant to be something you expose. In other words, if you expose your local instance to the public web, you probably have a different problem.

However, in your instance on `supabase.com`, you can easily change secrets and passwords. Go to **Project Settings** | **API**, and then within the **JWT Settings** section, you can force the creation of a new JWT secret by clicking the **Generate a new secret** dropdown and choosing **Generate a random secret**. This triggers the recreation of the Anonymous Key and the Service Role Key, as they are dependent on the JWT secret.

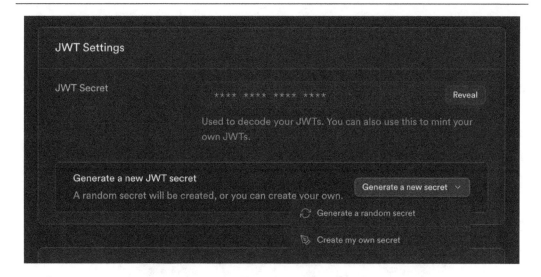

Figure 12.4: Generating a new JWT secret

Your new keys can then immediately be found in the **API Settings | Project API keys** section.

If you want to reset the database password, go to **Project Settings | Database**, find the **Database password** section, and click **Reset database password**:

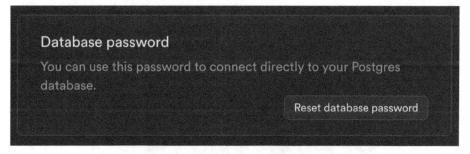

Figure 12.5: Resetting the database password

You're now aware of how to regenerate secrets.

Benefiting from Supabase Vault

Supabase **Vault** is an extension within the Supabase database (enabled by default) that allows you to easily manage and encrypt secrets. For instance, you can store API keys or environment variables inside of it, and then use them within your database with Edge Functions (which we look at in the next chapter), or even inside your application. Let me show you how it works.

Creating secrets in the Vault and reading them

To create a secret, simply call the `vault.create_secret()` function, as follows:

```
SELECT vault.create_secret(
  'THE_ACTUAL_SECRET',
  'A_SECRET_NAME',
  'Just some description'
);
```

Running this will confirm its creation by returning the specific UUID of the secret:

Figure 12.6: Running vault.create_secret

This key is encrypted and stored in the `vault` schema inside the `secrets` table. If you want to read it, you can do so using a special view:

```
SELECT decrypted_secret
FROM vault.decrypted_secrets
WHERE name = 'A_SECRET_NAME';
```

This will return **THE_ACTUAL_SECRET**, like so:

Figure 12.7: A decrypted secret

On `supabase.com`, Supabase Studio also provides a Vault UI where you can create secrets and list secrets. It's found at **Project Settings | Vault**:

Figure 12.8: The Vault UI

Now, you know how to get the secret inside of the database, but how does that help when you need it in business logic?

Using the secret in the business logic/within your application

In the next chapter, you'll see that the database is capable of handling scheduled tasks and even doing HTTP requests. That being said, you can have business logic directly within database functions. For such HTTP requests within the database, you already know how to retrieve Vault secrets – within the database – if you need to.

But the remaining question is how to get such a secret as part of code logic – for example, in your Next.js application. That's done by using what we already know about – RPCs. You can create a generic RPC, named `get_secret`, using **SECURITY INVOKER** and a `secretname` argument of type `text`, as follows:

```
BEGIN
  IF (auth.role() = 'service_role') THEN
    RETURN (SELECT decrypted_secret FROM vault.decrypted_secrets WHERE
    name = secretname);
  END IF;

  RAISE EXCEPTION 'no access';
END;
```

This function will only allow access to a Supabase admin client, as we check the roles in the `IF` block explicitly and raise an exception if the user (role) is not the `service_role`.

> **Note**
>
> If you also want to call this function within raw SQL queries (e.g., in the Supabase Studio SQL Editor), refer to the *Being careful with current_user usage and understanding auth.role ()* section to understand what additional conditions you'd have to add to the IF block of this function.

This means that this function cannot be called within a process started by the API where the client wasn't initialized with Service Role Key credentials.

With your application, you can then instantiate such an admin client and call supabase.rpc('get_sec', { secretname: 'A_SECRET_NAME'}) to retrieve any secret from the Vault.

Utilizing silent resets to avoid data manipulation

Within this book, we've written quite a few RLS policies and learned that access via RLS means access on a row. So, if I have **UPDATE** rights on a table, I can update whatever column I want within that row.

You can obviously extend an RLS expression to enforce certain limitations on columns, but have you ever thought about how to ensure that a created_at column stays the same and cannot be changed during an update? Have a think.

So, what's your solution for this? Maybe you were thinking about a way to somehow retrieve the existing created_at value and then compare it to the newly set value. However, trying to integrate such pseudo-column-level security within row-level security can be messy, extremely complex, or even simply impossible.

One solution to this problem is using actual column-level security, which we will discuss further in the next section, but sometimes, all you need is what I call a *silent reset*. A silent reset is a trigger that uses a function that ensures that certain values stick with previous values, even when a user tries to override them.

Doing this is very easy. Take a look at the following body of a trigger function:

```
BEGIN
  NEW.created_at = OLD.created_at;
  NEW.unchangeable_column = OLD.unchangeable_column;
  NEW.updated_at = now();

  RETURN NEW;
END;
```

Using this as a trigger function within a table's BEFORE UPDATE trigger will simply reset created_at and unchangeable_column, no matter which new value is provided.

Additionally, I added an update_at timestamp such that the user cannot manipulate the timestamp either; anytime the row is updated, the database sets the time when this happened.

This is a very easy but powerful utility to ensure data integrity. Now that you know how it works, let's move on to an alternative and complementary solution – column-level security.

Enabling column-level security/working with roles

In this section, I want to onboard you to **column-level security (CLS)**. This could probably fill a chapter on its own. While most people can live well without CLS, this section will allow you to decide whether you want to implement it.

Understanding CLS semantically is actually simple. You can define which database roles will be able to perform which operations on which columns. For instance, in a table such as service_users, you could define that a specific role can only read (SELECT) the id column. Trying to select anything else from that table will lead to failure then. CLS, being column-based, can be applied to SELECT, UPDATE, and INSERT (columns cannot be deleted; hence, DELETE always refers to a row that you want to control with RLS policies, not CLS).

Let's say we wanted the tickets table to only allow you to update the assignee column with CLS. First, we would have to revoke all UPDATE rights on that table to be able to give specific access to single columns instead. Revoking database operations such as UPDATE rights must be done using a database role. Our goal is to allow specific columns to be updated by signed-in users, and they have the authenticated database role, so we revoke the rights on the authenticated role:

```
REVOKE UPDATE ON TABLE public.tickets
FROM authenticated;
```

Next, you want to tell Postgres that the assignee column can be updated:

```
GRANT UPDATE (assignee)
ON TABLE public.tickets TO authenticated;
```

Now, when an update happens, the RLS determines which rows can be applied and the CLS prevents any modification from the user on any other column.

Note that granting CLS rights on multiple columns is done using GRANT UPDATE (col1,col2,col3,...).

This sounds great, and indeed, it works great, but it comes with a few caveats:

- Once CLS is enabled on a table and you've revoked specific columns, fetching rows with wildcard commands using plain select() or select('*') will fail. Instead, you must list the specific columns in the selection.

- You cannot have user-specific expressions and only expressions that grant or revoke column rights by role. This means that if you want user-specific CLS, you have to come up with more database roles, which you should only do if you have the required Postgres knowledge for it.

If you're interested in digging deeper into CLS, check out the documentation here: `https://supabase.com/docs/guides/database/postgres/column-level-security`.

There is also a UI that you can activate within your Supabase project to manage CLS visually. It's reachable with this link right within your project: `https://supabase.com/dashboard/project/_/database/column-privileges`. On your local instance, it's found at `http://localhost:54323/project/default/database/column-privileges`.

With this, you're all set with the basics of CLS to use it whenever you feel it helps you.

Understanding security on views and manually created tables

In this book, when creating tables, we used the Supabase Studio UI. Then, for those tables, we were able to create RLS policies. When no RLS policies are added, a table is safe, as it will prevent access to everyone but admins. However, some people assume that when creating a table with raw SQL, the same thing is true. That is not the case.

Here's an example of creating a `todos` table SQL:

```
CREATE TABLE todos (
   id SERIAL NOT NULL PRIMARY KEY,
   content TEXT,
   created_at TIMESTAMPTZ NOT NULL DEFAULT NOW(),
   updated_at TIMESTAMPTZ NOT NULL DEFAULT NOW(),
   completed_at TIMESTAMPTZ
);
```

Executing this will show the table in the Table Editor with an open lock icon, as well as state that RLS is disabled in the Table Editor view of Supabase Studio.

Figure 12.9: An unsafe table

To some, this might be obvious, but for many, it isn't. RLS is not activated by default on table creation. You can create tables with raw SQL, but you also then have to execute an additional statement to activate RLS on it, like so:

```
ALTER TABLE todos ENABLE ROW LEVEL SECURITY;
```

That covers RLS on tables. But in PostgreSQL, there is also the concept of a **view**, which, in essence, is just a filtered representation of an existing table. Given the fact that a view is a "live representation" of a table, most people assume it will inherit RLS from the table. It does not!

Let's have a look at what that means by using an example. You can create a view of the `todos` table to only show completed todos, like this:

```
CREATE VIEW completed_todos AS
    SELECT * FROM todos WHERE completed_at IS NOT NULL;
```

This view can now be used like a table, as well as in the Supabase client. However, you cannot write data to it; it's read-only. If you wanted to select completed todos with the Supabase client, simply use `supabase.from('completed_todos').select()`.

But as it is, the view doesn't care about RLS and is publicly exposed to everyone! This is because a view that is created from your admin user perspective (`postgres` user) has access to a table and will expose data (similar to a `SECURITY DEFINER` function). Therefore, you need to tell the view to respect RLS like this:

```
CREATE VIEW completed_todos WITH (security_invoker=on) AS
    SELECT * FROM todos WHERE completed_at IS NOT NULL;
```

Now, your view is secure and respects the same policies as the underlying table.

Changing the max_rows configuration

The `max_rows` configuration is a simple but effective strategy to avoid sending too many rows over the wire, either accidentally or maliciously. Its limit determines the maximum number of rows that can be returned by any selection request. By default, it is set to 1,000, but I usually tend to set it even lower, to something around 200.

The reason is simple – most UIs don't show 1,000 data points of anything, and if they do, they lazy-load chunks of it instead of loading all 1,000 data points at once.

So, my proposal is to start small and increase the data point value if needed. You can change it locally in `config.toml` and on `Supabase.com`; you'll find the setting within **Project Settings | API | API Settings** (further down the same screen shown in *Figure 12.3*).

Understanding safe-guarded API updates or deletion

In SQL, when you run a query such as `DELETE FROM table` or `UPDATE table`, it will obviously complete this for all rows if no `WHERE` clause is provided – you have admin rights. However, Supabase prevents you from using unlimited deletions or updates of rows when using the Supabase API (using the Supabase client, hence using PostgREST).

You can ensure that this setting is on by running this statement:

```
SELECT useconfig FROM pg_shadow WHERE usename = 'authenticator';
```

Inside the returned useconfig value, safeupdate should show up within the session_preload_libraries setting.

At this link, you can find out how to activate this extension for specific roles: https://supabase.com/docs/guides/database/extensions/pg-safeupdate#enable-the-extension. Now, I would love to activate this for the postgres admin role to avoid accidentally running DELETE expressions without a WHERE clause, but unfortunately, this cannot be enabled at the time of writing beyond the default. I've informed the Supabase team (because this was possible at some point), and they said they will track it at their end to work on it.

However, hypothetically, you should be able to enable this extension database-wide (for everyone) with the following SQL statement:

```
ALTER DATABASE postgres SET session_preload_libraries = 'safeupdate';
```

> **Note**
>
> I've created an issue on GitHub regarding this, where you can follow the issue's progress: https://github.com/supabase/supabase/issues/27455.

Adding middleware inside Postgres for each API request

This is one of the most unknown and underrated features that Supabase provides. The ability to add an API middleware will allow you to control the actual API request before it hits – for example, by implementing rate limiting. So, at its core, you have fine-grained control over API calls.

This consists of two parts:

1. Creating a public RPC function that returns tickets from a specific tenant
2. Safeguarding the RPC with PostgREST middleware

Let's say our ticket system has its own API for external developers to interact with tickets. For example, we want to allow developers to access tickets of a tenant at something like /api/tenant_id/tickets. Certainly, you can just add Route Handlers in the project and implement such an API with the usual code logic. But you can also use the existing PostgREST service to do that. This means there are no additional roundtrips on your server, and you don't need a backend application to build such an API. Sounds like magic!

> **Note**
>
> You'll find the following implementation in the Ch12_Security branch. Don't forget to run npx supabase db reset after starting it to keep all your content fresh.

To make this very simple, I'm going to show you how to implement an API on top of PostgREST to get tickets for a tenant, with special tenant-based API keys. For this, I've created a new schema called api (you can also do this in the public schema; I just like the separation) and created an api_keys table inside of it, like so:

```
CREATE SCHEMA api;

CREATE TABLE
  api.api_keys (
    api_key text not null,
    tenant text not null REFERENCES public.tenants (id),
    constraint api_key_id primary key (api_key),
    constraint api_key_tenant unique (tenant)
  );

ALTER TABLE api.api_keys ENABLE ROW LEVEL SECURITY;
```

This code will result in a table with two columns – the api_key being a text primary key and the tenant referencing an existing tenant. I also made the tenant unique, which means there can only be one API key per tenant (which makes things easier to explain).

Now, let's create an entry in the table, like so:

```
INSERT INTO
  api.api_keys (api_key, tenant)
  VALUES ('foobar', 'packt');
```

This means that foobar is the API key to access tickets for the packt tenant. You can see the key here:

Figure 12.10: The api_keys table

> **Note**
>
> Instead of saving `foobar` as cleartext, you can also use the Vault to only store the reference to an encrypted key.

Next, we want to create an RPC that, given a specific tenant, returns the most recent 10 tickets as JSON. So, within **Database functions**, create the `get_most_recent_tickets` function with **Return type** set to `json`. You would think that we also need a function argument for the tenant ID, but we don't, as we pass it via HTTP headers. Take a look at the following function logic to be used for the RPC:

```
BEGIN
  RETURN (
    SELECT json_agg(t.*) FROM (
      SELECT * FROM tickets
      WHERE

  tickets.tenant =
        current_setting('request.headers', true)::json->>'x-tenant'
      ORDER BY tickets.created_at DESC
      LIMIT 10
    ) t
  );
END;
```

Most of the function definition should be easy to understand – it selects tickets and converts the result into a JSON. But the WHERE condition I've highlighted is where the magic is. Each API request sets the current request information in the database for this specific transaction/call. What I'm doing in the WHERE clause is grabbing the HTTP headers as a JSON and accessing the `x-tenant` header. That means that we do not read a function argument but an argument from an HTTP header.

Before I can show you how to call that function, we must save it. As it will be called as an API from anywhere without being signed in, we need to make it run with **SECURITY DEFINER** to have access to all data (which is insecure, of course, but don't worry – we will safeguard this function in the *Adding middleware for PostgREST* subsection).

But how do we call that function now? At the time of writing, the `.rpc()` function unfortunately does not support adding custom headers. However, you can call the RPC with a `fetch` command using the PostgREST REST API instead (`https://postgrest.org/en/latest/references/api/functions.html`):

```
fetch('API_URL/rest/v1/rpc/get_most_recent_tickets', {
  method: 'POST',
  headers: {
    'apikey': 'ANON_KEY',
```

```
    'x-tenant': 'packt'
  }
})
.then(response => response.json())
.then(data => console.log(data))
.catch(error => console.error(error))
```

Here, you need to replace the highlighted values with your own Supabase values (remember that on your local instance, you can run `npx supabase status` to show those).

As you can see, we pass the `x-tenant` header with the `packt` value. Executing the given `fetch` command will then successfully give me 10 tickets from that tenant:

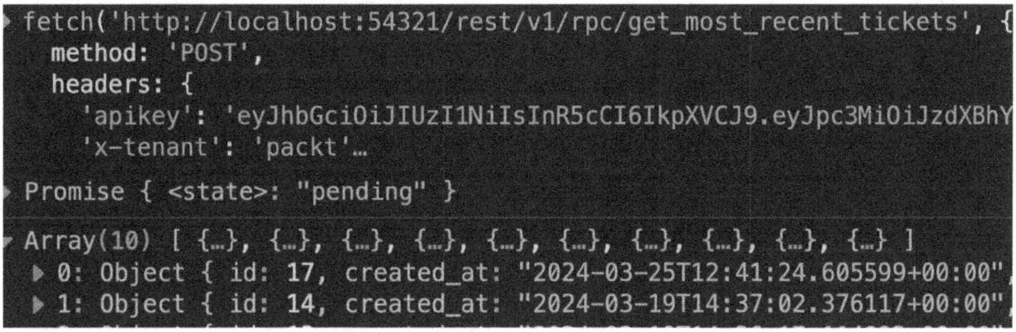

Figure 12.11: A sample result of calling the public RPC

But this is obviously super-insecure, as now everyone can get the most recent tickets from any tenant, especially since, by default, the OpenAPI feature exposes that this function exists! This is an obvious security hole, so let's fix that with middleware.

Adding middleware for PostgREST

In Supabase, you can define a function that runs before each PostgREST API request (no matter whether it's RPC or a table). In this function, which I call the PostgREST API middleware, we can now make use of the metadata we get with `current_setting` to also check the additional API key upfront. This will allow us to reject any invalid API keys immediately. Let's do it!

To achieve this `api-key` protection, we will add yet another function. This time, I'll give you the raw SQL definition for it; it's a rather long function but comparably simple to explain:

```sql
CREATE OR REPLACE FUNCTION public.my_api_middleware()
    RETURNS VOID
    LANGUAGE plpgsql
    SECURITY DEFINER
    as $$
```

```
DECLARE

given_api_key text := current_setting('request.headers', true)::json-
>>'x-tenant-api-key';
given_tenant text := current_setting('request.headers', true)::json-
>>'x-tenant';
given_request_path text := current_setting('request.path')::text;
grant_access boolean := false;

BEGIN
  IF given_request_path != '/rpc/get_most_recent_tickets' THEN
    RETURN; -- don't limit anything else
  ELSE
    grant_access := (EXISTS(
      SELECT FROM api.api_keys WHERE api_key = given_api_key AND
      tenant = given_tenant
    ));

    IF grant_access = true THEN
      RETURN; -- let the request pass through
    END IF;
  END IF;



  RAISE SQLSTATE 'PGRST' USING
  message = json_build_object(
    'code',    '123',
    'message', 'No permission found',
    'details', 'Wrong api key or tenant',
    'hint',    'Request your api key with your admin')::text,
    detail = json_build_object(
    'status',  403,
    'headers', json_build_object(
      'X-My-Custom-Response-header', 'Some data'))::text;
END;
$$;
```

In the `DECLARE` part, we define four variables that we can use within our function body:

- `given_api_key`, which contains the additional API key value from the header (`foobar`)
- `given_tenant`, which contains the tenant that has to match with the tenant's API key (`packt`).
- `given_request_path`, which contains the path of the API request – in our case, this will be `/rpc/get_most_recent_tickets`.
- `grant_access`, which is just a helper variable we fill to determine whether the tenant API key and the tenant match, initially set to `false`.

Next, in the function's body, we check whether the path is `/rpc/get_most_recent_tickets` or not; then, we simply want the function to finish by calling `RETURN;`, allowing us to continue the API request.

If, however, our middleware does match that specific function call, we define the `grant_access` boolean, based on whether the given values are inside of the `api.api_keys` table. If so, then the function returns (`RETURN;`) without throwing an error, which allows us to continue this request, as the API keys match.

However, if none of those cases were met, the function will continue executing and reach the state of running `RAISE SQLSTATE 'PGRST' ....`. This is similar to an exception/error, but it will make PostgREST (`PGRST`) respond with the given message data as a JSON object, and also allow us to modify the response using the `detail` value, such as setting the response status to `403 (Forbidden)`.

But wait! Don't be too impatient to test this. Creating this function now won't have any effect. We also need to tell Supabase to use this function as our middleware function. We can do so by executing the following code:

```
ALTER ROLE authenticator
  SET pgrst.db_pre_request = 'public.my_api_middleware';
```

With all of this in place, I can use the following headers in a `fetch` request (as we've done before) to get the most recent tickets:

```
headers: {
    'apikey': 'ANON_KEY,
    'x-tenant': 'packt',
    'x-tenant-api-key': 'foobar'
}
```

If I use the wrong headers (for example, set `x-tenant-api-key` to `snoobar`), I'll see a PostgREST error:

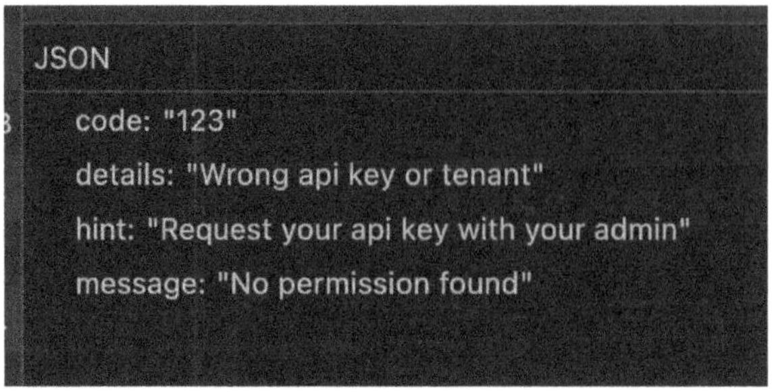

Figure 12.12: Calling the RPC with the wrong x-tenant-api-key

At this point, I hope you're as amazed as I am about the possibilities that we have within Supabase. However, although this is a fun demo, I want to remind you that this chapter is about security and this feature is here for a good reason – it allows you to precheck requests of any kind, away from actual function implementation. So, for instance, if you wanted to double-secure your public API with just another secret key, you now know how to do that!

> **Note**
>
> Supabase has a lovely example of how you can implement IP-based rate limiting. You can check it out here: `https://supabase.com/docs/guides/api/securing-your-api?queryGroups=database-method&database-method=sql&queryGroups=pre-request&pre-request=rate-limit-per-ip#examples`.

Using the Security Advisor

Supabase has a built-in system to detect potential threats within your database. You can find it within **Database | Security Advisor**. It is a list of advice, sorted by severity:

- **Error**: Something that should immediately have your attention
- **Warning**: Something that has less critical potential
- **Info**: About non-critical security knowledge.

How does it work though? For example, I have added a test RLS policy that uses the `auth.jwt()` `-> 'user_metadata'` expression. You've learned that the only metadata object on the user that is safe from being changed by the user is `app_metadata` and not `user_metadata`. The Security Advisor will detect that this is a security problem and list it in the **Errors** section:

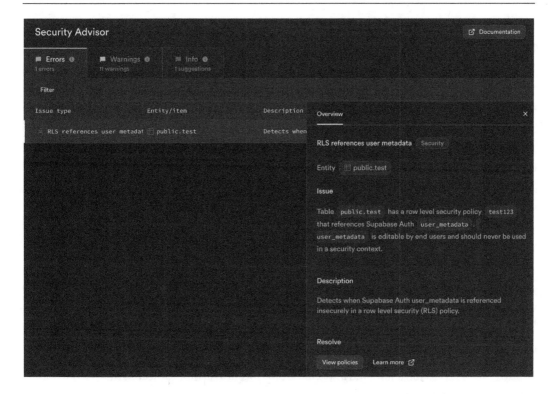

Figure 12.13: Supabase's Security Advisor

That being said, the Security Advisor is your best friend in regularly checking whether you have security-related issues in your database that you should work on. The Advisor UI really does explain itself.

Allowing a listing of IPs for database connections

With the PostgREST middleware I've shown you, you can block out any requests from specific IPs at the API level. Although this is covered in the link provided in the *Adding middleware for PostgREST* section's note box), let me quickly show you the necessary code to get the IP of a user who makes a request:

```
ip := SELECT split_part(
  current_setting(
    'request.headers', true
  )::json->>'x-forwarded-for',
',', 1);
```

However, since this is at the PostgREST level, this isn't useful when we want to restrict direct database connections. For that, we need network-level restrictions.

Say, for example, your server builds direct connections to the database of Supabase and your IPv4 server is just static, `123.123.123.1`. This means that your subnet for this IP is `123.123.123.1/32`, as you only have one IP (`https://mxtoolbox.com/subnetcalculator.aspx`). Likewise, your server will most likely have an IPv6 address network (e.g., `2001:0:0:0:0:0:0:0`). This means that its subnet notation is `2001:0:0:0:0:0:0:0/128`.

To make sure only your server can connect to the database, you can allow as many IPs as you want to be listed, using multiple `--db-allow-cidr` as follows:

```
npx supabase network-restrictions update \
  --db-allow-cidr 123.123.123.1/32 \
  --db-allow-cidr 2001:0:0:0:0:0:0:0/128 \
  --experimental
```

Once this is done, it will reject all connections from other addresses.

You can always check the status of such applied network restrictions by using the following command:

```
npx supabase network-restrictions get --experimental
```

On `supabase.com`, you'll also find a specific UI for this task within **Project Settings | Database | Network Restrictions**:

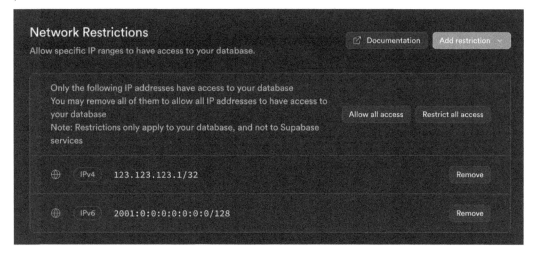

Figure 12.14: The Network Restrictions UI

Enforcing SSL on direct database connections

We've learned about connecting directly to the database in *Chapter 2*. This direct connection doesn't require SSL, which maximizes overall database client compatibility. However, most current clients do support SSL. You can enforce Supabase to only accept SSL-secured connections using the following command:

```
npx supabase ssl-enforcement --project-ref YOUR_PROJECT_ID update
--enable-db-ssl-enforcement --experimental
```

Here, I've used the `--project-ref` parameter (which is the ID of your `supabase.com` project in the API URL: `https://YOUR_PROJECT_ID.supabase.com`), but you can also use `--linked` if you've linked a project previously with `npx supabase link`.

Enforcing SSL for local development isn't useful, which is probably why it isn't possible with the CLI.

> **Note**
>
> You can find more information on enforcing SSL here: `https://supabase.com/docs/guides/platform/ssl-enforcement`.

Summary

In this chapter, we delved into the depths of Supabase security, equipping you with essential strategies to fortify your application. From concealing schema exposure to mastering role management and secret key regeneration, we left no stone unturned in safeguarding your data.

We explored the power of CLS and RLS to sculpt user permissions with precision. We unveiled the secrets of Supabase Vault for secure secret management and added battle-ready middleware to our PostgREST API for enhanced validation. Plus, with the Security Advisor by our side, we uncovered potential vulnerabilities and reinforced essential practices, such as IP allowlisting and SSL enforcement.

Armed with these tools and techniques, you're now ready to build and maintain a fortress of security around your Supabase application, ensuring that your data remains safe from unwanted manipulation and exposure.

In the final chapter, I'll supercharge your Supabase skills with advanced tips and tricks. Discover innovative features, optimize performance, and unlock the full potential of your application. Get ready for a fresh burst of inspiration and powerful enhancements to end your journey on a high note!

13

Adding Supabase Superpowers and Reviewing Production Hardening Tips

You've done a great job so far and you're so close to becoming a real expert in Supabase knowledge.

In this chapter, I'll give you some final tips to harden your application with the proper toolage within Supabase. I'll also show you some tools to improve your application workflow with Supabase, such as integrating Stripe directly as part of your database.

We'll not dive too deeply into the topics of this chapter, as it would call for another book. Instead, I want to provide you with the most necessary knowledge to help kickstart you into becoming a real expert in the field of Supabase.

Also, just as in the previous chapter, the following sections are not consecutive. I did, however, order them in a way that makes it easier to follow along but you're free to jump between them if you want.

So, in this chapter, we will cover the following topics:

- Making sense of `search_path`
- Familiarizing yourself with database extensions
- Adding an AI-based semantic ticket search
- Using anonymous sign-ins
- Transforming external APIs into tables with foreign data wrappers
- Using Webhooks

- Understanding Edge functions
- Using cronjobs to notify about due tickets
- Using `pg_jsonschema` for JSON data integrity
- Testing the database with pgTAP
- Setting `auth.storageKey` to avoid migration problems
- Extending `supabase.ts` with custom types
- Improving RLS and query performance
- Identifying database performance problems and bloat
- Working with complex table joins
- Reviewing the underestimated benefits of using an external database client
- Understanding migrations
- Utilizing database branching
- Disabling GraphQL or PostgREST (if you don't need it!)
- Using a dead-end built-in mailing setup
- Retrieving table data with the REST API and cURL

Technical requirements

The related code for this chapter can be found here: `https://github.com/PacktPublishing/Building-Production-Grade-Web-Applications-with-Supabase/tree/Ch13_Superpowers`.

Making sense of search_path

In *Chapter 12*, I showed you how to expose additional schemas to the API. At `supabase.com`, alongside the **Exposed schemas** configuration, you'll also find an **Extra search path** configuration:

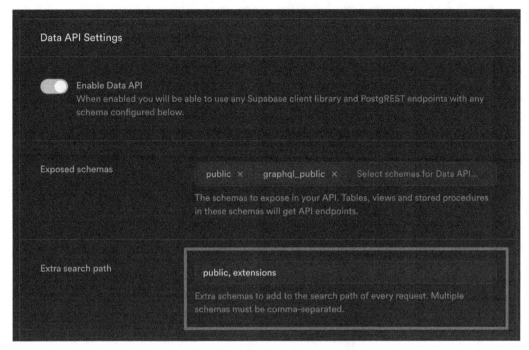

Figure 13.1: Extra search path (supabase.com)

The same configuration is found for your local instance in the `config.toml` file in the `[api]` section as follows:

```
extra_search_path = ["public", "extensions"]
```

For many, this opens the question, "What does this do and what is it good for?" To answer this question, I first want to discover what the search path does in Postgres and then explain the impact of this configuration.

Comprehending search path in Postgres

The **search path** is not Supabase-specific. It's a Postgres setting that is set for each Postgres database role and defines which schemas the database should look at to find data structures such as tables, functions, and views when there's no **fully qualified identifier** (from now on, I'll call it a *qualified identifier*).

A qualified identifier is a way to refer to a PostgreSQL resource, such as a table or a view, by specifying its complete path, including the schema name. For example, `public.tickets` explicitly identifies the `tickets` table in the `public` schema. This approach provides clarity and eliminates ambiguity, ensuring that the intended resource is accurately referenced.

While it's possible to use a shorter form such as `SELECT FROM tickets` without including the schema name, this can lead to potential confusion or errors. The database will search for the `tickets` table in the default search path, which includes the `public` schema by default. However, if there are multiple schemas with tables named `tickets`, the database may inadvertently access the wrong table.

Using qualified identifiers is a best practice because it leaves no room for ambiguity. Even though the shorter form may work in some cases, it's not an exact identification of the resource, and it relies on the default search path configuration, which may change or differ across environments.

Let's have a closer look at this configuration and what it means. If you move to the SQL editor and execute the `SHOW search_path;` command, it will output the following code:

```
"$user", public, extensions
```

> **Note**
>
> Don't get confused when the UI shows this value as `"\"$user\", public, extensions"`. Supabase adds quotes around the output of a SQL statement when the output is a string. If the output itself already contains quotes, it escapes those inner quotes with a `\`. The actual value you get when you copy it from the Supabase SQL Studio output won't include the additional quotes and without the escape characters.

As you can see, by default, for the `postgres` role, you can refer to data structures without needing to add the schema explicitly for the `public` and `extensions` schemas. But what does the `$user` schema refer to? It's a placeholder value referring to a schema with the same name as the user, which would be `postgres` for the `postgres` user role.

In other words, you can refer to resources by the names of these three schemas without needing to explicitly state the schema.

You can also configure this search path configuration. Let me show you how.

Setting search_path for a database role

Say, for example, you have a custom schema called `private` and you have a `private.get_secrets()` function defined. Now, if you don't want to write `SELECT private.get_secrets()`, but instead want the short-form `SELECT get_secrets()`, you must make sure that the search path is properly set as follows:

```
ALTER ROLE postgres SET search_path = "$user", public, extension,
private;
```

But in Postgres, you don't necessarily need to globally set `search_path`. Let's look at a temporary `search_path` solution.

Setting search_path temporarily with transactions and functions

You can set `search_path` for the time of a so-called database transaction only. A Postgres **transaction** is a sequence of database operations executed as a single unit, ensuring that either all operations complete successfully or none do; if one operation fails, everything is rolled back.

Setting the search path within a transaction looks like this:

```
BEGIN;
SET search_path = "$user", public, extension, private;
SELECT get_secrets(); -- => accesses private.get_secrets()
END;
```

If you execute this in the SQL editor, the search path is only set for everything between `BEGIN;` and `END;` and is not changed for anything outside of it.

> **Note**
>
> You can read more about transactions here, as we won't dive deeper into it (this is a Supabase book, not a Postgres one): `https://www.postgresqltutorial.com/postgresql-tutorial/postgresql-transaction/`.

Sometimes you may also want to do the same context-bound search path change when you're inside of a specific function. This is possible like so:

```
CREATE OR REPLACE FUNCTION public.my_awesome_function ()
RETURNS text
LANGUAGE plpgsql
SECURITY DEFINER
SET search_path TO 'public, private'
AS $$
BEGIN
   RETURN SELECT get_secrets(); -- => accesses private.get_secrets()
END;
$$;
```

`SET search_path TO 'comma, separated, schemas'` will set the search path to the given schemas for the context of the function execution. However, it's also recommended to set this to an empty string (`SET search_path TO''`) which will force you to use the full identifier (for example, `public.tickets`).

Now you're aware of search paths from a Postgres database perspective. Let's learn about the `extra_search_path` configuration of Supabase now.

Grasping the importance of extra_search_path

The `extra_search_path` configuration contains a list of schemas that are added to the search path of a request that is done with PostgREST (hence, when you use the Supabase client to fetch data).

When PostgREST executes database operations, it executes them as a single database transaction. This means that either all operations succeed, or none of them does. By default, PostgREST (and hence Postgres) can only access and operate on objects within the public schema during a transaction, as it is the only schema included in the search path.

However, you can extend the search path by using the `extra_search_path` option. This option allows you to specify additional schemas that should be included in the search path for the duration of the PostgREST transaction. With these extra schemas added to the search path, one can reference resources from those extra schemas without needing to use their fully qualified names. But when is that useful?

It is useful when, as part of the transaction, functions without the qualified identifiers are called. Let's say you have defined this simple `hello` function living in the `foobar` schema:

```
CREATE OR REPLACE FUNCTION foobar.hello()
RETURNS text
SECURITY INVOKER
LANGUAGE plpgsql
AS $$
BEGIN
  RETURN 'hello';
END $$;
```

Let's also say you have not exposed this schema to the API (as shown in *Chapter 12*). This means that this function cannot be called directly from the API.

However, you can call it within another function. For instance, say you define `public.call_foobar_hello()` as follows:

```
CREATE OR REPLACE FUNCTION public.call_foobar_hello()
RETURNS text
SECURITY INVOKER
LANGUAGE plpgsql
AS $$
BEGIN
  RETURN (SELECT hello());
END $$;
```

This now means `call_foobar_hello()` will be callable from the API as RPC – because it's in the `public` schema – however, the function it calls inside as part of the RETURN statement (`hello()`) is not specified with a qualified identifier. That means that the database can only execute it if `foobar` is in the search path.

From the previous section, you already know one option to add a search path on the function level. However, instead of setting the search path in the function itself, alternatively, you can set the schemas to be available globally for an API request. To do this, just change the `extra_search_path` configuration and set the schemas you want to be in the search path.

You now know that the `search_path` configuration in Postgres allows you to use unqualified identifiers and that the additional PostgREST configuration `extra_search_path`, adds schemas to the search path of API requests. Personally, I'm a big fan of using fully qualified identifiers for extra clarity. Hence, I never use unqualified identifiers for the reason of having less code to write.

However, there's one specific reason why adding a schema to `extra_search_path` can be required for database extensions. Let's take a look at that now.

Familiarizing yourself with database extensions

In the very first chapter of the book, I mentioned that the database (Postgres) within Supabase allows you to add extensions to become even more powerful. Such extensions are modular, self-contained units, which makes it easy to install and uninstall them. Each has its custom functionality built in and can expose functions in the schema they're installed; some, such as `pg_cron`, will also come with their own data structures (for example, tables). In this chapter, we will get to know a few such extensions, but for now, let's see how you can manage extensions generally.

Installing an extension in the default extensions schema

On the hosted instance on `supabase.com`, you can go to **Database | Extensions** to view a collection of available extensions, and enable or disable them by simply clicking the respective toggle button:

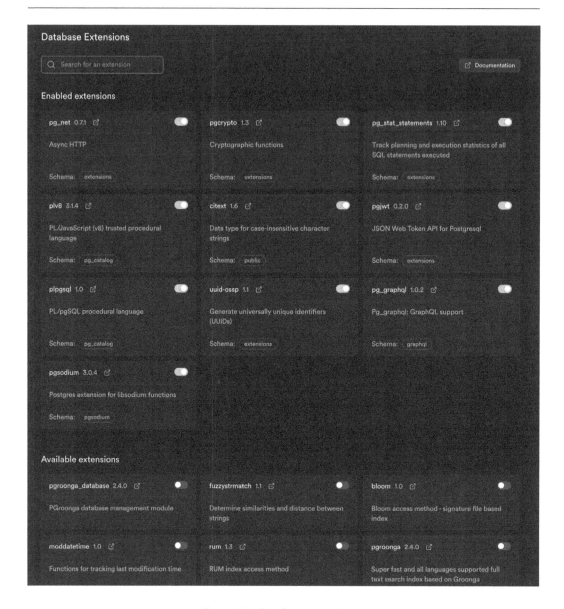

Figure 13.2: Database extensions

This also works the same way locally, or on a self-hosted version of Supabase, with no need to fiddle with the `config.toml`.

Besides using this Database Extensions UI to enable or disable extensions, you can also use raw SQL commands to enable a specific extension programmatically. Say, for example, you wanted to install the isn extension (`https://www.postgresql.org/docs/current/isn.html`), which is an extension for providing data types and functions for the international norm for product numbering **ISN**. You can install it programmatically by executing the following SQL:

```
CREATE EXTENSION "isn" WITH SCHEMA extensions;
```

This will return `Install the extension isn inside the extensions schema`.

> **Note**
>
> extensions isn't a special Postgres schema; it's just a prepared schema from Supabase for you to install extensions in to avoid installing extensions in the public schema. If you were to install an extension in the public schema and it adds functions there, those functions are exposed to the PostgREST API, which we definitely don't want.

Disabling an extension is then as easy as executing `DROP EXTENSION "isn";`.

Installing extensions in their own schema

When you install an extension, for example, isn, it will create whatever it needs in the selected provided schema. By default, that is the extension's schema. However, it is a recommended pattern to use a schema per extension. This is to avoid collisions between extensions and keep a clean structure. For isn, it would look like this:

```
CREATE SCHEMA IF NOT EXISTS isn;
CREATE EXTENSION "isn" WITH SCHEMA isn;
```

In the UI this is even easier. When you click the toggle button to enable the isn extension, you'll be shown a modal with a default selected (extensions schema). When you click the pre-selected schema, a dropdown opens and you can simply click **Create new schema "isn"** and it will create it along with installing the extension in it:

Figure 13.3: Creating a schema with extension installation

> **Note**
> There are a few rare extensions, such as `pg_cron`, that cannot be installed in a custom schema (they must be installed into the Postgres-internal `pg_catalog` schema). That's why I generally recommend activating such extensions in the UI over activating them programmatically as the UI will indicate such a situation and not allow you to select any other schema then.

For each extension installed in a custom schema you should do one more thing to ensure they can run smoothly: adding the custom schema to the search path of Postgres. Let me give you a quick sample. Later, in *Testing the database with pgTAP*, you'll find this line being executed:

```
SELECT plan(3)
```

It doesn't matter what this line does right now but the `plan` function is part of the `pgtap` extension. So, if you install the extension in its own custom same-named `pgtap` schema, you can either execute `SELECT pgtap.plan(3)` or add `pgtap` to the search path to omit the full identifier. However, even if you choose `SELECT pgtap.plan(3)`, the `pgtap` extension will internally execute the `_set` function (also coming from `pgtap`). But since it calls that internally without the fully qualified identifier, even running `SELECT pgtap.plan(3)` would fail due to it not being able to find `_set` internally (as the only default search paths are `public` and `extensions`).

That's why Supabase has the extensions schema selected as the place to install extensions by default, because it is already in the search path of both Postgres and the extra_search_path configuration.

You can easily set the search path to contain your custom extension schemas, though. Let's take the isn sample from before and add its schema to the search path as well:

```
CREATE SCHEMA IF NOT EXISTS isn;
CREATE EXTENSION "isn" WITH SCHEMA isn;

-- setting it for postgres (SQL editor)
ALTER ROLE postgres SET search_path = "$user", public, extension, isn;
```

The last line of the code block will extend search_path beyond the default path ("$user", public, extension) with the additional isn schema.

Additionally, you must add the custom extension schema (isn) to the extra_search_path configuration such that extensions can be properly executed when using the Supabase client.

Let me now tell you if and when using the programmatic approach vs using the UI for extension activation is useful.

Using the programmatic installation of extensions versus using the UI

Why would you choose the programmatic approach over the convenient UI-based approach for managing extensions? Let me elaborate. If you activate an extension in your local instance, someone else in your team won't have it activated very obviously because you did it on your computer. In *Chapter 5*, you learned about migrations and how they help you keep the same structure and setup of the database saved inside your repository.

Back then, we used this command to save the structure in migration files:

```
npx supabase db diff --local -f my_initial_structure
```

Now, when you activate an extension, no matter if in the extensions schema or a custom schema, this will diff all schemas and put the appropriate programmatic SQL commands into your migration files. Let's have a look at the flow:

1. I activate the isn extension with the custom isn schema in the UI.

2. I run npx supabase db diff --local -f added_isn_extension (more about this in this chapter in *Understanding migrations*).

3. The created migration file will automatically have the required programmatic SQL (CREATE EXTENSION IF NOT EXISTS "isn" WITH SCHEMA "extensions";), and by committing it in your repository, you're all set.

In other words, the programmatic approach is done for you. So, usually, if you don't specifically love writing the manual version over the UI, there's not much reason to do it.

However, as mentioned earlier, there are very few special cases, such as `pg_cron`, which is installed in `pg_catalog`. This will not be affected by the `db diff` command. In that case, since you can't choose the schema of the extension anyway, you must use your own added migration file and add `CREATE EXTENSION IF NOT EXISTS "pg_cron";` to it manually; hence, you don't specify the schema for this one. You'll read more about migrations in *Understanding migrations*.

Adding an AI-based semantic ticket search

In this section, we'll add a semantic AI search to our tickets, such that there's no need for the search to match exact words anymore but simply to match with the meanings of words. To implement an AI-based content search, we store a so-called **embedding**, a mathematical vector representation of the content alongside the row.

Even though embeddings are far more complex than this, I'd like to give you an analogy to understand embeddings at their core. Think of two topics, dogs and cars. Let's say, for a given set of words, we define the amount of dog topics and car topics between 0 and 1, where 1 means "highest possible semantic meaning of topic" and 0 means "no semantic meaning of the topic."

Now let's take the sample phrase, "My dog likes going for a walk." How much "dog" is in this phrase semantically? There's no clear answer to this as this is an analogy, but as this is clearly about a dog walk, let's take the number 0.7; it's not just about a dog but also going for a walk and liking things, but the overall semantic meaning is strongly attached to the dog, so 0.7 feels right. And how much "car" is in this sentence? I'd say very close to 0. Hence, our simple embedding would look like this: $[0.7, 0]$ (a vector with 2 dimensions). This is an extreme simplification of embeddings, but it allows us to understand why/how embeddings work.

We will use an existing AI model to create such embeddings. In our case, we'll create such an embedding for the title of a ticket and store it within the database. Then we implement a search where we convert the search phrase also to an embedding and compare those two to get the best-matching results.

First, to store embeddings (vectors), we need to activate the `pgvector` extension. So, within **Database | Extensions**, search for `vector` and click the toggle next to `pgvector`:

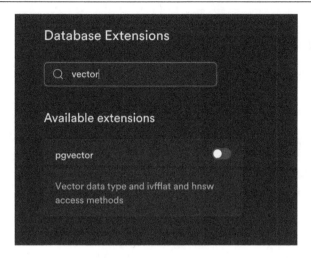

Figure 13.4: pgvector extension

Once the vector extension is enabled, let's look at what we need to do to achieve a semantic search:

1. We need an AI provider to create the actual embeddings based on text title and text content, for example, OpenAI, Claude, or Hugging Face.

2. We need to create a column within our tickets table that stores created embeddings.

3. We must fill the newly created column with embeddings.

4. We need to implement a search that takes a term, converts it to an embedding, and matches it with existing embeddings to find the best overall match.

So, let's get started.

Deciding on an embeddings provider

To create embeddings, you can use any provider that allows creating embeddings, for example, OpenAI, Claude, Gemini AI, or any openly accessible model on Hugging Face. The only important thing is to stick with one provider, as every provider and every model generates different embeddings and they're not interchangeable.

For example, Ollama (`https://ollama.com/`) is an awesome, open-source tool that runs AI models on your own system and also supports the creation of embeddings. Another model I love to use is the instructor-based model of the University of Hong Kong (`https://huggingface.co/hkunlp/instructor-base`). However, adjusting these open-source models to do what you expect them to do is usually more effort than using a generally well-established service provider. That's why I don't recommend starting with such OSS models in the beginning, and that's why we're using the OpenAI API instead.

More specifically, we'll be using the `text-embedding-3-small` model from OpenAI (`https://platform.openai.com/docs/guides/embeddings`).

Now let's move on to creating the embeddings column in our table.

Creating the embeddings column in the table

Creating an embeddings column cannot be done in the UI as the adjustable vector type will not be shown there. Instead, we will use the `ALTER` command to add an embeddings column to the `tickets` table. Here we use the `semantic_embedding` column name and set the embedding size (vector dimension) to `1536` (as our chosen model from OpenAI will give us vectors of this dimension):

```
ALTER TABLE tickets
ADD COLUMN semantic_embedding vector(1536);
```

You'll now see this newly added column in your table. Next, we fill it with actual embeddings from the API.

Creating embeddings with OpenAI

What we want to do now is to call this embeddings API for each ticket entry. So, for example, with the ticket title **Leave a good trace in this world**, we would want to create and store an embedding as follows (the following code block is pseudo-code to represent the logic we want!):

```
const embedding = getEmbeddingFromOpenAI('Leave a good trace in this
world');
storeEmbeddingInSupabaseDatabase(ticketId, embedding);
```

For the sake of generating the embeddings, we now create a `utils/openai.js` file that exports a function for generating the embedding:

```
const API_KEY = process.env.OPENAI_KEY;
const EMBEDDING_API_URL = "https://api.openai.com/v1/embeddings";
const EMBEDDING_MODEL = "text-embedding-3-small";

export async function getOpenAIEmbedding(input) {
  const { data } = await fetch(EMBEDDING_API_URL, {
    method: "POST",
    headers: {
      "Content-Type": "application/json",
      Authorization: `Bearer ${API_KEY}`,
    },
    body: JSON.stringify({
      model: EMBEDDING_MODEL,
      input,
```

```
    }),
  }).then((res) => res.json());

  return data[0].embedding;
}
```

> **Note**
>
> Make sure you add an `OPENAI_KEY=...` entry with your own OpenAI API key to your `.env.local` file for this to work.

Then, for this demo, I have prepared the following Route Handler that uses the Supabase admin client to fetch all tickets and then uses the newly created `getOpenAIEmbedding` function to store the returned embedding in our new table column. Add this code in the `app/[tenant]/embeddings-demo/route.js` file:

```
import { getSupabaseAdminClient } from "@/supabase-utils/adminClient";
import { getOpenAIEmbedding } from "@/utils/openai";
import { NextResponse } from "next/server";

export async function GET() {
  const adminSupabase = getSupabaseAdminClient();
  const { data } = await adminSupabase
    .from("tickets")
    .select("id, title, description");

  for (const ticket of data) {
    const embedding = await getOpenAIEmbedding(
      "Ticket title = " + ticket.title
    );

    adminSupabase
      .from("tickets")
      .update({
        semantic_embedding: embedding,
      })
      .eq("id", ticket.id)
      .then(console.log);
  }

  return NextResponse.json({ done: true });
}
```

You can then simply call this route at, for example, `packt.local:3000/embeddings-demo` (or even just `localhost:3000/embeddings-demo` if you have set `OVERRIDE_TENANT_DOMAIN` in the `.env` file).

Once executed, you'll see the generated embeddings in your `tickets` table in the database.

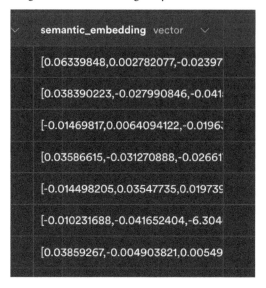

Figure 13.5: semantic_embedding column filled with embeddings

Now that we have the embeddings for each ticket stored in the database, let's create the last bit of the semantic search: the embeddings comparison.

Comparing embeddings to find matching search results

Given a specific search term, we have to convert that search term to an embedding, so that we can then search for similar embeddings. So, inside `TicketFilters.js`, adapt the placeholder first to become `placeholder="Search similiar tickets (AI)..."` (just so you don't forget you activated this feature).

Next, within `TicketList.js`, we currently use `countStatement` to get the tickets count and `ticketsStatement` to get the actual, paged tickets. Hence, the first intent is to adapt the filter conditions on both of those to use our embeddings for filtering and we're good to go. Unfortunately, this doesn't work, as none of the existing operators of PostgREST supports vector calculations (`https://postgrest.org/en/v12/references/api/tables_views.html#operators`).

This means we must build an RPC instead. Let's do that first, and then see how we can integrate it into our `TicketList.js`. Our RPC must take an existing embedding of a given search term and compare it against the ones we have in the `tickets` table. In Postgres, the so-called cosine distance can be calculated between two vectors using the `<=>` operator; the smaller the distance is, the more similar are the vectors. This leads us to the following RPC:

```
CREATE OR REPLACE FUNCTION match_tickets (
   search_embedding vector(1536),
   match_offset int,
   match_limit int
)
RETURNS SETOF RECORD
LANGUAGE SQL STABLE
as $$
   SELECT
   tickets.*,
   tickets.semantic_embedding <=> search_embedding as similiarity
   FROM tickets
   WHERE tickets.semantic_embedding <=> search_embedding < 0.45
   ORDER by tickets.semantic_embedding <=> search_embedding
   LIMIT match_limit OFFSET match_offset;
$$;
```

The given RPC takes a given embedding and selects only tickets from the database where the cosine distance is smaller than 0.45. A cosine distance is between 0 and 2 where 0 means they're identical and 2 means they're the total opposite. So, a value below 0.45 can be a good fit to find similar embeddings, but it always depends on the embeddings, hence the model and what you provided it. The results are then ordered by distance, and you can pass the offset and limit for the returned rows (e.g., to implement paging as part of your search).

For the sake of complying with our existing code, I've also added an additional RPC for the count only as I didn't want to change the existing `TicketList` code by overcomplicating both functions into one:

```
CREATE OR REPLACE FUNCTION match_tickets_count (
   search_embedding vector(1536)
)
RETURNS int8
LANGUAGE SQL STABLE
AS $$
   SELECT COUNT(tickets.id) FROM tickets
   WHERE tickets.semantic_embedding <=> search_embedding < 0.45;
$$;
```

Now, right before `const { count } = await countStatement;`, let's make sure to use the AI RPC search instead of the usual non-AI-based `select()` whenever we want the AI search to be active (when `useAiSearch` is set to `true`):

```
...
import { getOpenAIEmbedding } from "@/utils/openai";
...
const useAiSearch = true;
if (searchValue && useAiSearch) {
  const embedding = await getOpenAIEmbedding("Ticket title = " +
    searchValue);

  ticketsStatement = supabase.rpc("match_tickets", {
    search_embedding: embedding,
    match_offset: startingPoint,
    match_limit: 6,
  });

  countStatement = supabase.rpc("match_tickets_count", {
    search_embedding: embedding,
  });
}
```

If I now enter `i should relax` as the search term, it creates a vector representation of it using OpenAI and does a similarity search with our RPC in the database.

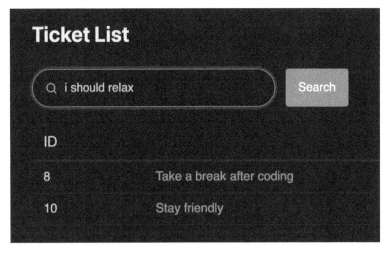

Figure 13.6: AI-powered ticket search results

Although the input content, being the title, is very small and does not contain "relax" at all, it seems to deliver a good semantic match – after all, "relax" can relate to taking a break from a hard task or taking a "relaxed" (friendly) approach to something. This is just a very basic sample to show you how embeddings can be integrated with Supabase.

Now you know how to use embeddings within `pgvector` and OpenAI, which also means you're ready for AI-powered application development. But before we finish, let me add a few final remarks:

- The way we implemented our AI search means we make a request to OpenAI each time the search is triggered. This can obviously be improved by caching results that were searched for before.

- In a simple search field like this, I'd usually not want to wait for an external API response as this can make your application feel unresponsive if the external API takes longer than expected. A way to avoid that would be to set a quick timeout for the response to cancel (e.g., fewer than three seconds).

- We have added an `embeddings-demo/route.js` file that sets embeddings for each ticket in the database. We did this as we wanted to create embeddings for all tickets that we already had in the database so that we're able to search those with embeddings. However, usually, you'll want to create an embedding along with the ticket creation instead, so one embedding at a time.

- As stated before, every model behaves differently, so you have to play around with the values such as the maximum cosine distance until it meets your expected outcome.

- When using other models than OpenAI embeddings, you must ensure that vectors are normalized (becoming unit vectors) before you store/use them; otherwise, you'll end up with values you cannot compare.

- We have only created embeddings for the title. If you create embeddings for better content than just my test content, you probably will see an increase in quality. Try playing around with creating embeddings for the ticket description as well.

- Supabase has cooperated with Hugging Face to support text embeddings without any additional API needs. A sample can be seen at the following link: `https://supabase.com/docs/guides/database/extensions/pgvector?queryGroups=database-method&queryGroups=database-method&database-method=sql`. However, there are some pitfalls in using it with Next.js, which are covered here: `https://huggingface.co/docs/transformers.js/tutorials/next`.

Using anonymous sign-ins

Nowadays, many apps provide the possibility to use an app like a signed-in user, even when not signed in, to be able to test its features and tempt users to sign up. In many cases, that's done by differentiating between signed-in users and non-signed-in users – this differentiation is a condition that adds complexity to code (as every condition does).

The trick to bypass this differentiation is to create fake users. For instance, using the fake address some-random-hash@user.yourdomain.com would still create a real account on the system. But then you'd have to make sure that this account is confirmed upfront, so long story short, all of this makes your app's code convoluted.

Supabase overcomes most of this complexity by allowing you to create users without needing an email. It creates a user with the internal information that this user is anonymous. Using supabase.auth.signInAnonymously().then(({ data: { user, session } }) => {}) will create a new, anonymous user and immediately sign in with that user (having a valid Supabase user session) and return the user object and session for that newly created anonymous user. At this point, the user of your app is authenticated with an unconfirmed, email-less, account.

After calling this, you'll also see such an **Anonymous** user listed in Supabase Studio:

Figure 13.7: Supabase Studio users

When you click on the user details of the anonymous user (which you can find through **Authentication | Users | View user info**), you'll see the user object with the is_anonymous attribute set to true:

```
{
  „instance_id": „00000000-0000-0000-0000-000000000000",
  „id": „23f5379b-bcf4-4db5-956e-f56d1e659201",
  "aud": "authenticated",
  "role": "authenticated",
  "email": null,
/** ... many more ... **/
  "is_anonymous": true
}
```

So, when you get a user object from Supabase (for example, via supabase.auth.getUser()), it will have this is_anonymous property.

This sounds very straightforward; however, there are a few important things to consider:

- This feature is disabled by default, as calling the signInAnonymously() function many times easily fills up your database with anonymous users.

- An anonymous user isn't running with the `anon` role but, just like a normal signed-in user, with the `authenticated` role. This is extremely important to acknowledge as enabling an anonymous user could change your policy behavior. For example, some people have very simple apps that only differentiate between a user being signed in or not. If someone is signed in, then that person gets access to certain blog articles, for example.

  Before enabling anonymous users, it was enough to create a `SELECT` policy with the RLS expression being `true` and the role set to `authenticated`, so that everyone who is signed in gets access. But as anonymous users also are authenticated, you'd need to expand the expression if you wanted to exclude anonymous users from being able to read the articles. That may look like this: `(auth.jwt() -> 'is_anonymous')::boolean = false`.

- Just calling `supabase.auth.signInAnonymously()` at page load is an antipattern as it will not check whether there's an existing user (even if there's already an anonymous one) that's signed in. Hence, it will just create another anonymous user. This not only needlessly fills the database but also effectively assigns the current user to a new (anonymous) Supabase user and whatever the user had done before will be lost. The solution here is to check the backend to see if someone is signed in, and if not, call `signInAnonymously()`. Alternatively, on the frontend, you can wait for the first initialization to happen and check whether the user is signed in or not (as shown in this video: `https://www.youtube.com/watch?v=Exiv8D1W5Mc`).

Now, imagine an anonymous user, for example, someone within an e-commerce application. The user wants to order a product, and by doing so, they create an account with all the items in the shopping cart being saved to the created account. What you want to do then is to convert the anonymous user to a real user. You can call `supabase.auth.updateUser({ email })` to set an email for the current anonymous user. With such a non-admin call of `updateUser({ email })`, usually, an existing user can update its email address and an email gets triggered to confirm that new email address: the `email_change` template is sent (see *Chapter 4* for more on built-in templates).

When this is being called for an anonymous user, however, not the `email_change` but the `confirmation` template is being sent to ask the user to confirm the created account. That means using `updateUser({email})` on an anonymous user triggers the same flow as calling `.signUp(..)`, sending a built-in confirmation email. In our project, we have worked with our custom emails instead of the built-in ones, so you'd need to make sure to send an activation link on your own (for example, by having a Route Handler calling `supabase.auth.admin.updateUser(..)`).

If you're using OAuth, as with Google Login, you'd trigger `supabase.auth.linkIdentity({ provider: 'google', options: {...}})`, which behaves the same way as `signInWithOAuth` (covered in *Chapter 7*) but converts the existing anonymous user into a real account using OAuth.

That's it. You're aware of the anonymous user feature and are capable of using it to improve whatever feature you have in mind.

Transforming external APIs into tables with foreign data wrappers

Imagine this – instead of querying an API to store data inside your database or to retrieve certain information, the only thing you have to do is select data from your database instead. This sounds like some kind of magic, and indeed it is.

Supabase supports so-called **foreign data wrappers** (**FDWs**). These FDWs act as if they're just another table in your database, which means you can query them like tables with the Supabase client, but they are essentially a proxy between the database and the API. That means you can connect an external source that will act as if it's just another table and you can hence query it as such.

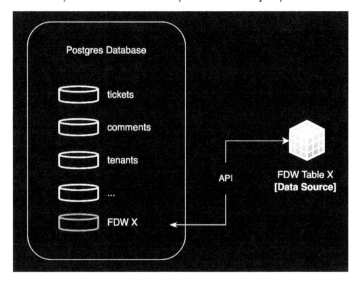

Figure 13.8: FDW visualization

This is best learned by using it, so let's do that. Within Supabase Studio, you can activate the FDW functionality by either activating the `postgres_fdw` extension in the **Database | Extensions** section or by clicking **Enable wrappers** in **Database | Wrappers** section:

Figure 13.9: Enabling FDWs

Once the wrappers are enabled, you can click on + **Create new wrapper** (or + **Add wrapper**) and you'll be asked for which API you want to create a wrapper. You'll see a list of supported APIs. In this sample, we're using the payment provider **Stripe**, so click on it!

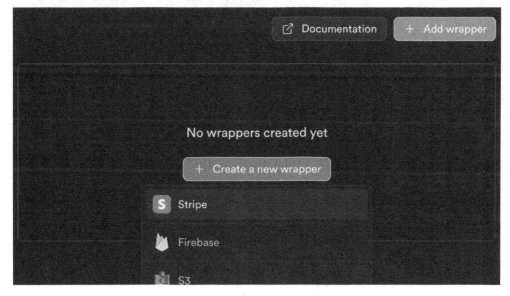

Figure 13.10: Button for adding a Stripe wrapper

> **Note**
> If you don't have a Stripe account, you can create a free Stripe account without providing business details. Then, after you've activated your account, go to `dashboard.stripe.com`.

Once you've selected **Stripe**, a mask opens (*Figure 13.11*) where you need to fill in **Wrapper Name** and **Stripe Secret Key**. For **Wrapper Name**, just enter `stripe`. Your Stripe secret key is found within the **Developers | API keys** area of your Stripe account (don't use **Publishable Key**, you need **Secret Key**). The **Stripe API URL** value for the wrapper is prefilled, so you can leave it as is.

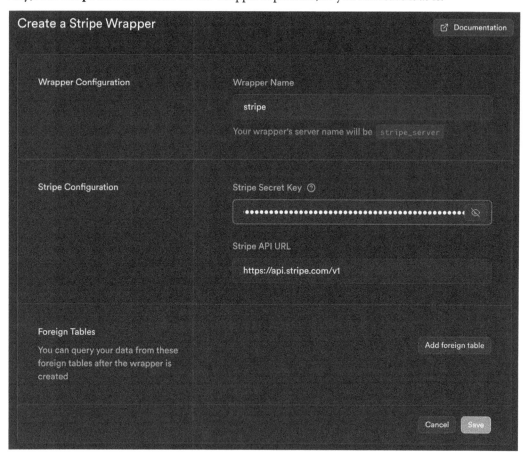

Figure 13.11: Stripe wrapper mask

Next, you have to choose data sources from Stripe to be reflected inside your database with the wrapper. To do that, click on **Add foreign table**. A mask will open where we will choose **Customers** as the data source from Stripe. For the schema, just leave it as `public`. Then choose a table name that will hold the Stripe data and that you can query for Stripe customers later. Simply enter `stripe_customers`. Then you can select the columns to fetch but, by default, all the columns are selected, so let's just click on **Save** to add this data source to our wrapper.

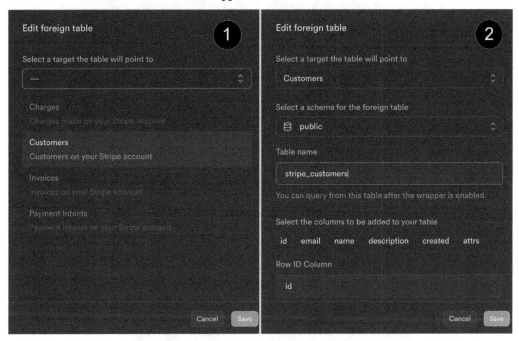

Figure 13.12: Selecting data from Stripe to reflect in our database

Then, click on **Save** on the wrapper mask to finally create the wrapper. Inside **Database | Wrappers**, you'll then see your wrapper listed (you can also edit it if needed):

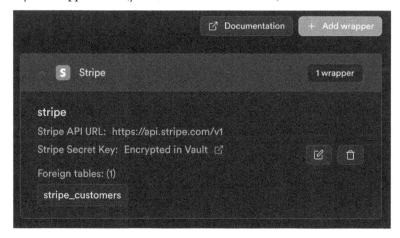

Figure 13.13: Stripe wrapper list

But this is just showing you that the wrapper is connected now. Far more interesting is experiencing how flawlessly this works. Go to the Table Editor and you'll find a highlighted `stripe_customers` table with an **F** icon indicating that its data is coming from an FDW:

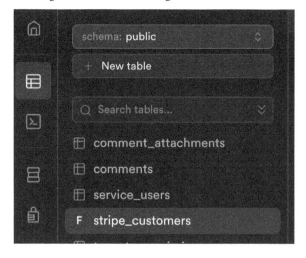

Figure 13.14: Stripe wrapper table in the Table Editor view

When you're using a fresh Stripe test account like I do, this table is naturally empty. So, let's get some data into it by signing in to your Stripe account, moving to the dashboard, and creating a few test customers. Immediately after creating four test customers in my Stripe account, I see them listed in `stripe_customers`. Isn't that magic? No more effort in doing Stripe API requests for stock data!

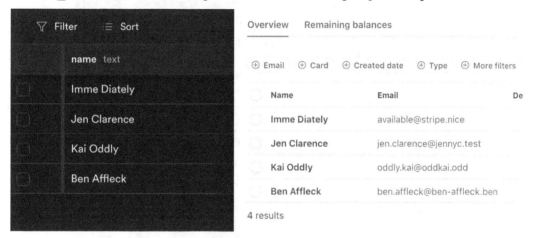

Figure 13.15: Supabase data (left) and Stripe data (right)

> **Note**
> The list of supported FDWs gets larger and larger over time, so if there's an FDW integration you're missing, just propose it in the GitHub discussions of Supabase: `https://github.com/orgs/supabase/discussions`.

Using webhooks

Webhooks are one of the most powerful and super-convenient features of Supabase. Under the hood, they're just database triggers that make HTTP requests with the `net` Postgres extension, but the webhook UI makes creating such triggers even simpler as it abstracts away complexity.

For example, when you want to trigger a specific URL whenever a new ticket is created, you can navigate to **Database | Webhooks**, click on **Create new hook**, then set **Table** as `tickets` and **Events** as `Insert`. Finally, all you need to do is to provide an endpoint URL for where the webhook request shall be sent (see the **HTTP Request** section in *Figure 13.16*).

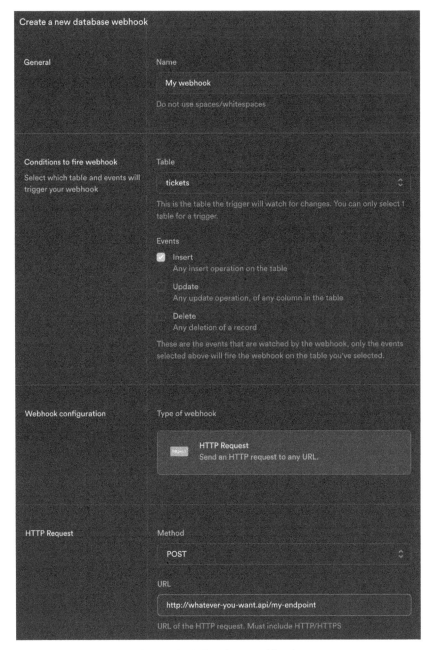

Figure 13.16: Creating a webhook

If you choose POST as the method of the request, Supabase will also internally make sure to send row data along the POST payload of the request. For example, for an **INSERT** happening on the tickets table, the payload on your endpoint will receive something like this:

```
{
  type: "INSERT",
  table: "tickets",
  record: {
    id: 1,
    created_at: ...,
    ...
  },
  schema: "public",
  old_record: null
}
```

The equivalent for creating this webhook via the webhook UI in raw SQL is as follows:

```
CREATE TRIGGER "my_webhook" AFTER INSERT
ON public.tickets FOR EACH ROW
EXECUTE FUNCTION supabase_functions.http_request(
    'http://whatever-you-want.api/my-endpoint',
    'POST',
    '{"Content-Type":"application/json"}',
    '{}',
    '1000'
);
```

If you look closely, you might think that the raw SQL statement will not send the row payload to the provided URL as all we pass is the URL itself, the method, a header object, an empty parameters object, and a timeout value. However, this is handled as part of the custom supabase_functions. http_request function. So, both variants will use this function and read the event data from the trigger and pass it to the provided URL in case of a POST request.

> **Note**
>
> It's extremely important to understand that your locally running Supabase will not be able to trigger webhooks on non-public-internet endpoints, so `http://localhost...` endpoints cannot be triggered from your local instance. This has nothing to do with Supabase but with the fact that the Webhook runner runs inside a Docker container. Inside the Docker container, localhost references the *container's* localhost, not your localhost! Usually, if you want to reach your host network from within a container, you can use the hostname `host.docker.internal`. So, if you have a webhook endpoint running on your system at `http://localhost:5000/my-webhook`, you must use `http://host.docker.internal:5000/my-webhook` when you add a webhook in your database.
>
> The upcoming subsection will also simplify this whole process of changing the URL hostname.

Next, let me show you how to generalize webhooks to the point that you don't have to recreate them for each environment (localhost, test, and production).

Creating webhooks with dynamic URLs per environment

The way webhooks are defined has one big downside – it doesn't respect which environment to run the webhook in (for example, a local server, a testing server, or the actual production server); it just triggers the URL you defined. But this isn't something that production web apps do. If we are on a test database, we don't want to run production hooks or vice versa. So, locally, we want to trigger `localhost`, but on production, we might want to trigger `my-production-system.domain`.

There's a trick for that. The idea is to store a base URL in each instance that is then used to construct the actual webhook URL. In your webhook definition, you then don't need to hardcode that URL but only provide the proper pathname of the URL (such as `/endpoint-abc`), and prefix it with the base URL. That way, you can change just the base URL in each environment and reuse the webhooks.

We've already learned about a lovely place to store secrets, the vault (we did this in *Chapter 12*). Even though the URL itself isn't security-sensitive, the vault is still a good place to store global key-value configurations. So, let's store our base URL there now:

```
SELECT vault.create_secret(
  'http://localhost:3000',
  'WEBHOOK_BASE_URL'
);
```

Now that we have this `WEBHOOK_BASE_URL` available, we want to create our own version of the existing `supabase_functions.http_request` trigger function, which only takes the URL pathname (`/my-endpoint`) and prefixes it with the base URL from the vault.

To create our own version of that custom Supabase `http_request` function, we first need to know the definition of the original one. You can get the full function definition with this statement:

```
SELECT
  pg_get_functiondef(
    'supabase_functions.http_request()'::regprocedure
  );
```

This gives you the source code of the original function. Then, copy the source code into the SQL editor and we'll adapt it to fulfill our needs of being environment-aware. All you have to do is to adapt the highlighted places, as shown in the following code block, accordingly (the rest of the code stays the same):

```
CREATE OR REPLACE FUNCTION public.http_request_with_base_url ()
 RETURNS trigger
 LANGUAGE plpgsql
 SECURITY DEFINER
 SET search_path TO 'supabase_functions'
AS $function$
    DECLARE
        request_id bigint;
        payload jsonb;
        url text := TG_ARGV[0]::text;
        ...
    BEGIN
        IF url IS NULL OR url = 'null' THEN
         RAISE EXCEPTION 'url argument is missing';
        END IF;

        IF (starts_with(url,'http://') OR starts_with(url, 'https://'))
        = false THEN
          url := CONCAT((SELECT decrypted_secret FROM
          vault.decrypted_secrets WHERE name = 'WEBHOOK_BASE_URL'),
          url);
        END IF;
    ....
  END;
$function$;
```

There's no need to scan through the rest of the function (it reads the trigger data and does the HTTP request). All we did was change the function name (using the `public` schema as we don't want to interfere with the one from Supabase) and adapt how the URL is resolved. This trigger function now checks whether a given Webhook URL starts with `http://` or `https://`, and if not, we prefix it with the value from the WEBHOOK_BASE_URL secret. This way, you stay flexible and can provide either a static URL or just a pathname that can change per environment.

You might ask yourself, why didn't we simply use the `public.get_secret(secretname)` function that we created in the previous chapter for accessing the `WEBHOOK_BASE_URL` vault secret? Why am I using `SELECT decrypted_secret ... instead`?

The reason is that we created the `public.get_secret` function with strong restrictions (check *Using the secret in the business logic / within your application* in *Chapter 12*). It will only work for the `service_role` key. That means if we used the function here and a ticket system user did something to trigger a webhook (such as creating a ticket) via the API, it would fail to execute `get_secret` with a normal, RLS-respecting Supabase client. This is because the trigger would be run as part of the insertion transaction of the database and hence `auth.role()` would be authenticated and `get_secret` would reject by throwing an exception. That's why I went for the raw SQL `SELECT` instead.

> **Note**
>
> If you don't like the raw `SELECT` in the previous function and want to use a more convenient `get_secret` function approach, I recommend creating a custom schema where you can put such a function. For instance, you can create a schema called `private` in which you define a `get_secret` function (as described in *Chapter 12*, specifically the *Using the secret in the business logic/within your application* section) but without the role checks. As it's in a custom schema, the PostgREST API naturally won't have access to the function and you can call it inside the database as you like.

Once you've executed the creation of your new `public.http_request_with_base_url` trigger function in the SQL editor, you can create environment-based webhooks as follows:

```
CREATE TRIGGER "my_tickets_webhook" AFTER INSERT
ON public.tickets FOR EACH ROW
EXECUTE FUNCTION public.http_request_with_base_url(
  '/your-webhook/endpoint',
  'POST',
  '{"Content-Type":"application/json"}',
  '{}',
  '1000'
);
```

This trigger would always make an HTTP request to `valueFromVault(WEBHOOK_BASE_URL)/your-webhook/endpoint` whenever a new ticket is created. So, you can use the same trigger on any environment and you just need to change the vault secret URL. So helpful!

Understanding Edge Functions

Edge is a term that many developers at least have heard of. Being on the "edge" refers to being as close to the user as possible with the help of a proper content delivery network that spreads data/code to multiple servers around the globe. This proximity reduces latency and improves performance by ensuring that data and code are as close to the end user as possible.

One practical application of this concept is seen in Supabase's **Edge Functions**. Edge Functions in Supabase are small pieces of TypeScript with the Deno runtime (`https://deno.com/`). These small pieces of code are executed right within the Supabase infrastructure, so you don't need to deploy them elsewhere. They're very good for isolated, small tasks that you can run independently from your application. But when should you use such Edge Functions specifically, where should you write them, and how should you use them? Let's have a look.

> **Note**
> The Deno runtime runs TypeScript files natively and can also make use of usual JavaScript packages (`https://jsr.io/docs/with/deno`).

Understanding when to use Edge Functions

It's important to know that Edge Functions can't do anything that you couldn't do with your application code. However, you can outsource small tasks that you don't want or need to have in your application logic. For instance, earlier in this chapter, we added embeddings to tickets, and I concluded that it would make sense to generate such embeddings per ticket when a ticket is created.

Hence, an Edge Function can be triggered when a ticket is created to exactly do that. This Edge Function then runs independently of your application code, goes to the Embeddings API, and updates the ticket with its embeddings data when done. All of that is independent from blocking your actual application code as it runs on the edge server, not on your Next.js application.

But how do we achieve that? Let's have a look.

Creating an Edge Function that runs for new rows

We will now create an Edge Function that runs for newly created tickets. To make it simple, I won't show OpenAI usage here; instead, we want our Edge Function to simply add the newly created ticket's title to a test table to confirm that our Edge Function works.

For this, execute the following statement in your SQL editor to create a new RLS-based table:

```
CREATE TABLE public.testtable (
  id int8 GENERATED BY DEFAULT AS IDENTITY,
  content text null,
  CONSTRAINT testtable_primary_key PRIMARY KEY (id)
);


ALTER TABLE public.testtable ENABLE ROW LEVEL SECURITY;
```

This creates a `testtable` instance with the usual incrementing `id` column, as well as an additional content column that we can fill with whatever text we like.

Next, we need to initialize a new Edge Function by running `npx supabase function new added_ticket`, where `added_ticket` is the function's name. A new file will appear, `supabase/functions/added_ticket/index.ts`, containing TypeScript code as it runs with Deno.

Within the created file, you'll find the following: `Deno.serve(async (req) => { ... })`. This is your edge request handler and is where your code logic must reside.

What we want to do is handle a request that gives us the ID of a newly inserted ticket as a payload, search for that ticket within the database with the Supabase client, and insert the ticket title and ticket ID as text into our `testtable`. To do this, replace your existing `Deno.serve(...)` code within `index.ts` with the following code:

```
import { createClient } from "https://esm.sh/@supabase/supabase-js@2";

Deno.serve(async (req) => {
  const data = await req.json();
  const ticketId = data.record.id;

  const supabaseAdmin = createClient(
    Deno.env.get("SUPABASE_URL"),
    Deno.env.get("SUPABASE_SERVICE_ROLE_KEY")
  );

  const { data: ticket, error } = await supabaseAdmin
    .from("tickets")
    .select("title")
    .eq("id", ticketId)
    .single();

  await supabaseAdmin.from("testtable").insert({
    content: `New ticket added: "${ticket.title}" with id:
${ticketId}`,
```

```
  });

  return new Response(JSON.stringify(ticket), {
    headers: { "Content-Type": "application/json" },
  });
});
```

This file imports `createClient` (the Deno way) to create a Supabase client. In the request handler, it expects the provided data to be a JSON (`await req.json()`) and assigns the passed `record`. `id` of the payload (`data`) to `ticketId`. This means that we expect our payload given to our Edge Function should be an object that contains `{ record: { id: TICKET_ID_HERE } }`. Then, the Supabase admin client is instantiated with environment variables that are automatically available via `Deno.env.get`.

> **Note**
>
> You cannot just use SUPABASE_URL, SUPABASE_ANON_KEY, or SUPABASE_SERVICE_
> ROLE_KEY with `Deno.env.get`. At the time of writing, the edge secrets are different than the vault secrets (you can list all of the secrets and manage the edge secrets with `npx supabase secrets`), but the Supabase team is aware of the confusion and plans to improve this.

We're then selecting `title` from the `tickets` table where the ticket ID is equal to `ticketId`. When that succeeds, we create an entry in our `test` table with `New ticket added: ....`

Now that the function is ready, we must ensure to trigger this Edge Function when a new database row is added.

> **Note**
>
> New or changed Edge Functions will only become active after restarting your local Supabase instance with `supabase stop` and `supabase start`, so do that now.

Triggering the Edge Function

Triggering the Edge Function can be done with an HTTP request (for example, with a webhook), but there are a few things to consider.

First and foremost, you have to know that *every* Edge Function is only executed if the Anonymous Key is also present within the HTTP header of the request (`Authorization: Bearer ANON_KEY`). An Edge Function always lives at `API_URL/functions/v1/function_name`. Hence, you can run a `fetch` command with JavaScript on the respective URL and set the additional `Authorization` header with ANON_KEY to trigger such an Edge Function.

This is the baseline, but there are a few further differences to respect between running them locally and on `supabase.com` to avoid running into pitfalls.

Triggering locally

In the *Using webhooks* section, I described how to use webhooks to trigger an HTTP request. I showed you that such a webhook will automatically receive the payload of a newly added row in the format we expect it in our Edge Function. So, we can use that for triggering an Edge Function.

For localhost, I've already noted (within *Using webhooks*) that you must use `host.docker.internal` inside the webhooks to reference the actual `localhost` on your system. That means that you can call an Edge Function named `added_ticket` locally by making a request to `http://host.docker.internal:54321/functions/v1/added_ticket`. However, you can also use `http://kong:8000/functions/v1/added_ticket`, which does the same thing locally. Let me explain why. Have a look at the following figure:

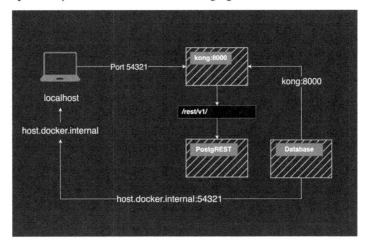

Figure 13.17: Local instance setup

Back in *Chapter 1*, you've already seen the big picture of how the Supabase services are connected. Now, in *Figure 13.17*, you see a simplified diagram where on the right side there are only Kong, PostgREST, and the database (Postgres) shown. On the left side, there is your computer (localhost). That means the database can either make an HTTP request to the Supabase API through your computer host, and by doing so hit Kong, which will forward that request to the PostgREST container; alternatively, the database can also just directly make a request to the proxy manager Kong via the internal Docker network. The container name is `kong` and the exposed container port of `kong` is `8000`.

This information might seem redundant in the context of the local instance but it can be very helpful when you decide to self-host, because then you might not have `host.docker.internal` available, and accessing the Kong container directly with `http://kong:8000` is straightforward.

> **Note**
>
> We don't cover self-hosting in this book as the demands for types of hosting (Docker, Kubernetes, and Terraform) are very different and could fill their own book. If you're interested in self-hosting Supabase, your starting point can be the official docs at `https://supabase.com/docs/guides/self-hosting`. On my YouTube channel, I've also shown how I'm using docker-compose to self-host Supabase, so you might want to have a look at that as well: `https://www.youtube.com/watch?v=wyUr_U6Cma4`.

For clarity, let's summarize once more that these two URLs, when requested from the database (for example, through a webhook) will point to the exact same target:

- `http://kong:8000/functions/v1/added_ticket`
- `http://host.docker.internal:54321/functions/v1/added_ticket`

I'd also like to repeat that `http://localhost:54321/functions/v1/added_ticket` will not work as localhost refers to the host where it's called, which is the database container.

Let's continue with our work to trigger the Edge Function with a webhook. You can obviously use our custom webhook function from *Creating webhooks with dynamic URLs per environment* section to create the webhook, but for this demo, I'll just use a hardcoded URL here with the webhook UI.

Under **Database | Webhooks**, create a webhook named `ticket_created_edge_trigger` for **Insert** events on the `tickets` table:

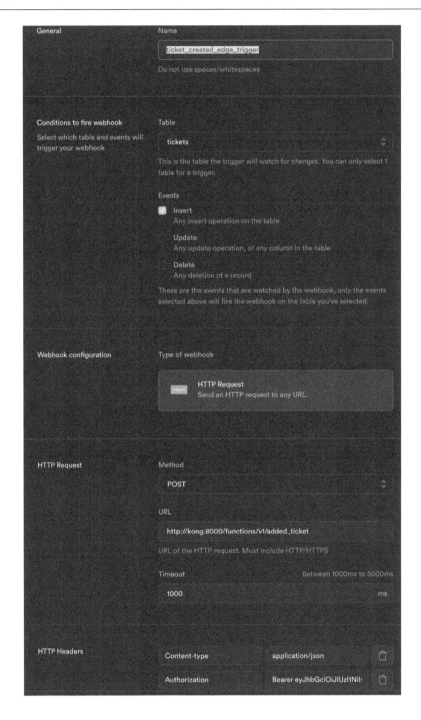

Figure 13.18: Webhook calling an Edge Function locally

As you can see, I'm using `http://kong:8000/functions/v1/added_ticket` as the webhook URL and adding the additional **Authorization** header in the **HTTP headers** section.

Once saved, this function will be triggered whenever a new ticket row is added. Go ahead and use the ticket system – create a new ticket and then come back to the Table Editor to see whether an entry was made to `testtable`. After adding a new ticket with the title `Hello i am a new ticket`, I see the successful insertion in `testtable`:

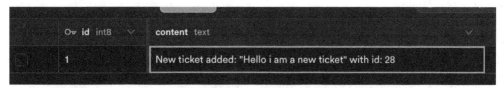

Figure 13.19: New row in testtable

> **Note**
>
> Webhook calls are also logged in the `supabase_functions.hooks` table. So, if something goes wrong, the first thing you can check is whether there's an entry there to see whether the hook triggered.

Lovely: we ran our first Edge Function with a webhook. Now, if you want to test your Edge Function, you can use any fetch tool to trigger the Edge Function manually. For example, when I want to trigger the function with ticket `id=14`, I can call this `curl` command on my system:

```
curl -X POST 'http://localhost:54321/functions/v1/added_ticket' \
    --header 'Authorization: Bearer
eyJhbGciOiJIUzI1NiIsInR5cCI6IkpXVCJ9.
eyJpc3MiOiJzdXBhYmFzZS1kZW1vIiwicm9sZSI6ImFub24iLCJleHAiOjE5ODM4MTI5
OTZ9.CRXP1A7WOeoJeXxjNni43kdQwgnWNReilDMblYTn_I0' \
    --header 'Content-Type: application/json' \
    --data '{"record":{"id": 14}}'
```

This will return the following:

```
{"title":"Leave a good trace in this world"}
```

That's how I can confirm the Edge Function works locally as it gives me the proper ticket title.

On `supabase.com`, calling Edge Functiosn and inspecting their invocations is even easier. Let's have a look.

Triggering Edge Functions on your supabase.com instance

To use our Edge Function on `supabase.com`, you must first deploy it to your instance by calling `npx supabase function deploy add_ticket`. This command will then ask you for the project instance on which to deploy it (many people have multiple Supabase instances).

Deploying takes a matter of seconds. Once completed, the Edge Function is listed in the **Edge Functions** area of Supabase Studio:

Figure 13.20: Edge Functions overview on supabase.com

In this overview, you can copy the URL of the Edge Function (which follows the exact URL convention described earlier). When you click on the function, you get a lovely little dashboard showing more details and providing an overview of invocations (i.e, every Edge Function call) and logs from the invocation (e.g., if you do `console.log` within the Deno code, it appears in the logs).

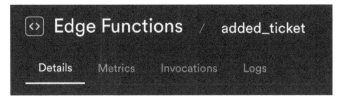

Figure 13.21: Edge Function details

To update a changed Edge Function, you simply call the `deploy` command again.

One last thing to mention about using Edge Functions with webhooks on `supabase.com` is that `supabase.com` allows you to create a webhook of the type **Supabase Edge Functions**, where you can choose the Edge Function to trigger, instead of the type **HTTP Request**.

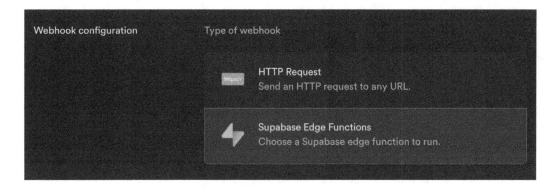

Figure 13.22: Webhook type

However, I would like to clarify that there is no difference. It's just a visual simplification and it will internally also just create a normal HTTP Request trigger.

With that, you now know how to create Edge Functions and how to execute and trigger them using webhooks, adding another powerful tool to your Supabase toolkit! Edge Functions can enhance your application by offloading small bits of logic to them.

However, it's important to note that Edge Functions are designed for small-scale operations and not for heavy computational loads. For instance, processing images is not suitable for Edge Functions. Instead, use them for lightweight tasks to ensure optimal performance.

Here's a link where you can read about current Edge Function limitations: `https://supabase.com/docs/guides/functions/debugging-limitations.sw`

As a final note for curiosity: Edge Functions on `supabase.com` are deployed on the Deno Deploy infrastructure (`https://deno.com/deploy`).

Using cronjobs to notify about due tickets

Many Supabase users may not be aware that Supabase supports the use of **cronjobs** through a dedicated database extension. This powerful feature allows you to automate and schedule various tasks within your Supabase environment. In this section, we'll demonstrate how to set up a "ticket due" notification system using cronjobs. This system will automatically send notifications when tickets are due, ensuring timely reminders without manual intervention.

> **Note**
>
> If you haven't read the *Understanding Edge Functions* section yet, please do, as we will use Edge Functions in this section. Also, we need the CREATE TABLE testtable statement from the same section as we will use that table for testing purposes in this section too.

We will implement something very simple to clarify the usage of cronjobs. When a ticket's state is still open after four weeks of existence, we will make an entry in `testtable` that says **Ticket #1234 is still open for more than 4 weeks** (to simulate notifying people via email).

The first thing you need to do is to activate the `pg_cron` extension in **Database | Extensions**. This will allow us to schedule cronjobs.

Next, we need to create an Edge Function that will fetch all tickets that are open and older than four weeks. To do so, initialize a new Edge Function with `npx supabase functions new notify_due_tickets`. Now fill the Edge Function (`supabase/functions/notify_due_tickets/index.ts`) with the following code logic:

```ts
import { createClient } from "https://esm.sh/@supabase/supabase-js@2";

Deno.serve(async (req) => {
  const supabaseAdmin = createClient(
    Deno.env.get("SUPABASE_URL"),
    Deno.env.get("SUPABASE_SERVICE_ROLE_KEY")
  );

  const fourWeeksAgo = new Date(
    Date.now() - 4 * 7 * 24 * 60 * 60 * 1000
  ).toISOString();

  const { data: tickets, error } = await supabaseAdmin
    .from("tickets")
    .select("*")
    .lte("created_at", fourWeeksAgo)
    .eq("status", "open");

  if (tickets && tickets.length > 0) {
    await supabaseAdmin.from("testtable").insert(
      tickets.map((ticket) => {
        return { content: `Ticket ${ticket.id} is due
        (${ticket.title})!` };
      })
    );
  }

  return new Response("processed tickets", {
    headers: { "Content-type": "text/plain" },
  });
});
```

The code shouldn't come with any surprises – we read all tickets with an `open` status and that were created more than four weeks ago. Then, we make a bulk-insert of those tickets to `testtable` by mapping them to the content column of `testtable` respectively.

Next, we need a cronjob that will call this Edge Function regularly. For the sake of testing it, let's say we want the cronjob to run every minute. The cronjob extension allows us to run SQL expressions like so:

```
SELECT cron.schedule('call-db-function', '*/5 * * * *', 'SELECT hello_
world()');
```

This sample would trigger a `hello_world` function every 5 minutes. However, we can also write function definitions directly, which is what we'll use for triggering our Edge Function now:

```
SELECT
  cron.schedule(
    'invoke-due-date-edge-function',
    '* * * * *',
    $$
    select
      net.http_post(
          url:='http://kong:8000/functions/v1/notify_due_tickets',
          headers:='{"Content-Type": "application/json",
          "Authorization": "Bearer YOUR_ANON_KEY"}'::jsonb,
          body:='{}'::jsonb
      ) as request_id;
    $$
  );
```

This will define a cronjob named `invoke-due-date-edge-function` and execute the code defined between $$ every minute. In this case, it's a POST request that is called every minute to trigger the `notify_due_tickets` Edge Function with the help of the `net.http_post` function (the function comes from the preinstalled `pg_net` extension). Make sure to set your Anonymous Key (`YOUR_ANON_KEY`) correctly. Then, execute this in the SQL editor to save this cronjob.

Once your cronjob is set up, you can monitor its execution and results. You can view all your defined cronjobs with the Table Editor in Supabase Studio within `cron.job` and see all cronjob invocations within `cron.job_run_details`. Shortly after defining this function, I'm seeing a lot of entries in `testtable` as many of them are older than four weeks and all of them are still open.

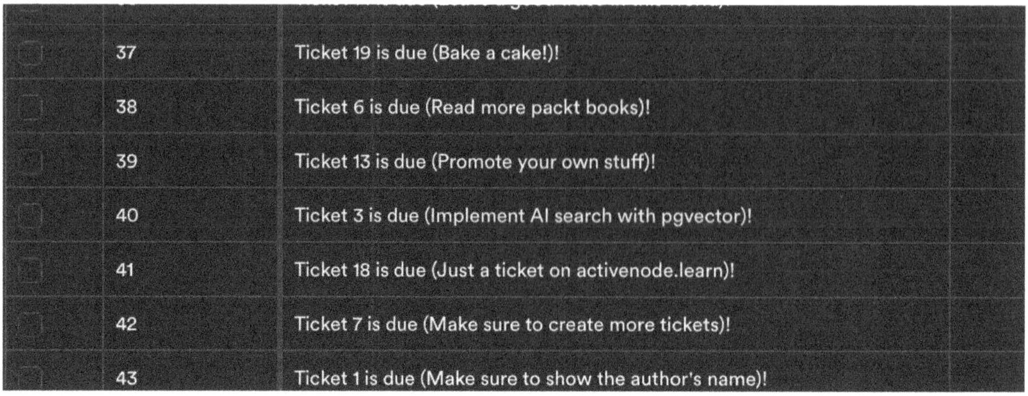

37	Ticket 19 is due (Bake a cake!)!
38	Ticket 6 is due (Read more packt books)!
39	Ticket 13 is due (Promote your own stuff)!
40	Ticket 3 is due (Implement AI search with pgvector)!
41	Ticket 18 is due (Just a ticket on activenode.learn)!
42	Ticket 7 is due (Make sure to create more tickets)!
43	Ticket 1 is due (Make sure to show the author's name)!

Figure 13.23: Testtable entries from the Edge Function triggered by the cronjob

This means the cronjob and the Edge Function work properly!

> **Note**
>
> I've not activated this demo cronjob in the source repository to avoid endlessly filling your database in the background. However, the `notify_due_tickets` Edge Function can be found in the repository of this chapter at `supabase/functions/notify_due_tickets/index.ts`.

Deleting a named cronjob is as simple as this:

```
SELECT cron.unschedule('invoke-due-date-edge-function');
```

With this crucial cron extension at hand, in combination with Edge Functions, you can process any kind of automated workflow in your Supabase instance. How cool is that?

> **Note**
>
> More cronjob examples can be found here: `https://supabase.com/docs/guides/database/extensions/pg_cron?queryGroups=database-method&database-method=sql#examples`.

Using pg_jsonschema for JSON data integrity

JSON Schema (`https://json-schema.org/`) is a well-known and widespread standard of defining and enforcing JSON data structure. Within Supabase, you can activate the `pg_jsonschema` database extension to be able to use utilities to confirm such JSON validation directly within Postgres. Now, I don't want to explain this too deeply, but I want to give you a heads-up on what this can do for you.

The following SQL expression, with `pg_jsonschema` extension activated, will return `true`:

```
SELECT json_matches_schema(
    '{
        "type": "object",
        "properties": {
            "foobar": {
                "type": "string"
            }
        }
    }',
    '{"foobar": "hello"}'
);
```

That's because in the first parameter, we provide a schema to validate against, and in the second parameter, we provide the value to validate, given the provided schema. Since the provided value is `{"foobar": "hello"}` and the given schema says it must be an object with a `foobar` property pointing to a string, we get the result `true`.

Now imagine you're building a **Remote Procedure Call** (**RPC**) and you need to validate an argument as a JSON object. Instead of complicating your code with numerous SQL `IF` statements, you can use the `pg_jsonschema` extension for this purpose. This extension simplifies the process by allowing you to validate JSON objects directly within your database.

Even more so, you can ensure data integrity within a table. You've already learned that you can add `CHECK` constraints to single columns (which we use to ensure a certain length in the ticket `title` for example). You can use `json_matches_schema` directly inside your column to check whether the column is a `json`/`jsonb` column.

With just this, you already have the crucial knowledge you need to know about the usage of this extension.

> **Note**
>
> Here is a more thorough resource from Supabase introducing the `pg_jsonschema` extension: `https://supabase.com/blog/pg-jsonschema-a-postgres-extension-for-json-validation`.

Testing the database with pgTAP

Most developers are aware of unit tests or integration tests/end-to-end tests. Just to drop a few tool names here: Jasmine, Cypress, Selenium – you name it. Often, these test tools will then implicitly test the database by setting up a fresh database for each test run, and then checking the effect/implications of the code on the database (are the correct rows inserted? Is rejecting row deletion working when I don't have permissions? And so on…).

However, in Postgres, within Supabase, you can run explicit database tests. For this, you need to enable the pgtap extension first.

Let's assume I am the database architect and I want to ensure that whoever will finally implement the database does so in the way I expect it. For example, I want to enforce that the tickets and comments tables are created and that they are related to each other. With pgtap, we can run a test like this:

```
BEGIN;
SELECT plan( 3 ); -- number of tests

SELECT has_table('tickets');
SELECT has_table('comments');
SELECT has_relation('comments', 'tickets');

SELECT * FROM finish();
ROLLBACK;
```

This code requests the test extension to run three tests (plan(3)) and finishes the test runs by calling SELECT * FROM finish().

With the ROLLBACK; command, we ensure that any execution between BEGIN; and ROLLBACK; will not be applied to the database. So, any operations done on the database, such as potential INSERT commands, had no impact on actual data – which is what we want because we only use the operations for testing.

To clarify: ROLLBACK is a feature of Postgres transactions, not a feature of the pgtap extension.

Beyond this, pgtap allows you to test RLS policies too; however, this section is here to give you the required starting point only. You can read more about database testing with pgtap here: https://supabase.com/docs/guides/database/extensions/pgtap?queryGroups=database-method&database-method=sql.

Setting the auth.storageKey to avoid migration problems

I've been a long-term user of Supabase, and as the platform has changed, so have some of the default values when using prebuilt packages. This sounds very generic, so let's look deeper. If you use any library, it comes with default settings. That also applies to the Supabase JavaScript libraries. In earlier versions of Supabase, the default cookie name was simply `supabase-auth-token`. Even back then, this cookie name was configurable, but this is what you got if you hadn't configured it.

So, if you kept the same version of the library in your project, all was fine. But once I updated my code to newer libraries, even though I did everything correctly, it didn't seem to work anymore in combination with another application we had in the product that was deployed on the same domain – this other application was still using the older library version. So, even though both were implemented correctly, we found that the reason they didn't work together was that the cookie name had changed; in newer versions of the Supabase library, a cookie starts with `sb-`.

However, you can determine how the session key is named by setting `storageKey` when creating a Supabase client. If you have multiple applications interacting with each other that might have different teams that use different library versions, I recommend setting this value. For instance, adapting `getSupabaseBrowserClient` to use the `my.supabase.auth.token` key can be done as follows:

```
export const getSupabaseBrowserClient = () =>
  createBrowserClient(
    process.env.NEXT_PUBLIC_SUPABASE_URL,
    process.env.NEXT_PUBLIC_SUPABASE_ANON_KEY,
    {
      auth: {
        storageKey: "my.supabase.supabase.auth.token",
      }
    }
  );
```

That way, you set the key and there will be no clashes anymore. Now you know how to ensure that `storageKey` stays the same, which helps you mitigate problems, especially when you have different projects with potentially different versions of libraries.

> **Note**
>
> One last thing – at the time of writing, there's a small issue with the `@supabase/ssr` library not respecting `storageKey` as the native `@supabase/supabase-js` does. It's tracked here: `https://github.com/supabase/ssr/issues/19`. If it still exists at the time you're reading this book, you can alternatively use the `cookieOptions.name` configuration instead of the `auth.storageKey` configuration for setting the cookie name.

Extending supabase.ts with custom typings

Most JavaScript-based projects use TypeScript as it prevents errors and makes application development much more resilient and stable. As stated in *Chapter 2*, Supabase can easily generate correct database types for you, derived from the table structure you have. For instance, a numeric primary key will be successfully derived as the TypeScript type `number`, whereas the database text type will be derived as `string`. Even a predefined enum will be properly derived as `columnName: 'Value 1' | 'Value 2' | '...';`.

However, for `json`/`jsonb` types, Supabase cannot derive a proper type as the database doesn't have rules for that. Any valid JSON will be allowed in a `json` field. That's why the type that is derived from Supabase for a `json` field is as generic as this:

```
export type Json =
  | string
  | number
  | boolean
  | null
  | { [key: string]: Json | undefined }
  | Json[]
```

The type is so generic that it's not helpful at all. But this is due to the fact that with JSON-based columns, it's possible to store all kinds of different, arbitrary JSON values. However, most of the time people store JSON data that follows a certain structure and not just random JSON values.

This is why I came up with a solution for this problem: extending the originally defined types with some TypeScript magic. Let me show you how.

Let's recall *Chapter 2*. When generating types, for instance, with `supabase gen types typescript > supabase.ts`, you'll find that `supabase.ts` will export a `Database` type, which you can pass as generic to the Supabase client (e.g., `createClient<Database>(...)`).

What we want to do now is to extend this `Database` type from Supabase, with more sophisticated JSON types. This is what I'm providing with the project found at `https://github.com/activenode/supabase-table-extensions-typed`. It is a small TypeScript file that reads the original `Database` type from `supabase.ts` and allows you to apply a subset of overrides to it.

Let's look at an example. You have a `blogpost` table with a `json` column called `metadata`. You know that `metadata` will always be of the same structural type `{ hash: string; originalUrl?: string, is_public: boolean }`. You take `supabase.extended.ts` from the mentioned GitHub project and adapt the `TableExtensions` type as follows:

```
type TableExtensions = {
  blogposts: {
    metadata: {
      hash: string;
```

```
      originalUrl?: string;
      is_public: boolean;
    };
  };
};
```

Then, all you have to do is import the `Database` type from this file instead of the original `supabase.ts` file and it will have the overrides in place. That's it. This way you can go beyond generic JSON types in Supabase with TypeScript.

Improving RLS and query performance

We implemented RLS optimizations in *Chapter 6* and learned that an RLS policy that selects data from another table will trigger the additional RLS policies on that table, and so on and so forth. We've also drastically reduced RLS complexity by checking for the tenant access directly with `auth.jwt()` and `app_metadata` instead of doing multiple sub-queries. However, there are a few more things that can come in handy to improve RLS performance, which we'll look at now.

Note

Before we go into RLS optimization specifics, I'd like to define the term **stable** (as in "stable query," "stable function", or "stable expression"). In Postgres, *stable* refers to something that is guaranteed to have no side effects and to return the same result for the same arguments for each row. It's basically the equivalent of the concept of *pure functions* within a programming context (`https://en.wikipedia.org/wiki/Pure_function`).

Stable functions/queries allow the database to apply internal optimizations as compared to unstable ones. For example, if you use a function as part of your `SELECT`, which returns the current milliseconds as a timestamp, the query will not be stable as this value is not stable for every row.

So, why is that important? Stable queries can be optimized by the database and are faster than non-optimized queries. This information will allow you to better understand some of the upcoming tips.

If you notice that some queries are slow or if you detect slow queries with the performance analyzing tools discussed in the next section, it's a good time to check whether your RLS policies are the culprit. The easiest way is to temporarily disable RLS, for example, by changing queries to `true OR actual_query` (do that only in test environments, not on production, obviously!). If the queries become suddenly much faster, the RLS is the culprit.

Now I'll give you a few tips to tackle RLS performance:

- If your policy uses a column of a table that doesn't have a natural database index (only primary keys and unique columns have indexes by default), you should add an index to the column of that table for faster access. For instance, when your RLS does this: `auth.uid() = tickets.assignee`, it compares `auth.uid()` to the `tickets.assignee` column; the `tickets.assignee` column, however, is neither a primary key nor is it unique. That means it has no index. You can tell the database to index this column by adding an index manually. This is done with the following command: `CREATE INDEX my_assignee_index ON public.tickets (assignee);`. This simple change has shown up to a 100x performance boost on large datasets.

> **Note**
> There's a great and concise video on how you can even create multi-column indexes from Jon Meyers from Supabase, which you can find here: `https://www.youtube.com/watch?v=fW-y-r7CgNI`.

- In our `service_users` table, we've used the RLS policy `supabase_user = auth.uid()`. This is simple and effective but is not the most efficient version. The performance-optimized variant is `supabase_user = (SELECT auth.uid())`. But why?

  I.   First of all, it's important to note that `auth.uid()` as well as `auth.jwt()` are stable. Hence, no matter how often they are called in the same transaction (for example, an API request), they will always return the same value.

  II. When you use `SELECT auth.uid()` instead of just `auth.uid()`, it will be similar to defining a variable before iteration in JavaScript. Instead of running the function for each iteration step, it will be evaluated once, and the returned value will be used for each iteration step.

  III. This is a very cheap and crucial optimization that you should always be doing (I've changed this specific `service_users` RLS expression in the respective repository of this chapter).

- Sometimes you don't need RLS to become active and you want to explicitly skip any sub-RLS checks. We've done that in *Chapter 11* when we created the `rls_helpers.file_exists_on_ticket` function. This is a function that is not publicly accessible (as it's in the `rls_helpers` schema) and it's running with `SECURITY DEFINER` rights. Hence, when called all queries that are running inside of it are skipping RLS checks (as they run with admin rights). So, using helper functions within your RLS expressions that run with admin rights is another way of improving performance.

- Always add a role to the policy definition. We already did that in all of our policies, but I haven't explained yet that it brings performance benefits. If you have an RLS policy on a table that applies to the **authenticated** role, and a non-authenticated user accesses that table, Postgres doesn't even try to execute the RLS policy as there is no policy with that role. Hence, it doesn't improve the performance for authenticated users, but it improves overall performance by not running useless processes.

- An extremely subtle but very helpful tip is to add as many filters to your selections as possible as the database is extremely good at optimizing such queries. For instance, in the `service_users` table, a user will only have access to its own service user row. You could just use the non-admin Supabase client and execute `supabase.select().from('service_users')`. This works, but since the user will be signed in at that point in time, you already know the user's `auth.users.id`. So, you can pre-optimize the request by adding a filter instead, like this:

```
const supabase_user = await supabase.auth.getSession().data?.
session?.user.id;


supabase
  .select()
  .from('service_users')
  .eq('supabase_user', supabase_user);
```

A quick analogy: it's faster to find a set of keys if you're told they are in a red box (the filter) than if I told you to search all boxes, of all colors, for the keys instead.

There is an awesome write-up that includes some of the given samples and beyond from Gary Austin of the Supabase team: `https://github.com/orgs/supabase/discussions/14576`.

Also, there is complementing information for RLS performance found in the docs: `https://supabase.com/docs/guides/database/postgres/row-level-security#rls-performance-recommendations`.

In the next section, we'll talk less about the actual optimization and instead focus on finding performance issues.

Identifying database performance problems and bloat

This section is about giving you direction on how to identify performance problems – hopefully before your users do. So, here are my tips for performance analysis within Supabase:

- Generally, in SQL, you can use the `EXPLAIN` statement to let the database tell you how expressions are resolved including explaining their performance impact. For example, when you run the following two statements independently, we can see that the latter will be less effective than the former, given the cost numbers the database returns:

```
EXPLAIN SELECT supabase_user FROM service_users s WHERE EXISTS (
  SELECT FROM tickets t WHERE t.created_by=s.id
```

```
);

EXPLAIN SELECT DISTINCT(s.supabase_user) FROM service_users s
INNER JOIN tickets t ON t.created_by = s.id;
```

Both statements will output a string with an explanation similar to the following figure:

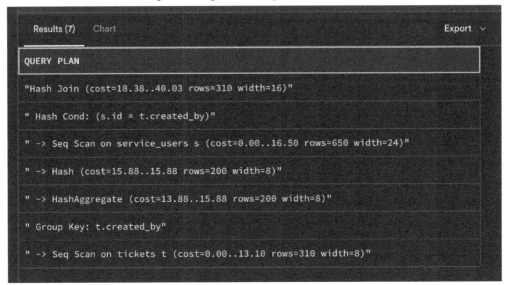

Figure 13.24: Running EXPLAIN in the SQL editor

The higher the cost factor, the worse the situation is. But even more helpful than just EXPLAIN is using those statements with both EXPLAIN and ANALYZE, which will show, for the executed statement, how long it took in milliseconds, so you have an even simpler number to compare with.

However, please note that EXPLAIN itself will be the same result for the same execution (as the costs are measured in unitless numbers), whereas the additional measurement given by ANALYZE can differ with each execution depending on how busy the server is. But as you get both factors with EXPLAIN ANALYZE, I always use EXPLAIN ANALYZE instead of just EXPLAIN.

- SQL's EXPLAIN is also built into the Supabase JavaScript client, and you can chain .explain({ analyze: true }) to the end of any .select() that you're making to retrieve performance data.

- You've seen the Security Advisor in the last chapter, but on supabase.com there's also **Database | Performance Advisor**, which works the same way – it's just with regards to performance tips instead of security.

- Complementary to **Performance Advisor**, there is also **Database | Query Performance**, which shows you your most time-consuming, most frequent, and slowest queries overall, based on the logs. So, these are good starting points for optimization.

- The database can bloat itself up over time. This process is well described in this blog post: `https://supabase.com/blog/postgres-bloat`. The post also shows how, with the help of the Supabase CLI (`supabase inspect db ...`), this bloat can be cleaned up.

- In general, the `supabase inspect db` command allows one to access the information that is provided on the **Query Performance** UI. Just call the command without additional parameters to see all the available features. For instance, if you're on localhost and you want to see long-running queries, you can do so with `npx supabase inspect db long-running-queries --local`.

These are my straightforward tips for quickly finding optimization needs.

Working with complex table joins

We've learned about joining related tables by using nested select strings (for example, fetching related comments in tickets), but with Supabase, you can do the following:

- Resolve ambiguous joins, to avoid errors, explicitly
- Rename joined tables on the fly
- Create inner joins to only return data where related data exists
- Combine the renaming feature with the inner join feature

Let me show you a few practical examples.

In our `tickets` table, we have two user references: `created_by -> public.service_users` and `assignee -> public.service_users`. I want to retrieve both user objects with a join. We do so by nesting the related table like so:

```
supabase.from("tickets").select("*, service_users(*)");
```

This will fail – since there are two pointers pointing to `service_users`, it will not know which one to resolve.

However, you can also resolve this another way, by declaring two explicit joins like this:

```
.select("id, assignee(*), created_by(*)")
```

Now it's not ambiguous anymore.

But PostgREST also has a renaming feature. So, if we wanted `created_by` to be returned as `author_user`, we could write the following:

```
.select(`
  id,
  assignee(*),
  author_user:service_users!created_by(*)
`);
```

The given syntax follows this scheme: `renamed_name:actual_referenced_table!actual_column_name`.

But what if I only wanted to have tickets resolved that actually have an assignee set? There are two good solutions. The first one isn't surprising: you simply check for the column not being `null`:

```
.not("assignee", "is", null);
```

However, in SQL, there's also the term `INNER JOIN`, which joins two tables but with the condition that there are results on both tables. In our case, if we do an inner join for the assignee column, it will also only return tickets that have an assignee. You can do that directly in `select` by adapting the selection as follows:

```
.select(`
  id,
  assignee!inner(*),
  author_user:service_users!created_by(*)
`);
```

Now you might be confused because previously we have used the exclamation mark to state the table name whilst renaming `created_by` to `author_user`. Does that mean we cannot rename the resolved `assignee` column now? You can combine both:

```
.select(`
  id,
  foobar_rename:service_users!assignee!inner(*),
  author_user:service_users!created_by(*)
`);
```

Now the related assignee object is resolved as `foobar_rename` and only rows are returned where an actual assignee exists as we use `!inner`.

Again, I would like to highlight that all of this is just PostgREST filtering, so if you ever need to search for even more features like this, I recommend not just searching in Supabase communities but also in PostgREST communities (and their documentation).

Reviewing the underestimated benefit of using an external database client

In this section, I quickly want to highlight the potential benefits of using a third-party application for database access.

Some people in the Supabase space have asked how they can give some people access to Supabase Studio but only allow reading data, not changing it, or even how to just allow access to single tables (for reading, or reading and writing). That's achievable with Supabase Studio on supabase.com but it comes with an additional price tag.

However, I'd like to highlight that rights management is built into any database and you can create as many additional database roles (hence users) as you want within the Postgres database of Supabase which will allow fine-grained data access. So, if you create a new user with your Postgres user and constrain the rights (I won't go into the details of Postgres role/rights management), you can just give people access with that user (plus the created password), the database URL and tell them to use an awesome database client such as DBeaver (https://dbeaver.io/). Problem solved!

Even more so, many of those clients allow to store snippets/scripts or even workflows much more conveniently than Studio can. Studio is a great abstraction, but it's a management tool for the whole Supabase stack and it will certainly have a hard time catching up with tailored database management tools.

Understanding migrations

You've already used migrations with supabase db diff and stored database data in seed.sql with supabase db dump. In fact, throughout this book, with every new chapter, I created an iteratively growing schema with supabase db diff. Whenever I changed something in the structure of the database, for example, when I added functions in the rls_helpers schema (or even the schema itself), I ran the following:

```
npx supabase db diff --local \
--schema rls_helpers
-f added_XYZ_in_rls_helpers
```

If you work like this when you develop your Supabase projects, you've already got iterative migration files.

> **Note**
>
> I wouldn't have to explicitly use --schema rls_helpers, but when I know which schema changed, I usually only pinpoint to that specific schema. It makes the process of creating the migration slightly more efficient.

Now, when I want to push only *new* changes to a production database (so not doing a full reset but only adding the new/changed structure), I can run `npx supabase db push --linked` (or instead of `--linked`). I can then provide a database connection string with `--db-url`) and it will then check which of the existing migrations were already applied and ask us whether we want to apply the ones that weren't applied yet.

You can also do a dry run to check what would be applied by running `npx supabase db push --linked --dry-run`. Additionally, you can list the applied migrations with `npx supabase migration list --linked`; this will show which migrations were only applied on the local instance and haven't been applied to the remote instance yet.

The most important part about these migrations, however, is the fact that `db push` will not actually compare the actual databases and check which migrations to apply and which not. Your remote instance holds a name-based list of migrations that it applied already and will check which new ones from the `migrations/` folder aren't applied yet. So, that's how it determines which to apply.

The migration flow to avoid problems should always be as follows: work locally, create migrations, and push them to the remote database with `db push`. If you change both the remote and the local instances individually, they logically diverge and you're getting yourself into potential problems when trying to apply them.

Hence, let me pin down the scenarios for using migrations with their respective solution to add clarity:

> **Note**
> For all commands, you must define the respective `--schema 'schemas,you,have, changed'` option if it's not just the `public` schema that is affected.

- You've already started working on an existing `supabase.com` project and now you want to start working locally? That's easy: you call `npx supabase db pull` to get your remote database into your local database. From here, you start working locally and create migrations that you'll then push to the remote instance.

- You've changed the structure locally and want it remote? Make sure to create a migration file as explained with `db diff`. Then, push these changes with `db push`.

- You've changed something locally and on the remote database? Quite honestly, this is calling for trouble and I don't want you to establish a habit of doing that. But if that happens, I recommend this link, which allows you to create migrations and mark them as applied so you don't get into trouble when trying to call `db push` again: `https://supabase.com/docs/reference/cli/supabase-migration-repair`.

As a last note on the topic of migrations, I'd like to say that we've only created migrations based on actual changes we did in the database and by calling `db diff` to automatically create the file containing the changes. However, you can also freely create migrations and put raw SQL statements into it.

This can be very useful, especially for SQL commands that are faster to manually write than changing something with Supabase Studio and running the `diff` command afterward to create a migration file.

To create a new empty migration file, simply run the following:

```
npx supabase migrations new my_migration_name
```

You can then fill this file with whatever change you want to apply:

```
ALTER TABLE my_table ADD COLUMN ...
```

Or you can activate an extension like so:

```
CREATE EXTENSION ...
```

To immediately apply the change locally, you'd run `npx supabase db reset` to create a new setup, based on your existing migrations and seed file. To push it to the remote database, use `db push`.

Utilizing database branching

Database branching is a feature from Supabase that deploys multiple versions of your database on your instance on supabase.com. Before I tell you how it works, let's answer the question: why is that useful?

- When working in a big team and developers implement features, they usually cannot give the repository to the manager and tell them, "Go check it out." Instead, they commit their feature and want it to be available as a preview.

- Many companies work with different staging deployments. For example, before something reaches production, it could be deployed to a `TEST.your-application.domain` URL that contains only test data to always check out the most recent version of the application in development. Then, many also use `STAGING.your-application.domain`, which usually is a version that is already agreed to be published soon, contains more realistic data (sometimes a copy of production), and when everything runs fine there, it goes to the primary domain.

So, how will branching work? Let's understand the branching mechanism step by step:

1. You configure Supabase for branching for specific Git branches (see the following link), and by doing so, you also select which branch is your primary branch (the production branch).

2. Whenever the branch changes, Supabase will deploy a database copy in your project based on the migration files in your `supabase/migrations` directory and then seed it with the `seed.sql` file data. Sounds familiar, right? It's basically the same process that happens when you call `npx supabase db reset` on your local instance – it's just that it's on Supabase.com and automated per branch.

3. When you merge an existing Git branch into the primary branch, Supabase will also apply new migrations (which haven't been applied yet) on the production database but it will very obviously not reset the data or load `seed.sql`.

I won't explain how branching exactly is configured as this is really best explained and up to date at `https://supabase.com/docs/guides/cli/managing-environments` and given any future changes, there's not much sense explaining this in this book.

However, one of the most common questions is, "How do I connect my application then to the branched database?" That's easy! Within Supabase Studio, after enabling and setting up branching, at the very top of the UI, you will be able to switch between those database branches. Hence, if you need the credentials for a specific branch (e.g., my `preview` branch as shown in *Figure 13.25*), then you just select that branch and move to API settings where the specific credentials for that branch are shown.

Figure 13.25: Branch selection in the top of the UI

Now you know what branching is and can activate it whenever you need it.

Disabling GraphQL or PostgREST (if you don't need it)

By default, any Supabase project comes with GraphQL support. This is good for those who want it but I'd rather not have it enabled by default as Supabase already provides a good way of data handling without GraphQL with the Supabase client.

So, if you're like me, not in need of GraphQL, I highly recommend disabling this exposed API. This means one less API, one less attack option, and in the best case, you're saving resources.

As GraphQL is also just a database extension (`pg_graphql`), you can easily disable it in the **Database | Extensions** area of Studio. And, as shown in *Understanding migrations*, you can also pre-deactivate it as part of a migration file.

Sometimes people use Supabase for connecting to the database only and using its Auth system. Then, exposing/using PostgREST at all doesn't make sense. In that case, you can completely disable it. In your local instance you can set the `enabled = false` within the `[api]` section of the `config.toml` to deactivate PostgREST (despite the config section being named `[api]`, it will not disable all APIs, only PostgREST).

On `supabase.com`, you find the option to disable PostgREST within **Project Settings | API | Data API**, as shown here:

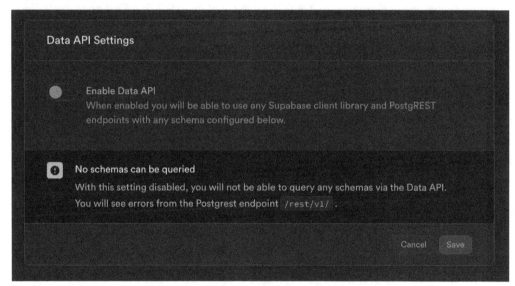

Figure 13.26: Disabled API on supabase.com

Deactivating unused APIs: done!

Using a dead-end built-in mailing setup

At the moment, Supabase does not allow us to switch off the built-in mailing system (this is an open GitHub issue: `https://github.com/orgs/supabase/discussions/18528`). However, with apps like our ticket system, we really don't want or need this feature.

To overcome this, I have a simple but effective solution. Inside Supabase, change all of the email templates so that they are empty. That way, if someone maliciously triggers a built-in function that triggers an email, there will be nothing but an empty email in the inbox.

You can also obviously use an SMTP server that will just throw away all emails to avoid an email being sent at all, but setting up empty HTML templates is certainly less effort.

Retrieving table data with the REST API and cURL

Throughout this book, we've used the Supabase client to communicate and manage data with our Supabase instance. That's great, and, as you've learned in *Chapter 2*, there are many other libraries for Supabase for other languages.

At this point of the book, I want to remind you that, ultimately, it's all just API requests. We've already dealt with such when using webhooks, but now I want to specifically point out once more how easy it is and take away the fear of playing around with the REST API directly. Why? First and foremost, it's because remembering that, with Supabase, you're independent of language-specific implementations is uplifting; then, secondly, it allows you to run Supabase requests in the command line.

Say you're on a system where you only need one simple HTTP request (for example, an automation system) and you want to retrieve some data from your Supabase project. Many people using Supabase start by building an internal backend endpoint that then uses the Supabase client to retrieve the result. But often, a simple REST API request can solve it. If you wanted to get the latest three tickets of your system, the request (using the `curl` CLI to trigger it) would look like this:

```
curl 'http://localhost:54321/rest/v1/tickets ?order=created_
at.desc&limit=3' \
-H "apikey: ANON_KEY" \
-H "Authorization: Bearer ANON_OR_SERVICE_KEY" \
-H "Accept-Profile: public"
```

Here we tell the API to go to the tickets table (`/tickets`) and provide options to only return three rows (`limit=3`) and order them by descending creation date (`order=created_at.desc`).

In our project, I'd obviously need to use the Service Role Key as the Anonymous Key wouldn't give me access to any ticket without an additional valid session header.

Sometimes people wonder where they would find the specific Supabase documentation for the REST API for the database. However, Supabase uses PostgREST as a database API, so there is no specific "Supabase Database API" – instead, there is simply the PostgREST API. Everything found at `YOUR_SUPABASE_API_URL/rest/v1/` is just the PostgREST API. This means that the documentation isn't found on `supabase.com`, but on `https://postgrest.org/`. More specifically, to see the documentation about fetching data from tables, you can check out this link: `https://postgrest.org/en/v12/references/api/tables_views.html`.

Now, the REST API for data fetching is demystified and you're ready for any adventure, even if no official library exists.

Summary

As we conclude this journey through the expansive world of Supabase, it's impossible not to admire how far you've traveled in your quest for expertise. In this final chapter, you've been given a toolbox brimming with advanced techniques and insights that, combined with the rest of the book, stretch extremely far into the galaxy of Supabase, preparing you to tackle complex challenges and enhance your applications in ways you hadn't imagined when starting this book.

You've dipped your toes into the powerful currents of database extensions, explored AI-based features such as semantic ticket search, and embraced the flexibility of anonymous sign-ins. Explorations into FDWs have shown you how to extend your database beyond its conventional boundaries, while delving into the practicalities of using webhooks and understanding Edge Functions has brought the reactive capabilities of the web right to your database's doorstep.

The introduction of cronjobs, `pg_jsonschema`, and database branching illuminated paths to automate, validate, and manage multiple database environments effectively, ensuring your projects remain agile and error-free. By setting the `auth.storageKey` type and extending the `Supabase.ts` type, you've tailored Supabase to better fit your project's unique requirements.

Your foray into enhancing RLS and query performance, together with tackling complex table joining and migrations, has undoubtedly sharpened your database schema design skills. Moreover, disabling GraphQL and leveraging Supabase's built-in mailing setup exposed you to the vast capabilities within Supabase to build tailored, efficient applications.

This chapter – and indeed this book – was designed not just to educate but to empower: to provide you with a foundation that's both broad and deep, upon which you can build your ideas, solve problems, and create applications that are as secure as they are innovative.

As we turn the last page of this guide, remember that each topic touched upon opens a door to further exploration and learning. The world of Supabase is dynamic, and you are now equipped with the knowledge and skills to navigate it confidently.

Thank you for embarking on this journey through Supabase with us. The road doesn't end here; it's merely the beginning of what you can achieve with the tools and insights you've gathered. Here's to your future projects and successes – may they be as fulfilling and enlightening as the process of learning itself. Celebrate your newfound expertise, and never stop exploring the possibilities that Supabase and the broader world of web development offer.

> **Note**
>
> If you have any questions or face any issues, feel free to reach me using **@activenode** on most social media platforms. You can also contact me through the book's Discord channel, which can be found here: `supa.guide/discord`.

Index

A

advanced storage restrictions 396-399

AI-based semantic ticket search
  adding 442, 443
  embeddings column, creating in table 444
  embeddings, comparing to find
    matching search results 446-449
  embeddings, creating with OpenAI 444-446
  embeddings provider, deciding 443

Anonymous Key 34, 123
  generating 412

anonymous sign-ins
  using 449-451

API URL 33

app_metadata
  auth.jwt() checks, using in RLS
    expressions 197, 198
  auth.jwt() function 194-197
  extending, manually 193, 194
  extending, with tenant permissions 192, 193
  tenants value, setting with
    auth hook 198-201

App Router 41

authentication process tenant-based
  forbidden tenant URLs, rejecting 206, 207
  invalid tenant URLs, rejecting 206, 207

  magic link login, preventing
    for foreign tenant 205
  making 203
  password login, preventing on
    foreign tenant 204, 205

authentication protection, with Supabase
  access, protecting to Ticket
    Management system 101-103
  adding 92
  login functionality, implementing
    in app 99-101
  middleware, preparing 95-98
  users, creating 93-95

authentication service user 15

auth.jwt() checks
  using, in RLS expressions 197, 198

auth.jwt() function 194-197

auth.role() 410, 411

Auth Service
  OTP, requesting from 125

auth.storageKey
  setting, to avoid migration problems 477

auth.uid() function 182

B

backend
used, for adding Log out button 107, 108
Backend as a Service (BaaS) 3
Blob
reference link 393
brew
reference link 31
buckets
options 372, 373
RLS policies, writing directly on 395, 396
built-in templates
adapting 134, 136
bulk insert 383

C

client ID 246
client secret 246
column-level security (CLS) 417
enabling 417, 418
comment creation
implementing 344, 345
comments table
creating 332-335
complex table joins
working with 483, 484
containers 8
content delivery network (CDN) 62, 369
controlled forms 68
reference link 68
convenient wrapper function
reference link 301
cosine distance 447
createBrowserClient
utilizing, on frontend 42, 43
cronjobs 471
using, to notify due tickets 471-474

CSS Modules
reference link 78
cURL
used, for retrieving table data 490
current_user 409, 410
custom access token auth hook 198
problems 201, 202
custom claim 190
syncing, with table data 202, 203
used, for minimizing RLS
complexity 190-192
custom email content
used, for implementing server-
only magic link flow 123
custom-made email
used, for sending crafted link 126-128

D

database
testing, with pgTAP 476
database branching
utilizing 487, 488
database extensions 437
installing, in default extensions
schema 437, 438
installing, in own schema 439-441
programmatic approach, versus UI 441, 442
database, for multi-tenancy
designing 139
permission structure, designing 152-155
planning 139
tenant access mechanism, defining 140
tenant, defining 139
tenants table, creating 141-146
user, defining 140
users table, designing 146-152
database function
defining 311-315

database passwords
  generating 412
database performance problems
  identifying 481-483
database roles
  working with 417, 418
database state
  committing 155-157
DBeaver
  reference link 485
DB URL 33
dead-end built-in mailing setup
  using 489
Deno 24
Deno runtime
  reference link 463
Docker volume 156
document.cookie
  reference link 40
dotenv
  reference link 37
dynamic URLs per environment
  used, for creating webhooks 460-462

E

Edge Function limitations
  reference link 471
Edge Functions 463
  creating 463-465
  need for 463
  optimization stack, completing 23, 24
  reference link 23
  triggering 465
  triggering, locally 466-469
  triggering, on supabase.com
    instance 470, 471

Elixir Lang
  reference link 21
embedding 442
enum 263
exchange code 122
existing comments
  listing, from server 346-349
external APIs
  transforming, into tables with foreign
    data wrappers (FDWs) 452-457
external database client
  benefit, reviewing 485
extra_search_path 436, 437

F

file attachments on comments 376
  connected files, displaying 385-390
  files, uploading to storage 377-380
  RLS storage policy, creating for
    comment uploads 380-382
  UI, preparing with file upload
    possibility 376, 377
  uploaded files, connecting with
    written comment 382-385
files
  exploring, within bucket
    programmatically 369
Fine Grained Authorization (FGA) 10
flash of content 121
foreign data wrappers (FDWs) 452
  used, for transforming external
    APIs into tables 452-457
frontend
  used, for adding Log out button 104, 105
fully dynamic domain mapping 217, 218
fully qualified identifier 433

G

getServerSideProps
Supabase, using within 51, 53
getStaticProps method 51
Supabase, using within 51
Google Cloud console
reference link 242
GoTrue 18
GraphQL
disabling 488, 489
GraphQL API
for database 19, 20
GraphQL URL 33

I

image attachments
serving, directly in UI 390
Image Proxy service 22
Image Transformations 390
using 390, 391
impersonated real-time updates
triggering, with Table Editor 354-358
Inbucket URL 34
infinite loading
versus paging 290
invalid user registration
dealing with 254
IPs for database connections
listing 427, 428

J

JSON Web Token (JWT) 34, 411
reference link 412
JWT secret 34

K

Kong
service orchestrator, overarching 25

L

Load Balancer 369
localhost
with mapped domains 218, 219
log out button
adding 103
adding, with backend 107, 108
adding, with frontend 104
log out button, with frontend
event-driven redirection, after
logging out 105-107
explicitly, redirecting after signOut() 105

M

magic links
authenticating with 114
sending, with signInWithOtp()
on frontend 114-119
signInWithOtp(), usage
considerations 119, 120
magic link verification route
adding 128-130
mapped domains
localhost 218, 219
max_rows configuration
changing 419
migration files 157
migrations 485, 486
multiple local Supabase instances
managing 34

managing, with start-stop technique 35

ports, changing 35

multi-tenant application 138, 139

N

Next.js

Pico.css, setting up with 62, 63

Next.js application

making, tenant-aware 158, 159

Next.js application, making tenant-aware 158, 159

error page and magic thanks page 162

Login.js component 164

login, making tenant-based 165-168

middleware, enhancing to safeguard routes 159-161

Nav.js component 163, 164

next.config.js 162

Route Handlers 165

static routes, fixing 161, 162

TicketList.js component 164

Next.js project

creating 31

non-existing tenant

reference link 178

Nuxt 3 57

reference link 58

Supabase, connecting to 57

O

OAuth/Sign-in, enabling with Google 242

crypto/HTTPS security problem, solving 249

Google OAuth credentials, obtaining 242- 246

reference link 242

Sign in with Google option, adding 247, 248

Supabase instance, configuring 247

verification route, building 250-253

OAuth standard

wikipedia link 242

objects table

RLS policies, writing directly on 395, 396

One Time Password (OTP) 115

OpenFGA

reference link 12

P

Pages Router 41

paging

fresh data, enforcing 294, 295

logic, implementing 290-294

versus infinite loading 290

password login

enhancing 110-114

password recovery

adding 130, 131

pg_jsonschema

reference link 475

using, for JSON data integrity 475

pg-meta

internal helper service, for database 24

pgTAP

used, for testing database 476

PHP

reference link 109

Pico.css

reference link 62

setting up, with Next.js 62, 63

Platform as a Service (PaaS) 3

Postgres

search_path, comprehending 433, 434

Postgres for API request
middleware, adding 420-426
PostgREST 19, 20
disabling 488, 489
PostgREST OpenAPI schema 404, 405
exposing, to API 409
exposure, preventing 407, 408
removing, from usage via API 408
PostgREST roles
anon role 411
authenticated role 411
reference link 412
service_role 411
private bucket 372, 373
pseudo-CDN, building 391-393
selecting 374, 375
production-grade ticket system
   project 25-28
proxying 25
pseudo-CDN
building, for private buckets 391-393
using, in UI 393-395
public bucket 368
examining 367-369
selecting 374, 375
pure functions
reference link 479
Python
reference link 59
Supabase, connecting to 58, 59

Q

query performance
implementing 479-481

R

Realtime comments
enabling 350-353
implementing 350
subscribing 350-353
UI, updating with 353, 354
real-time communication
reference link 21
Realtime service
additional insights, embracing 359-361
pitfalls 359-361
redirect URLs 132, 133
configuring 133, 134
registration page
implementing 224-227
registration, with Route Handler
account creation, handling 232, 233
activation email, sending 236-239
form data, reading 228-230
form data, validating 228-230
permission rows, adding 234, 235
processing 228
rejecting 230, 231
service user, adding 234, 235
user, redirecting to success page 240, 241
relations 13
remote procedure calls (RPCs) 307, 331, 475
function, using 316, 317
table structure, enhancing 308, 309
used, for adding user list 308
used, for fetching users 309
REST API
for database 19, 20
RLS complexity
app_metadata, extending with
   tenant permissions 192
minimizing, with custom claim 190-192

RLS helper functions
creating 338-342
RLS optimization
implementing 479-481
RLS policies
adding 338
adding, to bucket 370-372
creating 342-344
optimizing 338
writing, directly on buckets 395, 396
writing, directly on objects table 395, 396
RLS storage policy
creating, for comment uploads 380-382
**Role / Relationship Based Access
Control (RBAC)** 10
Route Handlers 49
Router Cache 294
row-level security (RLS) 15, 169
implications, learning 189, 190
implications, solving 185, 187
permission-based RLS policy,
creating 178-184
policies, defining to access tenants 173-178
policies, re-evaluating 187, 188
tenant data, fetching with restrictive
Supabase client 171-173
working 170, 171

S

SaaS multi-tenancy solutions
types 138
safe-guarded API deletion 419, 420
safe-guarded API updates 419, 420
schema 19
scoop
reference link 31

search_path 433
comprehending, in Postgres 433, 434
extra_search_path 436, 437
making 432, 433
setting, for database role 434
setting, with function 435
setting, with transaction 435
Security Advisor
using 426
security views 418
Server actions 50
server authentication 109, 110
server-based authentication
adding, reasons 109
server components 49
server-only magic link flow 121-123
implementing, with custom
email content 123
**server-only magic link flow, with
custom email content**
crafted link, sending with custom-
made email 126-128
link with OTP, crafting 126
magic link verification route,
adding 128-130
OTP, requesting from Auth Service 125
Superadmin client, preparing 123, 124
service role key 34
generating 412
services 8
sharp
reference link 394
signed URL 373, 374
signInWithOtp()
usage, considerations 119, 120
used, for sending magic links
on frontend 114-119

signups

disabling, generally 223

impact of disabling 222, 223

specific signup methods, disabling 223, 224

silent resets

utilizing, to avoid data manipulation 416

Simple Storage Service (S3) 21

site URL 132, 133

configuring 133, 134

Software as a Service (SaaS) 5, 103

SQL expression

writing, for same tenant users 310, 311

SSL

enforcing, on direct database
    connections 429

stable 479

start-stop technique 35

Storage buckets 364

creating 364-367

files, exploring programmatically 369

public buckets, examining 367-369

RLS policy, adding to 370-372

Storage Image Transformations

reference link 390

Studio URL 33

Supabase

connecting, to Nuxt 3 57

connecting, to other frameworks 57

connecting, to Python 58, 59

database connection, building 54, 55

significance 5-8

using, within getServerSideProps
    or API routes 51-53

using, within getStaticProps 51

using, with TypeScript 55-57

Supabase Auth (GoTrue)

authentication handler 18, 19

Supabase CLI

installing 31

Supabase instance

initializing 32

running, on machine 32

starting 32-34

Supabase JavaScript client 36

base 40

base, testing within Next.js 36-39

createBrowserClient, utilizing
    on frontend 42, 43

createServerClient, using with
    Pages Router 50

createServerClient, utilizing
    on backend 43-45

creating, for middleware 45-47

initializing 36

request/response client, using
    in middleware 47

Supabase backend clients, creating
    with App Router 48-50

Supabase backend clients, creating
    with Pages Router 50

used, for connecting to Supabase 35, 36

using, with App Router 40, 41

using, with Pages Router 40, 41

Supabase Realtime Server

user experience, elevating 21

Supabase services

Kong 25

Supabase Storage

simple and scalable object storage 21, 22

Supabase Studio

convenient web dashboard 17, 18

reference link 485

supabase.ts

extending, with custom typings 478

Supabase Vault 413

secret, creating in Vault 414, 415

secret, using in business logic 415

Supabase, with Postgres

access control, handling 13, 14

access system under hood,
considerations 14-17

inner works, demystifying 8, 9

logic, accessing as central service 10-13

logic, accessing within route 9, 10

Superadmin Key 123

T

table data

retrieving, with cURL 490

retrieving, with REST API 490

Table Editor

used, for triggering impersonated
real-time updates 354-358

tenant per domain

custom domains, adding via
hosts file 208, 209

domains, mapping 209

matching 207

tenant problem 103

text-embedding-3-small model

reference link 444

ticket assignee

adding, at ticket creation 322

adding, to ticket creation 323, 324

assignee columns, adding in
ticket table 318, 319

displaying 324-326

editing 317

select component, creating 322, 323

setting 317

trigger function, creating to
cache name 319-322

UPDATE policy limitations 328

updating 326-328

TicketComments component 346

ticket creation logic

implementing 267-272

ticket details

author name, caching with trigger 284-287

date and status view, improving 287, 288

viewing 281-283

ticket management system

comments section, adding to
ticket details 78-83

login form, building 63-68

navigation component, enhancing 86-88

page, implementing to create ticket 83, 84

shared UI layout, creating with
navigation elements 69-72

Ticket Details page, constructing 75-78

Ticket List page, designing 72-75

UI, visualizing 68, 69

user overview, implementing 84-86

Ticket Management system

access, protecting with 101-103

tickets

creating 266

deleting 303-305

filter, creating 297-302

filtering 289

listing 288, 289

paging, enabling 290

sorting 296

tickets table

creating, in database 260-265

transactions

reference link 435

trigger

  adding, to set tenant automatically 335-337

  checks, enforcing on database
      columns 279, 280

  loading behavior, improving 279

  user ID, setting 273-278

  using 266, 267

  using, to derive 273-278

TypeScript

  Supabase, using with 55-57

U

UI

  image attachments, serving directly 390

  preparing, with file upload
      possibility 376, 377

  pseudo-CDN, using 393-395

  updating, with Realtime data 353, 354

**ultra-high-performing domain
      mapping 209-213**

  links, fixing 214, 215

  static TENANT_MAP dynamic, making 216

users

  fetching, with remote procedure
      calls (RPCs) 309

W

Warrant

  reference link 12

webhooks

  creating, with dynamic URLs per
      environment 460-462

  using 457-460

wildcard 133

  reference link 134

Y

Y Combinator Accelerator Program

  reference link 6

Z

Zod validation library

  reference link 230

www.packtpub.com

Subscribe to our online digital library for full access to over 7,000 books and videos, as well as industry leading tools to help you plan your personal development and advance your career. For more information, please visit our website.

Why subscribe?

- Spend less time learning and more time coding with practical eBooks and Videos from over 4,000 industry professionals

- Improve your learning with Skill Plans built especially for you

- Get a free eBook or video every month

- Fully searchable for easy access to vital information

- Copy and paste, print, and bookmark content

Did you know that Packt offers eBook versions of every book published, with PDF and ePub files available? You can upgrade to the eBook version at packtpub.com and as a print book customer, you are entitled to a discount on the eBook copy. Get in touch with us at customercare@packtpub.com for more details.

At www.packtpub.com, you can also read a collection of free technical articles, sign up for a range of free newsletters, and receive exclusive discounts and offers on Packt books and eBooks.

Other Books You May Enjoy

If you enjoyed this book, you may be interested in these other books by Packt:

Full Stack Development with Spring Boot 3 and React

Juha Hinkula

ISBN: 978-1-80512-246-3

- Make fast and RESTful web services powered by Spring Data REST
- Create and manage databases using ORM, JPA, Hibernate, and more
- Explore the use of unit tests and JWTs with Spring Security
- Employ React Hooks, props, states, and more to create your frontend
- Harness the Material UI component library to customize your frontend
- Use the fetch API, Axios, and React Query for networking
- Add CRUD functionality to your apps
- Deploy your apps using AWS and Docker

React Application Architecture for Production

Alan Alickovic

ISBN: 978-1-80107-053-9

- Use a good project structure that scales well with your application
- Create beautiful UIs with Chakra UI and emotion
- Configure a base Next.js app with static code analysis and Git hooks
- Learn to mock API endpoints for prototyping, local development and testing
- Choose an optimal rendering strategy in Next.js based on the page needs
- Learn to choose the best state management solution for given problem
- Write unit tests, integration tests and e2e tests in your React Application
- Deploy your React applications on Vercel

Web App Development Made Simple with Streamlit

Rosario Moscato

ISBN: 978-1-83508-631-5

- Develop interactive web apps with Streamlit and deploy them seamlessly on the cloud
- Acquire in-depth theoretical and practical expertise in using Streamlit for app development
- Use themes and customization for visually appealing web apps tailored to specific needs
- Implement advanced features including secure login, signup processes, file uploaders, and database connections
- Build a catalog of scripts and routines to efficiently implement new web apps
- Attain autonomy in adopting new Streamlit features rapidly and effectively

Packt is searching for authors like you

If you're interested in becoming an author for Packt, please visit authors.packtpub.com and apply today. We have worked with thousands of developers and tech professionals, just like you, to help them share their insight with the global tech community. You can make a general application, apply for a specific hot topic that we are recruiting an author for, or submit your own idea.

Share Your Thoughts

Now you've finished *Building Production-Grade Web Applications with Supabase*, we'd love to hear your thoughts! Scan the QR code below to go straight to the Amazon review page for this book and share your feedback or leave a review on the site that you purchased it from.

https://packt.link/r/1-837-63068-2

Your review is important to us and the tech community and will help us make sure we're delivering excellent quality content.

Download a free PDF copy of this book

Thanks for purchasing this book!

Do you like to read on the go but are unable to carry your print books everywhere?

Is your eBook purchase not compatible with the device of your choice?

Don't worry, now with every Packt book you get a DRM-free PDF version of that book at no cost.

Read anywhere, any place, on any device. Search, copy, and paste code from your favorite technical books directly into your application.

The perks don't stop there, you can get exclusive access to discounts, newsletters, and great free content in your inbox daily

Follow these simple steps to get the benefits:

1. Scan the QR code or visit the link below

https://packt.link/free-ebook/9781837630684

2. Submit your proof of purchase

3. That's it! We'll send your free PDF and other benefits to your email directly

www.ingramcontent.com/pod-product-compliance
Lightning Source LLC
LaVergne TN
LVHW080110070326
832902LV00015B/2498

* 9 7 8 1 8 3 7 6 3 0 6 8 4 *